Advance Fluid Catalytic Cracking

Testing, Characterization, and Environmental Regulations

Edited by
Mario L. Occelli
MLO Consulting, Atlanta, Georgia, USA

CRC Press
Taylor & Francis Group
Boca Raton London New York

CRC Press is an imprint of the
Taylor & Francis Group, an **informa** business

CHEMICAL INDUSTRIES

A Series of Reference Books and Textbooks

Founding Editor

HEINZ HEINEMANN
Berkeley, California

Series Editor

JAMES G. SPEIGHT
CD & W, Inc.
Laramie, Wyoming

Advances in Fluid Catalytic Cracking

Testing, Characterization, and Environmental Regulations

Edited by

Mario L. Occelli

MLO Consulting, Atlanta, Georgia, USA

CRC Press
Taylor & Francis Group
Boca Raton London New York

CRC Press is an imprint of the
Taylor & Francis Group, an **informa** business

CRC Press
Taylor & Francis Group
6000 Broken Sound Parkway NW, Suite 300
Boca Raton, FL 33487-2742

First issued in paperback 2017

© 2010 by Taylor and Francis Group, LLC
CRC Press is an imprint of Taylor & Francis Group, an Informa business

No claim to original U.S. Government works

ISBN-13: 978-1-4200-6254-0 (hbk)
ISBN-13: 978-1-138-11635-1 (pbk)

Library of Congress Cataloging-in-Publication Data

Advances in fluid catalytic cracking : testing, characterization, and environmental regulations / editor, Mario L. Occelli.
    p. cm. -- (Chemical industries)
    Summary: "Since 1987, the Petroleum Division of the American Chemical Society (ACS) has sponsored at three years intervals an international symposium on fluid cracking catalysts technology. Papers presented at these symposia have been published in book form in seven separate volumes. The recent global economic downturn together with the H1N1 flu scare, have limited participation and contributions to the recent 238th ACS meeting in Washington DC, August 2009. As a result the present volume contains, in addition to research presented at the symposium, several invited papers"-- Provided by publisher.
    Includes bibliographical references and index.
    ISBN 978-1-4200-6254-0 (hardback)
    1. Catalytic cracking--Congresses. I. Occelli, Mario L., 1942-

TP690.4.A38 2010
665.5'33--dc22
                                                    2010040232

Visit the Taylor & Francis Web site at
http://www.taylorandfrancis.com

and the CRC Press Web site at
http://www.crcpress.com

Contents

Preface

Since 1987, the Petroleum Division of the American Chemical Society (ACS) has sponsored an international symposium on fluid cracking catalysts technology at three-year intervals. Papers presented at these symposia have been published in book form in seven separate volumes. The recent global economic downturn together with the H1N1 flu scare have limited participation and contributions to the recent 238th ACS meeting in Washington, DC in August 2009. As a result the present volume contains, in addition to research presented at the symposium, several invited papers.

To refiners, changes and challenges are everyday occurrences. After overcoming oil supply limitations from Middle East politics and the obstacles of fuel reformulations and rising crude prices, the industry is now facing an ever-growing number of mandates by governmental bodies worldwide at a time when there is a decline in demand for transportation fuels based on traditional fossil feedstocks. As a result, feeds, processes, and therefore catalysts will have to change.

The refiners' efforts to conform to ever stringent environmental laws and use of fuels derived from renewable sources are evident in chapters reporting FCC emission reduction technologies. Today, modern spectroscopic techniques continue to be essential to the understanding of catalysts performance and feedstock properties. This volume contains a detailed review in the use of adsorption microcalorimetry to measure acidity, acid site density, and strength of the strongest acid sites in heterogenous catalysts as well as a discussion in the use of ^{1}H-NMR to characterize the properties of a FCCU feedstock. In addition, several chapters have been dedicated to pilot plant testing of catalysts and nontraditional feedstocks, to maximizing and improving LCO (heating oil) production and quality, and to the improvement of FCCU operations.

The Clean Air Act (CAA), passed in 1970, created a national program to control the damaging effects of air pollution. The CAA Amendments of 1990 protect and enhance the quality of the nation's air by regulating stationary and mobile sources of air emissions. The EPA has identified the refining industry as a targeted enforcement area. As a result, a "Refining Initiative" was commissioned in 2000 with the expressed goal to have 80% of the refining industry enter into voluntary consent decrees by 2005.

The negotiation of a consent decree for a given refinery is a complex process driven by the strength and severity of the CAA and the refinery's desire to avoid litigation. Consent decree negotiation and FCC emissions (SO_x, NO_x, CO, PM) reduction technologies through consent decrees implementation are discussed in Chapters 14 through 18 of this volume.

The views and conclusions expressed herein are those of the chapters' authors, whom I thank for their time and effort in presenting their research and for preparing their manuscripts for this volume.

Mario L. Occelli, PhD
MLO Consulting
Atlanta, Georgia
mloccell@mindspring.com

Contributors

Marco Antonio Santos Abreu
Petrobras R&D Center
Rio de Janeiro, Brazil

Luis Almanza
Instituto Colombiano del Petróleo,
 ECOPETROL
Santander, Colombia

Sven-Ingvar Andersson
Chalmers University of Technology
Department of Chemical and
 Biological Engineering
Applied Surface Chemistry
Göteborg, Sweden

Aline Auroux
Institut de Recherches sur la
 Catalyse et l'Environnement
 de Lyon
Villeurbanne, France

Andy Batachari
Shaw Energy & Chemicals
Houston, Texas

Colin Bowen
Shaw Energy & Chemicals
Houston, Texas

Wu-Cheng Cheng
W. R. Grace & Co.-Conn.
Columbia, Maryland

Ray Cocco
Particulate Solid Research, Inc.
Chicago, Illinois

Dilip Dharia
Shaw Energy & Chemicals
Houston, Texas

Diana Duarte
Instituto Colombiano del Petróleo,
 ECOPETROL
Santander, Colombia

William Gaona
Instituto Colombiano del Petróleo,
 ECOPETROL
Santander, Colombia

William Gilbert
Petrobras R&D Center
Rio de Janeiro, Brazil

Song Haitao
Research Institute of Petroleum
 Processing, SINOPEC
Beijing, People's Republic of China

Roy Hays
Particulate Solid Research, Inc.
Chicago, Illinois

Ruizhong Hu
W. R. Grace & Co.-Conn.
Columbia, Maryland

Tian Huiping
Research Institute of Petroleum
 Processing, SINOPEC
Beijing, People's Republic of China

David Hunt
W. R. Grace & Co.-Conn.
Houston, Texas

E. F. Iliopoulou
Laboratory of Environmental Fuels and
    Hydrocarbons (LEFH)
Aristotle University of Thessaloniki
Thessaloniki, Greece

Long Jun
Research Institute of Petroleum
    Processing, SINOPEC
Beijing, People's Republic of China

S. B. Reddy Karri
Particulate Solid Research, Inc.
Chicago, Illinois

Ted M. Knowlton
Particulate Solid Research, Inc.
Chicago, Illinois

Larry Langan
W.R. Grace & Co.-Conn.
Columbia, Maryland

A. A. Lappas
Laboratory of Environmental Fuels
    and Hydrocarbons (LEFH) Aristotle
    University of Thessaloniki
Thessaloniki, Greece

Warren Letzsch
Shaw Energy & Chemicals
Houston, Texas

Hongbo Ma
W. R. Grace & Co.-Conn.
Columbia, Maryland

Carlos Medina
Instituto Colombiano del Petróleo,
    ECOPETROL
Santander, Colombia

Edisson Morgado Jr.
Petrobras R&D Center
Rio de Janeiro, Brazil

Trond Myrstad
Statoil
Oil and Gas Processing
Trondheim, Norway

Prashant Naik
Shaw Energy & Chemicals
Houston, Texas

Uriel Navarro
W.R. Grace & Co.-Conn.
Columbia, Maryland

Michelle Ni
W.R. Grace & Co.-Conn.
Columbia, Maryland

Dariusz S. Orlicki
W.R. Grace & Co.-Conn.
Columbia, Maryland

Georgeta Postole
Institut de Recherches sur la Catalyse et
    l'Environnement de Lyon
Villeurbanne, France

A. C. Psarras
Laboratory of Environmental
    Fuels and Hydrocarbons
    (LEFH) Aristotle University of
    Thessaloniki
Thessaloniki, Greece

Chris Santner
Shaw Energy & Chemicals
Houston, Texas

Jeffrey A. Sexton
Marathon Oil Company
Findlay, Ohio

Frank Shaffer
National Energy Technology
    Laboratory
Pittsburgh, Pennsylvania

Steve Tragesser
Shaw Energy & Chemicals
Houston, Texas

Jack R. Wilcox
Albemarle Corporation
Houston, Texas

Zhu Yuxia
Research Institute of Petroleum
    Processing, SINOPEC
Beijing, People's Republic of China

Da Zhijian
Research Institute of Petroleum
    Processing, SINOPEC
Beijing, People's Republic of China

1 Maximizing FCC Light Cycle Oil by Heavy Cycle Oil Recycle

Hongbo Ma, Ruizhong Hu, Larry Langan, David Hunt, and Wu-Cheng Cheng

CONTENTS

1.1   INTRODUCTION

Recent years have seen an increasing interest for diesel due to the energy demand and new regulations on energy efficiency [1]. Refiners are looking for technologies to raise the production of light cycle oil (LCO) from their fluid catalytic cracking unit (FCCU) to take advantage of the significant value of diesel relative to gasoline. LCO, like gasoline, is an intermediate product whose yield increases with conversion at very low conversion levels, eventually reaching an overcracking point. Past the over-cracking point, LCO yield declines with increasing conversion [2,11]. Figure 1.1 shows how LCO and bottoms oil yields shift with conversion. FCCU traditionally operates at high conversion and feed rate to produce gasoline, C4s and C3s, which is referred to as Max Gasoline Mode. To increase LCO yield, refiners can change the FCCU operating conditions and use catalysts with lower activity to shift the opera-tion away from Max Gasoline Mode toward the lower conversion regime. However, this shift also increases the yield of undesired bottoms oil. Maximizing LCO in the FCCU at reduced conversion without producing incremental bottoms oil presents

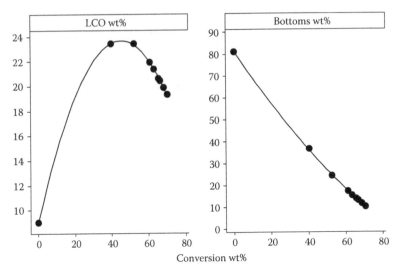

FIGURE 1.1 Yields of LCO and bottoms vs. conversion. Bottoms oil yield monotonically decreases as conversion, while LCO yield experiences a peak (overcracking point). Recycling operation enables FCCU to run at conversion close to LCO yield peak with little or no bottoms penalty compared to max gasoline mode.

the true challenge. Recycling is eventually required to minimize bottoms oil production as the refinery reduces conversion to reach an optimal LCO yield. However, the refining industry has removed recycling from the FCCU since the 1970s largely due to the introduction of the zeolite catalyst and improved equipment technology. As a result, knowledge on the recycle streams and their effect on FCC yields using modern catalyst systems and equipment is very limited.

In this paper, we developed a lab-scale method to evaluate the recycling operation, and investigated ways to optimize the operation in terms of recycle stream, recycle ratio, and conversion level. In Section 1.2, a two-pass experimental scheme to simulate the recycling operation is introduced. Experimental results are discussed in Section 1.3. Recycling of two typical FCC feeds, vacuum gas oil (VGO) and resid are compared in Section 1.3.1, using resid feed. The effect of a recycle stream boiling point range is investigated in Section 1.3.2. In Section 1.3.3, optimization of overall yields based on the experiment data is discussed. The effect of first-pass conversion level is presented in Section 1.3.4. In Section 1.4, we summarize the experimental findings and provide recommendations for refiners who want to adapt a recycle operation in their FCCU.

1.2 EXPERIMENTS

In steady-state FCC operation with heavy cycle oil (HCO) recycling, it is conceivable that some hydrocarbon molecules could go through the riser multiple times. We developed a two-pass scheme that combines the Davison circulation riser (DCR) and advanced cracking evaluation (ACE) unit to simulate the recycling operation.

First, a feedstock is cracked in DCR. Sufficient HCO or bottoms oil is collected and blended back with the original feed. Then the blend is fed into the ACE unit to be cracked again. Because of the low-recycle ratio, this two-pass cracking is expected to be close to the steady-state recycling operation. This steady-state approximation will be discussed further in the data processing section below.

1.2.1 DCR PILOT PLANT RUNS AND PREPARATION OF RECYCLE STREAMS

A commercially available MIDAS catalyst was deactivated, without Ni or V, at 1465°F for 20 hours, using the advanced cyclic propylene steam protocol described by Wallenstein et al. [3]. After deactivation, the catalyst had a 94 m²/g zeolite surface area, an 83 m²/g matrix surface area, and a unit cell size of 24.30 Å. The deactivated catalyst was charged in the DCR pilot plant [4], where cracking of VGO and residual (hereafter referred to as resid) feedstock were conducted. Properties of VGO and resid used in the study are listed in Table 1.1. Reaction severity was varied by adjusting the temperature set points of the riser top, regenerator, and feed preheater. We obtained 55% conversion by weight for VGO feed, and conversion levels of 54%, 58%, 68%, and 75% for resid feed. Ideally, a 55% conversion run of resid should be used to compare with VGO. However, accurate control of conversion in DCR and ACE is difficult. To overcome this problem, we always interpolate yields to 55% conversion before making the comparison. The DCR conditions and product yields are listed in Table 1.2. The C4 and lighter products were analyzed by gas chromatograph, while C5 and above liquid products were analyzed by simulated distillation and expressed as gasoline (C5-430°F), LCO (430°F–650°F) and bottoms (650°F+). The detailed boiling point distribution of the bottoms fraction is also provided in Table 1.2. These results provide the amount of hydrocarbon in a given boiling point range when an ideal distillation is achieved, which were used as a basis to determine the maximum available quantity of each recycle stream.

Liquid product from each DCR run was first separated by atmospheric distillation on a modified Hempel still (ASTM D295) to obtain the 650°F+ fraction. Each 650°F+ fraction was further separated by vacuum distillation (ASTM 1160) to obtain fractions with a desired boiling point range. The properties of the various fractions are shown in Figures 1.2 and 1.3. These fractions are referred to as recycle streams later.

1.2.2 ACE CRACKING OF THE RECYCLE BLENDS

Each of the recycle streams from DCR runs was blended back with its starting parent feedstock for cracking in ACE unit to simulate the recycling operation in FCCU. The percentage of recycle stream in each blend was determined based on simulated distillation listed in Table 1.2. The recycle streams were blended at two recycle ratios to demonstrate the sensitivity and reproducibility of yield changes. These feed blends, listed in Table 1.3, can be separated into three groups. The first group are the recycle streams with boiling point range of 650°F–750°F obtained from VGO

TABLE 1.1

Properties of VGO and Resid Used in the Study

Feed Name	Vacuum Gas Oil	Resid
API gravity at 60°F	25.5	20.6
Specific gravity at 60°F	0.9012	0.9303
Refractive index	1.5026	1.5222
K factor	11.94	11.76
Aniline point, °F	196	196
Average molecular weight	406	445
Paraffinic ring carbons, wt.%	63.6	59.2
Naphthenic ring carbons, wt.%	17.4	15.4
Aromatic ring carbons, wt.%	18.9	25.4
Sulfur, wt.%	0.369	0.416
Total nitrogen, wt.%	0.12	0.18
Basic nitrogen, wt.%	0.05	0.069
Conradson carbon, wt.%	0.68	5.1
Ni, ppm	0.4	6.6
V, ppm	0.2	16.5
Fe, ppm	4	6.1
Na, ppm	1.2	0
Simulated Distillation, vol.%, °F		
IBP	307	455
5	513	597
10	607	653
20	691	734
30	740	793
40	782	844
50	818	894
60	859	950
70	904	1017
80	959	1107
90	1034	1265
95	1103	1295
FBP	1257	1324

at 54% conversion and resid at 55% conversion. The ACE yield from the later will be interpolated to 55% conversion to make the comparison with VGO. The second group consists of recycle streams with boiling point ranges of 650°F–750°F, 650°F–800°F, 650°F–850°F, 650°F+, and 750°F+ obtained from resid at 54% conversion. The results of this group help us determine the best recycle stream. The last group consists of recycles with one boiling range, 650°F–750°F, but obtained at various first pass conversion levels of 54%, 5%, 68%, and 75% from the resid.

The ACE runs [5] used the same laboratory deactivated MIDAS catalyst as in the DCR runs above. All ACE testing were conducted at a reactor temperature of 930°F

TABLE 1.2

Product Yields and Conditions of Cracking VGO and Resid in DCR

Feed	VGO	Resid	Resid	Resid	Resid
Reactor exit temperature, °F	930	950	950	971	970
Regenerator temperature, °F	1250	1350	1350	1270	1270
Feed temp, °F	700	701	574	700	299
Conversion, wt%	55	54	58	68	75
C/O ratio	5.78	4.32	5.01	5.92	9.37
Dry gas, wt%	1.41	2.01	1.93	2.63	2.24
LPG, wt%	9.54	8.21	8.88	11.39	13.34
Gasoline, wt%	42.8	38.38	41.99	47.97	51.89
LCO, wt%	24.84	22.15	21.71	19.18	16.73
Bottoms, wt%	20.48	23.97	20.04	12.84	8.55
Coke, wt%	2.22	5.15	5.34	5.89	7.13
Boiling Point Distribution of 650°F+ Bottoms, wt%					
650°F–700°F	5.5	5.31	4.75	3.45	2.47
700°F–750°F	5.03	4.82	4.25	2.87	1.95
750°F–800°F	4.25	4.30	3.61	2.18	1.43
800°F–850°F	2.75	3.57	2.85	1.62	1.05
850°F–900°F	1.53	2.51	1.95	1.17	0.73
900°F–950°F	0.85	1.63	1.24	0.77	0.49
950°F–1000°F	0.57	1.02	0.78	0.80	0.44
1000°F–1050°F	0.00	0.83	0.60	0.00	0.00

for VGO type feed and 950°F for resid type feed, using the same amount feedstock of 1.5 g and a constant feedstock delivery rate of 3.0 g per minute. In order to achieve desired conversion, catalyst to oil ratio was varied by changing the amount of catalyst charged in the reactor in each run. As in the DCR run, gas and liquid products were analyzed by gas chromatography and simulated distillation. Coke on catalyst was measured using a LECO analyzer.

1.2.3 Data Processing

In the DCR-ACE scheme, the steady-state yields are approximated by the yields from two-pass cracking. The validity of this approximation can be checked by tracking the path of a feedstock element. Consider 100 grams of oil, which is fed into the FCC unit and cracked into various products, of which the bottoms oil is partially recycled. For example, 10 grams of bottoms oil is recycled and fed into the unit again to crack further. Additional products are obtained, and some of the resulting bottoms oil (e.g., 1 gram) is recycled and cracked again in the next pass, and so on. The whole process is shown in Figure 1.4. R is the recycle ratio, defined as the fraction of the recycle stream in the total feed into the unit. R is equal to 0.1 in the above example. By accumulating the products along the path of this 100 gram feedstock, we can get the product yields as weight percentage on the 100 grams fresh feed basis. Using this

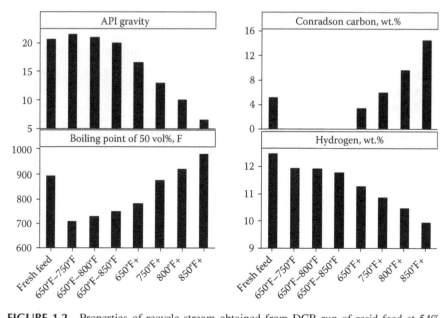

FIGURE 1.2 Properties of recycle stream obtained from DCR run of resid feed at 54% conversion. Conradson carbon and 50 vol% boiling point increases with boiling point range, while API gravity and hydrogen content decreases.

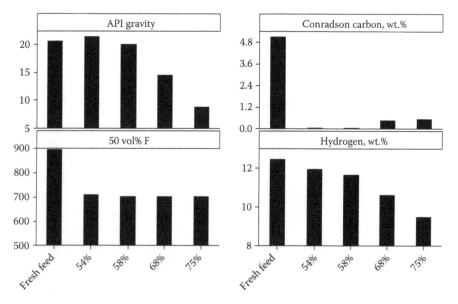

FIGURE 1.3 Properties of 650°F–750°F recycle stream obtained from DCR runs of resid feed at 54%, 58%, 68%, and 75% conversion. Conradson carbon increases with conversion level, while API gravity and hydrogen content decreases.

TABLE 1.3

Combined Feeds Used in ACE Cracking

Original Feed	Recycle Stream	First Pass Conversion (wt%)	Blend Ratio (wt%)	Original Feed (wt%)	API
VGO	—	—	—	100	25.50
Resid	—	—	—	100	20.60
VGO	650°F–750°F	55	9.5	90.5	25.03
VGO	650°F–750°F	55	11.5	88.5	25.00
Resid	650°F–750°F	54	8.3	91.7	20.42
Resid	650°F–750°F	54	6.3	93.7	20.39
Resid	650°F–800°F	54	11.7	88.3	20.37
Resid	650°F–800°F	54	9.7	90.3	20.38
Resid	650°F–850°F	54	13.4	86.6	20.29
Resid	650°F–850°F	54	11.4	88.6	20.30
Resid	650°F+	54	14.7	85.3	19.83
Resid	750°F+	54	7.1	92.9	19.87
Resid	650°F–750°F	58	8.3	91.7	20.29
Resid	650°F–750°F	58	6.3	93.7	20.29
Resid	650°F–750°F	68	7.3	92.7	19.95
Resid	650°F–750°F	68	5.3	94.7	20.03
Resid	650°F–750°F	75	5.4	94.6	19.72
Resid	650°F–750°F	75	3.4	96.6	19.93

Note: Each recycle stream was blended with its original feed at two different ratios.

method, the yield of any product on a fresh feed basis can be calculated as in the following:

$$Y = Y_1 + R \times Y_2 + R^2 \times Y_3 + \cdot + R^{i-1} \times Y_i, \qquad (1.1)$$

$$\text{Bot} = (\text{Bot}_1 - R) + R \times (\text{Bot}_2 - R) + R^2 \times (\text{Bot}_3 - R) + \cdot + R^{i-1} \times (\text{Bot}_i - R), \quad (1.2)$$

where Y_i is the yield of the ith pass cracking of the recycle stream from the $(i-1)$th pass except bottoms oil. Bottoms oil yield needs to be calculated differently from other yields because of the recycling. If the recycling ratio R is small, the second- and higher-order terms of R could be ignored. In this work, the maximum R is 0.15; so, the third term on the right-hand side of Equations 1.1 and 1.2 is negligible, only about 2.25% of the first term. Therefore, if we can get Y_2, the yield of the recycle stream in the second-pass cracking, a reasonable estimate for Y, the yield on a fresh feed basis can be obtained. The total feed in the second pass consists of $(1-R)$ fresh feed and R recycled stream from the first-pass cracking by weight fraction. This second pass corresponds to the ACE study in our DCR + ACE scheme. Analog to the partial molar properties in thermodynamics, we define the incremental yield of recycling stream as the change in normalized yield due to the addition of the

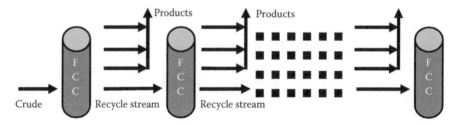

FIGURE 1.4 Schematic diagram of the cracking path of a feed element in FCCU recycling operation.

recycling stream into the base feed. Note that the interactions between the molecules from the recycling stream and those from the base feed during cracking complicate the interpretation of the incremental yield, and make it recycling ratio dependent. For simplicity and because of the small range of recycling ratio in this study, we ignore those interactions and assume a linear addition of the yield from recycling stream and that from base feed. This approach was proposed in an earlier paper and proven to be insightful [6]. Given that, the incremental yield of the recycling stream can be readily calculated as in the following:

$$Y_2 = Y_1 + (Y' - Y_1)/R, \tag{1.3}$$

$$\text{Bot}_2 = \text{Bot}_1 + (\text{Bot}' - \text{Bot}_1)/R, \tag{1.4}$$

where Y' is the yield of any product for the combined feed of recycle stream and base feed, Y_2 is the incremental yield from the recycle stream, Y_1 is the yield from the base feed, and R is the recycling ratio. Accordingly, bottoms oil yield is calculated as in Equation 1.4.

Substitute Equations 1.3 and 1.4 into Equations 1.1 and 1.2), the yields on fresh feed basis are

$$Y = R \times Y_1 + Y', \tag{1.5}$$

$$\text{Bot} = R \times \text{Bot}_1 + \text{Bot}' - R. \tag{1.6}$$

Second-order and above terms have been ignored. The incremental yield of a specific product can be deduced from Equation 1.5. For example, the LCO yield on fresh feed basis can be calculated as

$$\text{LCO} = R \times \text{LCO}_1 + \text{LCO}'. \tag{1.7}$$

1.3   RESULTS AND DISCUSSION

1.3.1   Effect of Feed Type

Resid and VGO are two typical types of feed processed in FCCU. It is of interest to know which feed can benefit more from the recycle operation. Generally speaking,

VGO contains more saturate and less aromatic hydrocarbons than resid. VGO also has a lower boiling point range. As a result, VGO is easier to crack and generates less bottoms oil. In this section, the yield structure of VGO will be compared to that of resid, and the differences are discussed.

Using the method described in Section 1.2, one could calculate the incremental yields derived from the second-pass cracking of each of the recycle streams. The incremental yields of the recycle streams with boiling point range of 650°F–750°F from VGO and resid at interpolated to 55% conversion are shown in Table 1.4. The recycled streams are less crackable than the base feed, as indicated by the much higher catalyst to oil ratios (C/O ratio) required to achieve the same conversion. This is expected, as the easy to crack material of the base feed has been cracked in the first pass. The crackability of the recycle streams increases with the API gravity. The gasoline yield of the 650°F–750°F recycle stream from VGO is higher than that from resid. Surprisingly, although the base VGO feed generates more LCO and less bottoms oil than resid, the recycle stream of 650°F–750°F from VGO gives the opposite results: less LCO and more bottoms oil. Another observation is that the yields of total C4s and total C1s and C2s are almost doubled for the recycle stream from resid than that from VGO, while the hydrogen yield only increases slightly. These striking differences need to be explained by the details of the molecular composition instead of the boiling point distributions [6], which are indistinguishable as shown in Figure 1.5. We used GC mass spectrometry to quantify the different hydrocarbon types in the 650°F–750°F recycle streams from

TABLE 1.4

Yields Structure of VGO, Resid and 650°F–750°F Recycle Streams from Them at 50% Conversion

Original Feed		VGO		Resid
Boiling range	—	650°F–750°F	—	650°F–750°F
Recycle ratio, wt%	0	10.5	0	7.3
Cat-to-oil ratio	3.29	3.49	3.43	4.80
Hydrogen, wt%	0.04	0.05	0.09	0.07
Total C1s & C2s, wt%	0.65	0.69	0.99	1.06
Propylene, wt%	2.25	2.39	2.08	2.75
Total C3s, wt%	2.53	2.69	2.38	3.28
Total C4s, wt%	3.80	3.88	3.95	5.46
Total C4s, wt%	5.86	6.20	5.59	10.07
C5+ gasoline, wt%	44.01	43.58	40.63	39.79
LCO, wt%	26.04	26.32	24.72	36.97
Bottoms, wt%	18.96	18.68	20.28	8.03
Coke, wt%	1.90	1.81	5.59	5.45

Note: The yields for the recycle streams are the incremental yields. Recycle stream from resid gives much higher LCO and lower bottoms oil yields. Furthermore, yields of total C4 and total C1 and C2 are almost doubled.

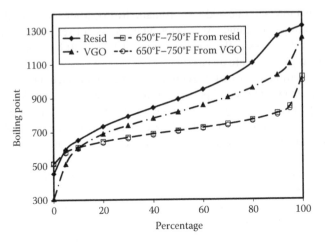

FIGURE 1.5 Boiling point distribution of the resid and VGO, and that of the 650°F–750°F recycle streams obtained from them, at 55% and 54% conversion, respectively. The boiling point distribution of 650°F–750°F recycle streams from VGO and resid are identical.

TABLE 1.5
Weight Distribution of Different Hydrocarbons in the Recycle Streams from Resid and VGO Obtained by GC Mass Spectrometry

Recycle Stream	650°F–750°F from VGO	750°F+ from VGO	650°F–750°F from Resid	750°F+ from Resid
Total saturates	30.3	25.2	20.4	22.4
Total aromatics	55.4	64.1	65.3	67.4
Mono-aromatics	4.8	5.1	8.3	9.5
Di-aromatics	22.7	10.8	39.9	13.8
Tri-aromatics	27.3	36.6	17.1	34.3
Tetra-aromatics	0.6	11.6	0.0	9.9

Note: The 650°F–750°F recycle stream from resid has more di-aromatics and less tri-aromatics.

VGO and resid. Weight percentage of different hydrocarbons in the two recycle streams obtained from GC mass spectrometry is presented in Table 1.5 and visualized in Figure 1.6.

As shown in Table 1.5, the recycle stream of 650°F–750°F from VGO contains more total saturates and less total aromatics than the recycle stream from resid. After cracking, the fragments of the saturates contribute to gasoline, which explains the higher gasoline yield of stream from VGO. A close examination of the composition of the aromatic hydrocarbons reveals that although the recycle stream of 650°F–750°F from resid has more total aromatics than that from VGO, the weight fraction of the aromatics are not across the board higher: slightly higher (higher but accounting for a small fraction of total hydrocarbon molecules) mono-aromatics

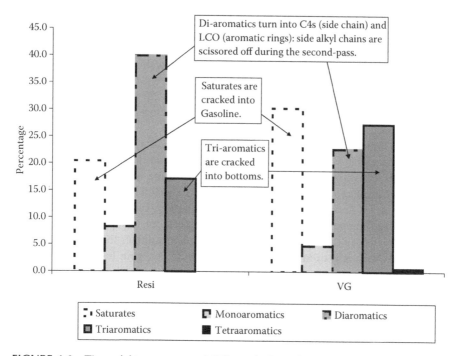

FIGURE 1.6 The weight percentage of different hydrocarbons in the 650°F–750°F recycling streams from resid and VGO, at 54% and 55% conversion, respectively. Recycling from VGO has more saturates that are cracked into gasoline. Recycling stream from resid has more di-aromatics that are cracked into LCO and light gas. It also has less tri-aromatics that go to the bottoms oil after cracking.

(8.3% vs. 4.8%), significantly higher di-aromatics (39.9% vs. 22.7%) and lower tri-aromatics (17.1% vs. 27.3%). Tetra-aromatics are low in both streams of 650°F–750°F. They mostly reside in boiling point range higher than 750°F, which is evident in Table 1.5. The molecular structure of all the aromatic hydrocarbons contains an aromatic core and some saturate side chains/rings. The aromatic nucleus cannot be cracked in the FCC condition, and cracking reaction generally happens to the saturate side chains. After cracking, most of the saturate side chains end up as wet gas (C4 and minus hydrocarbons) with heavier ones possibly going to gasoline. So the higher level of total parent aromatics in the recycle stream from the resid should give more wet gas. This is confirmed by our experimental data (almost doubled total C4s and total C1s and C2s, higher total C3s). The aromatic cores, however, can follow very different paths depending on the complexity of their structure. Mono-aromatic cores (with residual side chains) usually have less carbon atoms and a boiling point less than 430°F, so they are the precursors of gasoline [7]. Di-aromatic cores from the cracking, on the other hand, contain more carbon atoms and feature a higher boiling point. These aromatic structures fall into the LCO boiling point range [7,8]. Therefore, the higher di-aromatics level in the 650°F–750°F recycle stream of the resid explains its higher LCO yield (37.0% vs. 26%). Tri-aromatic cores are even heavier. Most of them enter into the bottoms oil. The 650°F750°F recycle stream from VGO has more tri-aromatics (27% vs. 17%) and thus produces more bottoms (18% vs. 8%). Lastly, tetra-aromatics are extremely heavy and inactive toward cracking. They are prone to dehydrogenation and tend to turn into coke. In the recycle streams of 650°F–750°F from VGO and resid, tetra-aromatics level is less than 1% by weight, so their contribution to the yields is negligible. In summary, the detail distribution of the aromatics in the 650°F–750°F recycle streams from VGO and Resid determines their different yield structure, which, at first glance, seems to contradict to the general perception.

In FCCU operation, recycling is typically adapted at a lower conversion level. The differences between VGO and resid suggest that in the first pass at low conversion, paraffins are selectively cracked into gasoline and LCO, which favors VGO. After product separation, HCO is recycled and fed into the unit again. For feeds with more di-aromatics like resid, the following cracking pass efficiently upgrades di-aromatics in HCO into LCO. Furthermore, comparing to gasoline mode, the recycling operation with low conversion minimizes the overcracking of LCO. Feeds with less di-aromatics like VGO, however, gain less LCO from the recycling operation.

1.3.2 EFFECT OF RECYCLE STREAMS

In the previous section, we showed that resid feed takes more advantages of HCO recycling. In this section, we look at the effect of the boiling point range of recycle stream on the cracking yields. In Section 1.3.4, the effect of conversion level will be discussed. The data presented in these two sections are from resid feed.

Table 1.6 shows the interpolated yields of the original resid feed at 70 and 55% conversion, as well as the yields of the combined feeds of recycle stream and original resid feed at 55% conversion. As a base case, 70% conversion of the resid feed represents the typical maximum gasoline mode. The yields are expressed as weight percentage of the total feed amount. To better illustrate the contribution of each recycle stream, the yields of LCO, bottoms, coke, and gasoline, as a function of the recycle ratio, are plotted in Figure 1.7. With the exception of the 750°F+ recycle feed, all the combined feeds made higher LCO and lower bottoms than the original resid feed. With the exception of the 650°F–750°F recycle feed, all the combined feeds made higher coke and lower gasoline than the original resid feed. The data quality confirms that the ACE testing has the sensitivity to measure the yield contribution of the recycle streams at the desired range of recycle ratio.

Among these recycle streams, the 650°F–750°F one made the lowest coke, the most LCO and gasoline at a given conversion. The trends in LCO and gasoline yields from the lightest stream (650°F–750°F) to the heaviest stream (750°F+) appear to be continuous and consistent with the trend in the API gravity (see Figure 1.2). However, the increase of coke is very gradual up to the 650°F–850°F stream and becomes stepwise higher for the 650°F+ and 750°F+ streams. As shown in Figure 1.8, the coke yield trends very closely to the Conradson carbon, which is concentrated in the 850°F+ range (see Figure 1.2). These results suggest that during first-pass cracking, coke precursors in the boiling range of 850°F+ are formed. These molecules are responsible for coke production during second-pass cracking. The recycling

TABLE 1.6
Interpolated Yields of Base and Combined Feeds at 70% and 55% Conversion

Conversion, wt%	70	55	55	55	55	55	55	55	55	55
Recycle Boiling Range	—	—	650°F–750°F	650°F–750°F	650°F–800°F	650°F–800°F	650°F–850°F	650°F–850°F	650°F+	750°F+
Recycle ratio	0.0%	0.0%	8.3%	6.3%	11.7%	9.7%	13.4%	11.4%	14.7%	7.1%
Cat-to-oil ratio	6.05	3.43	3.52	3.54	3.56	3.60	3.74	3.57	3.71	3.59
Hydrogen, wt%	0.11	0.09	0.08	0.09	0.09	0.09	0.10	0.09	0.10	0.09
Total C1s & C2s, wt%	1.37	0.99	1.02	0.98	1.08	1.07	1.08	1.11	1.04	1.06
Propylene, wt%	3.34	2.08	2.16	2.10	2.17	2.19	2.24	2.21	2.10	2.01
Total C3s, wt%	3.88	2.38	2.48	2.41	2.49	2.51	2.57	2.55	2.44	2.36
Total C4s, wt%	5.11	3.95	3.91	4.17	3.82	3.90	3.96	3.97	3.87	3.81
Total C4s, wt%	8.50	5.59	5.74	6.03	5.61	5.71	5.87	5.80	5.81	5.51
C5+ gasoline, wt%	49.44	40.63	40.82	40.38	40.15	40.12	39.84	39.80	39.88	40.03
RON	89.61	89.48	89.32	89.55	89.50	89.48	89.45	89.48	89.50	89.26
MON	78.66	77.55	77.65	77.72	77.73	77.74	77.79	77.72	77.81	77.65
LCO, wt%	20.49	24.72	25.75	25.47	25.64	25.45	25.13	24.99	25.01	24.67
Bottoms, wt%	9.51	20.28	19.25	19.53	19.36	19.55	19.87	20.01	19.99	20.33
Coke, wt%	6.73	5.59	5.60	5.57	5.69	5.62	5.70	5.67	6.07	6.09
Relative Combined Feed Rate										
Coke burn limited	1.00	1.20	1.20	1.21	1.18	1.20	1.18	1.19	1.11	1.10
Wet gas limited	1.00	1.53	1.49	1.46	1.49	1.48	1.44	1.45	1.48	1.54
Catalyst circulation limited	1.00	1.76	1.72	1.71	1.70	1.68	1.62	1.69	1.63	1.68

(Continued)

TABLE 1.6 (CONTINUED)
Interpolated Yields of Base and Combined Feeds at 70% and 55% Conversion

Conversion, wt%	70	55	55	55	55	55	55	55	55	55
Recycle Boiling Range	—	—	650°F–750°F	650°F–750°F	650°F–800°F	650°F–800°F	650°F–850°F	650°F–850°F	650°F+	750°F+
			Relative Fresh Feed Rate							
Coke burn limited	1.00	1.20	1.10	1.13	1.04	1.08	1.02	1.05	0.95	1.03
Wet gas limited	1.00	1.53	1.36	1.37	1.32	1.33	1.25	1.29	1.26	1.43
Catalyst circulation limited	1.00	1.76	1.58	1.60	1.50	1.52	1.40	1.50	1.39	1.56

Note: The yields are expressed as weight percentage of the total feed amount. Seventy percent conversion run is considered as a typical maximum gasoline mode. The feed rate relative to maximum gasoline mode subjected to the same operation constraints are listed as well.

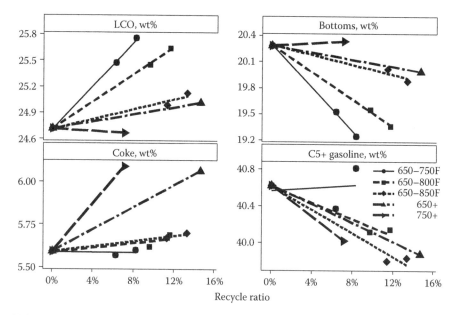

FIGURE 1.7   Interpolated yields at 55% conversion vs. recycle ratio. The yields are expressed as weight percentage of the total feed amount.

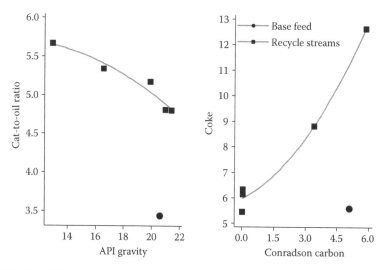

FIGURE 1.8   Effect of API gravity and Conradson carbon on catalyst to oil ratio and coke yield at 55% conversion.

of 650°F+ bottoms made lower LCO and higher bottoms oil than the recycling of 650°F–800°F and 650°F–850°F HCO. Thus, it is advantageous to recycle HCO rather than bottoms. Gasoline yields on fresh feed basis for all the recycling streams are about 4%–6% higher than that of the case without recycling, which corroborates the results reported by Fernandez et al. [12].

While the 750°F+ stream is not a practical recycle stream, it does provide valuable insight on the negative impact of recycling heavy bottoms oil. This stream made more than double the coke yield of the base feed. A close examination of the hydrocarbon compounds by GC mass spectrometry (see Table 1.5) shows that the 750°F+ stream contains higher aromatic compounds, and in particular tetra-aromatic compounds, than the 650°F–750°F+ stream. This result suggests that the coke precursors formed during the first-pass cracking are the tetra-aromatic compounds. We noticed that Ye and Wang [9] reported slightly less coke formation (0.6%) with recycling of high aromatics bottoms in FCC unit. However, their recycling ratio was much lower, only 1.5%.

1.3.3 Modeling Overall Yields

Table 1.6 also reveals the impact of recycling on cracking throughput. Compared to the yields at 70% conversion, the LCO yield at 55% conversion is higher while the yields of wet gas and coke are much lower and the C/O ratio is lower. If the unit changes from maximum gasoline (70% conversion) to maximum LCO (55% conversion) operation, one should be able to increase total feed rate until the unit reaches coke burn, wet gas compressor, or catalyst circulation constraint, assuming there is no other limitation. The results in Table 1.6 suggest the coke burn constraint will be reached much sooner than the wet gas or catalyst circulation constraint, which could be a limit at reduced catalyst activity. At coke burn limit, the combined feed rate of the maximum LCO operation is 10%–20% higher than the maximum gasoline operation.

The data analyses so far have been confined to yields with the selected recycle ratios. The following examples demonstrate how to use this data to determine the recycle stream and optimize the recycle ratio to maximize the LCO production under various constraints. We will examine a maximum recycle case and a constant bottoms case.

Case 1: Maximum Recycle

The goal of the calculation was to maximize recycle ratio of each recycle stream until the coke yield of the base feed at 70% conversion was reached. The calculated hydrocarbon yields, on the fresh feed basis are shown in Table 1.7. In the cases of the 650°F–750°F and 650°F–800°F streams, the maximum available recycle levels, based on SIMDIST (Table 1.2), were reached before the coke limit was hit; therefore, the maximum available recycle ratio was used.

The highest LCO yield of 30.2% was achieved with maximum available recycle (14.4%) of the 650°F–800°F HCO stream. The next highest LCO yield of 29.9% was achieved with 15.6% recycle of 650°F–850°F HCO stream. Even though the 650°F–750°F stream had the best incremental yields as shown in Section 1.3.2, the combined feed with 650°F–750°F stream made only 28.9% LCO and much higher bottoms because this stream was limited to a maximum available recycle ratio of 10.1%. In the case of the 650°F+ bottoms recycle, due to coke limitation, only 15% out of the available 24% recycle stream could be recycled. The lower coke yield allows the feeds with HCO recycle to be processed at higher rate. The relative feed rate with the same coke production rate as maximum gasoline mode was calculated and shown in Table 1.7. These results suggest that the selection of recycle stream, recycle ratio, and feed rate need to be balanced in order to optimize the recycling operation.

TABLE 1.7

Yields and Feed Rate at Maximum Recycle Subject to Coke Burn Limit; Yields Are on Fresh Feed Basis

	Max Gasoline	Base No Recycle	650°F–750°F	650°F–800°F	650°F–850°F	650°F+
Conversion	70.0	55.0	61.2	64.2	65.2	64.7
Recycle ratio	0	0	0.101	0.144	0.156	0.150
Maximum recycle available			0.10	0.14	0.18	0.24
Cat-to-oil ratio	6.05	3.43	3.48	3.56	3.59	3.60
Hydrogen, wt%	0.11	0.09	0.10	0.11	0.12	0.12
Total C1s & C2s, wt%	1.4	1.0	1.1	1.3	1.4	1.4
Propylene, wt%	3.3	2.1	2.4	2.6	2.7	2.7
Total C3s, wt%	3.9	2.4	2.7	2.9	3.1	3.1
Total C4s, wt%	5.1	3.9	4.5	4.5	4.7	4.8
Total C4s, wt%	8.5	5.6	6.6	6.6	6.9	7.0
C5+ gasoline, wt%	49.4	40.5	44.6	46.8	47.0	46.4
RON	89.6	89.2	89.4	89.5	89.5	89.7
MON	78.6	77.3	77.7	77.8	77.7	77.9
LCO, wt%	20.5	24.7	28.9	30.2	29.9	29.3
Bottoms	9.5	20.2	9.9	5.6	5.0	6.0
Coke, wt%	6.7	5.6	6.1	6.5	6.7	6.7
Relative combined feed rate with constant coke	1.00	1.20	1.23	1.21	1.18	1.18
Relative fresh feed rate with constant coke	1.00	1.20	1.10	1.04	1.00	1.00
Relative coke production rate	6.7	6.7	6.7	6.7	6.7	6.7

Case 2: Constant Bottoms

The goal of this calculation was to adjust the recycle ratio of each recycle stream until the bottoms yield of the base feed at 70% conversion was reached. Again, in the case of the 650°F–750°F stream, the maximum available recycle level, based on SIMDIST (Table 1.2), was reached before the target bottoms oil yield was hit; therefore, the maximum available recycle ratio was used. The hydrocarbon yields, on the fresh feed basis, are shown in Table 1.8. In this case, all the combined feeds with HCO recycle had higher LCO selectivity than bottoms (650°F+) recycle. The difference on the coke yield also allows higher throughput. The relative feed rate with the same bottoms yield by weight percentage of the fresh feed and coke production rate as maximum gasoline mode was calculated and shown in Table 1.8.

1.3.4 EFFECT OF CONVERSION LEVEL

The objectives of this section are to determine how the composition of the HCO stream changes with conversion, and how recycling HCO obtained at varying conversion levels affects the LCO yield. As described earlier, DCR liquid products

TABLE 1.8

Yields and Feed Rate at Constant Bottoms Oil Yield Recycle; Yields Are on Fresh Feed Basis

	Max Gasoline	Base No Recycle	650°F–750°F	650°F–800°F	650°F–850°F	650°F+
Conversion	70.0	55.0	61.2	62.0	62.4	62.5
Recycle ratio	0.00	0.00	0.101	0.112	0.118	0.120
Maximum recycle available			0.10	0.14	0.18	0.24
Cat-to-oil Ratio	6.0	3.4	3.48	3.53	3.56	3.56
Hydrogen, wt%	0.1	0.1	0.10	0.10	0.11	0.11
Total C1s & C2s, wt%	1.4	1.0	1.1	1.2	1.3	1.3
Propylene, wt%	3.3	2.1	2.4	2.5	2.5	2.6
Total C3s, wt%	3.9	2.4	2.7	2.8	2.9	3.0
Total C4s, wt%	5.1	3.9	4.5	4.4	4.5	4.6
Total C4s, wt%	8.5	5.6	6.6	6.4	6.6	6.8
C5+ gasoline, wt%	49.4	40.5	44.6	45.1	45.1	44.9
RON	89.6	89.2	89.4	89.5	89.5	89.6
MON	78.6	77.3	77.7	77.7	77.7	77.8
LCO, wt%	20.5	24.7	28.89	28.60	28.19	28.01
Bottoms, wt%	9.5	20.2	9.9	9.5	9.4	9.5
Coke, wt%	6.7	5.6	6.1	6.2	6.4	6.4
Relative combined feed rate with constant coke	1.00	1.20	1.23	1.21	1.19	1.19
Relative fresh feed rate with constant coke	1.00	1.20	1.10	1.08	1.05	1.04
Relative coke production rate	6.7	6.7	6.7	6.7	6.7	6.7

obtained at 54%, 58%, 68%, and 75% conversion levels were distilled, and the 650°F–750°F fraction of each liquid product was collected and analyzed (Figure 1.3). Those 650°F–750°F fractions were blended with their original feeds and tested in the ACE unit.

Based on the ACE test results, the incremental yields of the second-pass cracking are calculated using the method introduced in Section 1.2. The difference in the yields of gasoline, LCO, and coke between the recycle stream and the fresh feed is shown in Figure 1.9. The maximum recycle ratio at each conversion based on SIMDIST is also plotted in Figure 1.9. At lower conversion, there is more 650°F–750°F fraction available for recycle. The low-conversion recycle stream made much higher LCO than the fresh feed, while making about the same gasoline and coke. At higher conversion there is less 650°F–750°F fraction available. The high-conversion recycle stream made much lower gasoline, similar LCO and much higher coke. These results can be explained by examining the properties of the recycle streams in Figure 1.3. Generally speaking, higher cracking severity in FCC units leads to more gasoline, but leaves a much higher concentration

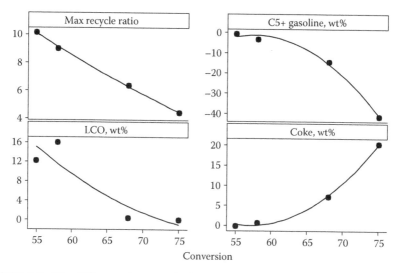

FIGURE 1.9 The difference between the incremental yields of recycling stream and the yield of fresh feed versus conversion.

TABLE 1.9

Weight Distribution of Different Hydrocarbons from GC Mass Spectrometry in the Recycle Streams from Resid at 54 and 68% Conversion

Recycle Stream	54% Conversion from Resid	68% Conversion from Resid
Total saturates	20.4	11.0
Total aromatics	65.3	70.1
Mono-aromatics	8.3	3.5
Di-aromatics	39.9	43.4
Tri-aromatics	17.1	23.1
Tetra-aromatics	0.0	0.0

of condensed aromatics in the bottoms oil [10]. Although the 50 vol% boiling points are about the same for each stream, the API gravity and hydrogen content decrease with increasing conversion. This is consistent with the GC mass spectrometry data in Table 1.9, which shows the di-aromatics and tri-aromatics of the 650°F–750°F stream, obtained at 68% conversion, is much higher than that at 55% conversion.

Figure 1.10 shows the yields of gasoline, LCO, bottoms, and coke as a function of conversion for cracking of the base feed (the first-pass cracking). The same figure also shows the corresponding yields, normalized to the fresh feed basis, for cracking of the combined feed (base feed + maximum recycle of the 650°F–750°F stream at

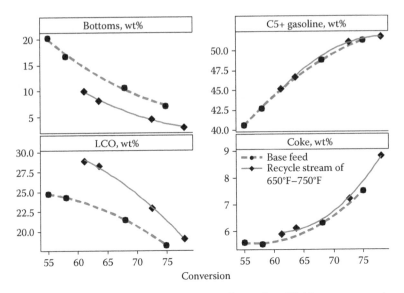

FIGURE 1.10 Conversion effects on the yields of recycling. Yields are expressed on the fresh feed basis.

each conversion level). As the figure suggests, by recycling the 650°F–750°F fraction one can lower bottoms and increase LCO without sacrificing gasoline and with only a minor penalty on coke. The LCO boost and bottoms oil reduction are more profound at lower conversion.

1.4   CONCLUSION

Maximizing LCO yield is largely a bottoms management process. Recycle can be employed to fully maximize LCO at reduced conversion, while maintaining bottoms equal to that of a traditional maximum gasoline operation. Due to the lower conversion, coke yield is also reduced. Feed type, conversion level, recycle stream need to be chosen carefully to fully optimize the recycle operation.

The comparison of VGO and resid feed shows that di-aromatics in HCO can be converted into LCO by recycling at lower conversion without overcracking the gasoline and LCO products. Feeds with more di-aromatic hydrocarbons could benefit from the recycle more than those with less. The crackability and LCO yield produced by a particular recycle stream are consistent with its API gravity and hydrogen content. The 650°F–750°F stream, when recycled, produces the most LCO, gasoline and the lowest coke for a given conversion in terms of incremental yields. However, it is not produced with sufficient quantity to fully maximize LCO and reduce the bottoms oil. High-Conradson carbon level due to more tetra-aromatic and other heavier compounds limits the yield of LCO when 650°F+ or 750°F+ streams are recycled. The comparison on different conversion levels shows that the lower conversion level, the more and the better quality of HCO for recycling.

REFERENCES

1. Energy Information Administration (EIA). January, 2009. Short-Term Outlook. http://www.eia.doe.gov/emeu/steo/pub/contents.html
2. Ritter, R. E., and Greighton, J. E. Producing Light Cycle Oil in the Cat Cracker. *Davison Catalagram* 69 (1984): 5.
3. Wallenstein, D., Harding, R. H., Nee, J. R. D., and Boock, L. T. "Recent Advances in the Deactivation of FCC Catalysts by Cyclic Propylene Steaming (CPS) in the Presence and Absence of Metals." *Applied Catalysis A: General* 204 (2000): 89–106.
4. Young, G. W., and Weatherbee, G. D. *FCCU Studies with an Adiabatic Circulating Pilot Unit.* Paper presented at the Annual Meeting of the American Institute of Chemical Engineers, San Francisco, 1989.
5. Kayser, J. C. *Versatile Fluidized Bed Reactor.* U.S. Patent 6,069,012, 05/23/1997.
6. Harding, R. H., Zhao, X., Qian, K., Rajagopalan, K., and Cheng, W.-C. Fluid Catalytic Cracking Selectivities of Gas Oil Boiling Point and Hydrocarbon Fractions. *Industrial & Engineering Chemistry Research* 35 (1996): 2561–69.
7. Fisher, I. P. Effect of Feedstock Variability on Catalytic Cracking Yields. *Applied Catalysis* 65 (1990): 189–210.
8. Mariaca-Dom, E., Rodriguez-Salomon, S., and Yescas, R. M. Reactive Hydrogen Content: A Tool to Predict FCC Yields. *International Journal of Chemical Reactor Engineering* 1 (2003): A46.
9. Ye, A., and Wang, W. Cracking Performance Improvement of FCC Feedstock by Adding Recycle Stock or Slurry. *Lianyou Jishu Yu Gongcheng* 34 (2004): 5–6.
10. Venugopal, R., Selvavathy, V., Lavanya, M., and Balu, K. Additional Feedstock for Fluid Catalytic Cracking Unit. *Petroleum Science and Technology* 26 (2008): 436–45.
11. Corma, A., and Sauvanaud, L. Increasing LCO Yield and Quality in the FCC: Cracking Pathways Analysis. In *Fluid Catalytic Cracking VII: Materials, Methods and Process Innovations.* Edited by M. L. Occelli, 41–54. Amsterdam: Elsevier, 2006.
12. Fernandez, M. L., Lacalle, A., Bilbao, J., Arandes, J. M., de la Puente, G., and Sedran, U. Recycling Hydrocarbon Cuts into FCC Units. *Energy & Fuels* 16 (2002): 615–21.

REFERENCES

2 A New Catalytic Process Approach for Low Aromatic LCO

William Gilbert, Edisson Morgado Jr., and Marco Antonio Santos Abreu

CONTENTS

2.1 INTRODUCTION

In a refinery the fluidized catalytic cracking (FCC) process remains the major process to convert high boiling range vacuum gasoil (VGO) and other heavy hydrocarbon product intermediates from other refinery processes into higher value lighter hydrocarbons. FCC will produce a high yield (40–50 wt%) of cracked naphtha boiling in the 35–221°C range that will require relatively simple adjustments to meet motor gasoline specifications. As long as there is a strong market for motor gasoline, FCC will remain a very profitable process. Recent changes in the fuel market, however, have weakened the demand for gasoline and are eroding FCC profitability.

In the Otto or gasoline engine, ignition temperature of the air–fuel mixture must be high to avoid knocking and the fuel must have a high octane number, which is favored by thermodynamically stable aromatic hydrocarbons. In the diesel engine, the ignition temperature must be less than the final temperature of the compression stroke and the fuel must have a high cetane number, properties that are negatively affected by aromatic hydrocarbons [6]. The diesel engine has a compression ratio that is two to three times as large as that of the gasoline engine, which translates into 30% higher

23

fuel efficiency. The higher fuel efficiency of diesel means that it is a better choice for fuel economy and for lower CO_2 emissions measured in grams/km.

The fuel chemical requirements of the gasoline engine are easier to meet than those of the diesel engine and have allowed the market to consider several economically viable alternatives to gasoline. In Brazil, sugar cane ethanol and compressed natural gas have captured a large fraction of the gasoline market, whereas biodiesel, the most important alternative to diesel, still has to overcome a series of technical and economical hurdles before it can capture a significant fraction of this market [5].

In Brazil, the lack of alternatives for diesel and the displacement of gasoline by other fuels have created an imbalance in the market that forced the country to import, in 2008, 14% of the diesel consumed and to export 11% of the gasoline produced. In addition to the gasoline/diesel market problem there was a strong reduction in fuel oil consumption creating a surplus of 46% of the fuel oil produced that had to be exported [2]. The shrinking fuel oil market creates a strong pressure on Brazilian refineries that must increase their conversion capacity and simultaneously change the diesel/gasoline production ratios to meet the market requirements. Similar diesel/gasoline market imbalances are seen in several other parts of the world, most notably in the EU where sales of diesel cars had already surpassed gasoline cars in 2007.

The two most important active ingredients of the FCC catalyst, the Y-zeolite and the active matrix, both have strong acid character and promote cracking reactions that favor the production of iso-paraffins, iso-olefins, and aromatics, hydrocarbons that are very good for gasoline octane number maximization, but very bad for the diesel cetane number [11]. Thermal cracking reactions, on the other hand, are far more efficient in preserving longer chained hydrocarbons in the mid-distillate range, while producing less aromatics. The greatest challenge in FCC catalyst design is exactly how to modulate its acidity and how to introduce a new functionality that will promote cracking reactions similar to those of thermal cracking.

2.2 EXPERIMENTAL

This paper describes the historical sequence of experiments performed during the development of a new FCC process designed for low aromatic mid-distillate production.

Catalyst hydrothermal deactivation was carried out in two different equipments: a 100g capacity fixed bed steamer was used for the advanced cracking evaluation (ACE) unit tests and a 5 kg capacity fluidized bed steamer was used for the other testing protocols. Steaming conditions in the two cases were the same: 788°C for 5 hours under 100% steam flow. Although conditions were similar, higher pressure buildup in the fixed bed steamer led to lower surface area retentions.

Several different reactor types were used for catalyst evaluation, including a DCR pilot riser [3] an ACE fixed fluidized bed (FFB) reactor [7], a Riser simulator [4,9], and a specially designed extended residence time circulating pilot unit. The reaction conditions of each of the reactors will be reported in the sections dealing with the specific reactor type. Different grades of Brazilian Campos Basin derived VGOs were used in the experiments. Feed properties are presented in Table 2.1.

TABLE 2.1

Feedstock Properties

Feed	VGO-A	VGO-B	VGO-C	VGO-D
°API gravity	18.7	18.1	18.8	19.6
CCR, wt%	0.88	—	0.43	0.43
Basic nitrogen, mg/kg	1338	1200	1204	1014
Distillation 10 wt%, °C	373	369	368	392
Distillation 50 wt%, °C	463	423	454	470
Distillation 90 wt%, °C	545	536	537	543
Aniline point, °C	80	—	76.2	83.6
Saturates, wt%	52.5	53.9	50.3	47.4
Mono-aromatics, wt%	17.3	16.9	18.8	18.3
Di-aromatics, wt%	17.8	17.8	19.5	21.0
Tri-aromatics, wt%	7.9	7.1	7.8	8.6
Poli-aromatics, wt%	4.6	4.3	3.5	4.8

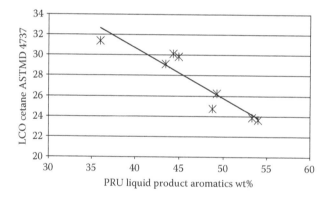

FIGURE 2.1 Cetane index correlation with total liquid product aromatics of LCO samples obtained from TBP distillation of pilot riser liquid products.

Standard cut points from ASTMD 2887 simulated distillation were used for liquid product yield calculation: gasoline was defined as the C5-221°C fraction, light cycle oil (LCO) as the 221°C–343°C fraction and slurry oil as the 343°C fraction. Gasoline quality was determined by gas chromatography using PONA column and a variation of the methodology described by Anderson [1].

LCO quality was determined by two different methods. Before 2-D chromatography was available, liquid product aromatics were measured for all samples using a variation of the ASTMD 5186-96 method. A selected number of samples were cut between 221°C and 343°C using a TBP column to produce an LCO fraction for direct ASTMD 3747 cetane index determination. The cetane values of the LCO cuts were then correlated to the total liquid product aromatics (Figure 2.1) and the correlation was used to estimate the LCO quality of the other samples.

With the development of 2-D chromatography, direct hydrocarbon speciation in the LCO range for synthetic crudes produced in FCC laboratory reactors became possible. The new method in addition to a greater understanding of the mid-distillate chemical composition avoided the effect of variations in light naphtha condensation efficiency on total aromatics. The C5 + fraction lost to the gas phase will concentrate aromatics in the liquid phase and numerical compensation by adding the gas phase C5s back to the liquid phase and is subject to errors because of the low precision of C5 + determination in the gas phase.

In the second part of the work, 2-D chromatography using a Zoex system 2-D chromatography kit with a 15 m × 0.25 mm × 1 µm nonpolar DB-1 column coupled to a 1.5 m × 0.25 mm × 1 µm polar Carbowax column was used for LCO speciation. Figures 2.2 and 2.3 show 2-D chromatograms of FCC products from two catalysts compared at similar slurry oil yields. Figure 2.2 is from the product of the novel

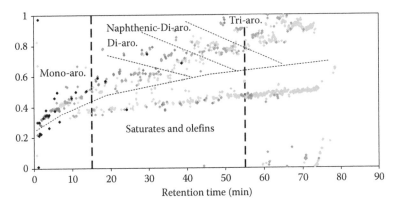

FIGURE 2.2 2-D chromatogram from a MAB catalyst product showing higher concentration of peaks in the saturates region of the LCO range (area between retention time 15 and 55 min).

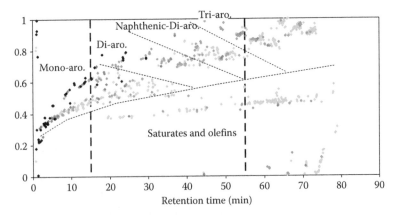

FIGURE 2.3 2-D chromatogram from the LZM catalyst product showing lower concentration of peaks in the saturates region of the LCO range.

low aromatic catalyst described in the following sections, Figure 2.3 is from the low zeolite to matrix ratio (Z/M) catalyst used in the pilot riser study.

2.3　RESULTS AND DISCUSSION

2.3.1　CONVENTIONAL APPROACH FOR FCC MIDDLE DISTILLATE MAXIMIZATION

The classical way for mid-distillate maximization in the FCC is to decrease reaction temperature and reduce catalyst activity as well as Z/M [10]. This kind of approach will always be limited by a trade-off between conversion (slurry oil yield) and LCO quality (aromatics). There is also an operational constraint that limits the minimum reaction temperature, below which there is hydrocarbon carryover from the stripper to the regenerator, leading to a sharp temperature increase in that equipment. The success of reducing reaction temperature and catalyst zeolite content will be dependent on the feed quality. Aromatic feeds are more demanding and have a narrower operational window than paraffinic feeds, requiring higher reaction temperatures and catalyst-to-oil ratios (CTO) to achieve minimum acceptable conversion levels. These conditions will also enhance aromatic production in the FCC, leading to worse LCO cetane. The gains in LCO quality are therefore much more limited for difficult feeds, even in cases where there is room for some loss in conversion. Brazilian VGOs in addition to moderately high levels of aromatics also have high levels of basic nitrogen and require catalyst formulations with high intake of active ingredients to compensate for nitrogen poisoning effects.

A DCR pilot riser unit (PRU) study was performed to evaluate the potential improvement of LCO yield and quality using Brazilian VGO-A (Table 2.1) and standard catalyst technologies. Two catalyst formulations were compared, representing a maximum gasoline and a maximum middle distillate grades. Catalyst deactivation was carried in the fluidized bed large scale unit. Catalyst properties are displayed in Table 2.2 and the test results are presented in Figures 2.4 through 2.6.

The change from a high Z/M (HZM) catalyst to a low Z/M (LZM) catalyst produced a 15-point drop in conversion at constant CTO (Figure 2.4) and a 1-point increase in coke yield at constant conversion (Figure 2.5). These differences were in agreement with expectations for the two kinds of catalyst formulation. LCO cetane improvement at constant slurry oil yield, however, was very modest (Figure 2.6) and was not large enough to compensate for the very significant drawbacks of the LZM catalyst. Figure 2.6 shows that most of the LCO cetane improvement produced by

TABLE 2.2
Catalyst Properties

Catalyst	Fresh Z/M Ratio	Fresh BET SA m²/g	Steamed BET SA m²/g
HZM	3.5	332	232
LZM	0.25	293	190

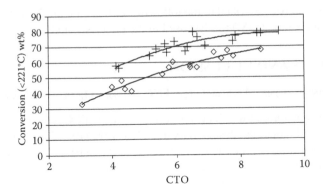

FIGURE 2.4 Conversion as a function of pilot riser CTO for catalyst with HZM (+) and LZM (◇).

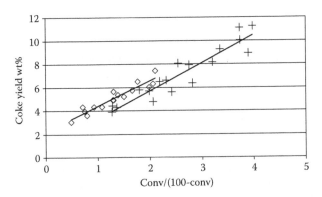

FIGURE 2.5 Coke yield as a function of second-order conversion for the catalysts with HZM (+) and LZM (◇).

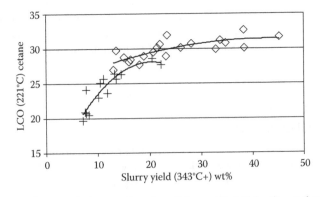

FIGURE 2.6 LCO cetane index as a function of slurry oil yield for the catalysts with HZM (+) and LZM (◇).

using the LZM catalyst could be gained with a HZM catalyst by reducing cracking severity (CTO) and giving up bottoms conversion.

2.3.2 A NEW FCC CATALYST FOR MID-DISTILLATES

In an exploratory experiment, 13 different powder materials were tested in a FFB ACE unit. Most of the results were unremarkable except for three catalysts: a low Z/M commercial maximum distillate catalyst (the same LZM catalyst used in the pilot riser experiment), a spray dried low surface area silica (inert) and the minimum aromatics breakthrough (MAB) catalyst. The inert material was included in the study to represent thermal cracking. The catalysts were steam deactivated in the fixed bed steamer prior to testing. Catalysts and the VGO-B feed properties are displayed in Tables 2.3 and 2.1, respectively. LCO aromatics were measured with 2D GC. Figures 2.7 through 2.9 illustrate the main results.

The conversion levels of the three catalysts in the ACE unit, displayed in Figure 2.7, were very different. In agreement with expectations, the commercial LZM catalyst was by far the best. Its superiority in coke selectivity was also outstanding compared to the other materials, which fell along a common trend in Figure 2.8. The advantages of the MAB catalyst were only apparent when product aromatics were considered. In Figure 2.9, most of the catalysts would fall along a line linking the LZM

TABLE 2.3
ACE Unit Study Catalyst Properties

Catalyst	Fresh BET SA m²/g	Steamed BET SA m²/g
LZM	293	164
MAB	295	207
Inert	—	19

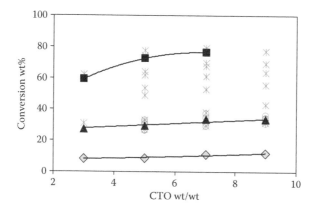

FIGURE 2.7   ACE unit conversion (<221°C) as a function of CTO. (✳) New materials—high to moderate acidity, (○) new materials—low acidity, (◇) inert, (■) LZM catalyst, (▲) MAB.

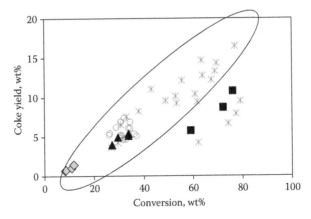

FIGURE 2.8 ACE unit coke yield versus conversion (<221°C). Catalysts within the ellipse region produced roughly twice as much coke as the LZM catalyst at the same conversion. (✳) New materials—high to moderate acidity, (○) new materials—low acidity, (◇) inert, (■) LZM catalyst, (▲) MAB.

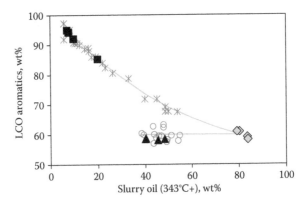

FIGURE 2.9 ACE unit LCO aromatics versus slurry oil yield (>343°C). (✳) New materials—high to moderate acidity, (○) new materials—low acidity, (◇) inert, (■) LZM catalyst, (▲) MAB.

catalyst to the inert material. As was the case with the pilot riser study (Section 2.3.1), catalysts in this trend would show very little improvement in LCO quality over a standard maximum gasoline catalyst when compared at the same bottoms conversion. The points of the group of low acidity new materials, however, departed from the main trend in Figure 2.9. With these catalysts, it was possible to reduce slurry oil yield relative to the inert material thermal cracking at the same LCO aromatics level. Within this group, the winning catalyst was MAB, which in addition to the good combination of activity and selectivity had the best hydrothermal stability and very good physical properties in the group, which made it a strong candidate for a commercially viable catalyst.

In a follow-up study, the MAB catalyst was submitted to a metals resistance test using Mitchell impregnation [8]. Two levels of metal contamination, 500 ppm and

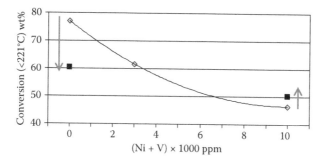

FIGURE 2.10   Comparing MAB catalyst with a standard residual catalyst at extreme metal contaminant levels with atmospheric residue in the ACE unit. (◇) LZM resid catalyst, (■) MAB.

1500 ppm each of nickel and vanadium, were compared to the metals free catalyst after steam deactivation. BET surface area retention of the three catalysts was the same, equal to 65%. Subsequent testing in an ACE unit compared favorably with the LZM catalyst. When exploring extreme metal contamination levels, the MAB catalyst showed a loss in conversion at constant coke of 1.6 points for every 1000 ppm nickel plus vanadium increase. The LZM catalyst showed a loss of 3.5 points in conversion for the same increase in metal contamination. Although the LZM catalyst starts at a much higher level of conversion than the MAB catalyst, at low metal levels, the difference becomes progressively less with the increase in vanadium concentration (Figure 2.10) up to a point where there is a ranking reversal in the activity of the two catalysts. This effect may be explained by the destruction of zeolite and the loss of the superior coke selectivity of the LZM catalyst.

2.3.3   A New Catalytic Conversion Process for Middle Distillates

The results from the ACE FFB unit showed clearly that the low activity of the MAB catalyst would produce high slurry oil yields in present day FCC configurations, optimized for Y-zeolite based catalysts. Some adjustments of the FCC hardware would certainly be necessary to fully exploit the new catalyst selectivity potential.

To evaluate the new process conditions that would allow the MAB catalyst to produce lower slurry oil yields than those obtained in the FFB unit, a series of experiments were carried out using a Riser Simulator—built and perfected at the Institute of Petrochemistry and Catalysis Research (INCAPE/FIQ-UNL) in Santa Fe, Argentina (Figure 2.11).

In the Riser Simulator, an impeller rotating at very high speed on the top of the reaction chamber keeps the catalyst fluidized between two metal porous plates, inducing the internal circulation of the reacting mixture in an upward direction through the chamber. When the reactor is at the desired experimental conditions, the reactant is fed through an injection port, and immediately after the set reaction time is attained, products are evacuated and analyzed by gas chromatography. Of particular importance to the experiments performed was the ability to extend reaction

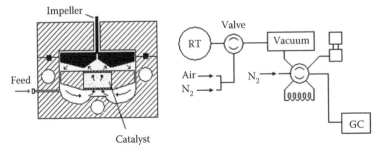

FIGURE 2.11 Riser simulator reactor schematics, showing reaction chamber at left and peripherals at right, including a preheated vacuum cylinder linked to a gas chromatograph to drain, quantify and identify reaction products at the end of a run.

TABLE 2.4
Riser Simulator Test Conditions

Catalyst	Reactor T (°C)	Catalyst/Oil Ratio	Reaction Time (s)
LZM	480	9.8	0–60
MAB	540	9.8	0–90
Inert	540	9.8	0–70

time well beyond the typical values used to simulate riser cracking in the laboratory. In the Riser Simulator, the full vapor product evacuated from the reactor is injected without condensation into a gas chromatograph equipped with a PONA column, capable of discriminating the full hydrocarbon composition from C1 all the way to the gasoline endpoint (up to 221°C). Direct determination of aromatic composition in the LCO range was not possible with this system and was estimated based on the product yields of C10 and C11 aromatics in the heavy naphtha fraction of the gasoline.

Tables 2.1 and 2.4 show the VGO-C feed quality properties and the test conditions of the Riser Simulator experiments. The three catalysts tested were the same ones used in the FFB reactor experiments. Temperature for the LZM catalyst was lower than for the other catalysts to reproduce typical conditions used for mid-distillate maximization in commercial units.

In the Riser Simulator it was possible for the MAB catalyst to reach slurry oil yield levels compatible with the conventional LZM catalyst operating at typical conditions for maximum mid-distillate. However, the yields of heavy naphtha range aromatics were half of those obtained with the LZM catalyst compared at the same slurry oil yield (Figure 2.12). The MAB C10–C11 aromatics trend as a function of slurry oil yield was a continuation of the inert catalyst trend, an indication that a similar reaction mechanism could be taking place. The minimum slurry oil yield for the inert catalyst, even at maximum severity, was still above 40 wt%;

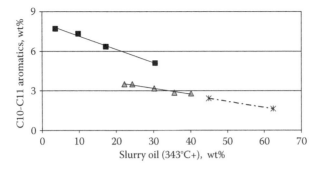

FIGURE 2.12   Riser simulator results: Heavy naphtha aromatics (C10–C11) as a function of slurry oil yield (343°C+): (✳) Inert, (■) LZM catalyst, (▲) MAB catalyst.

TABLE 2.5
Riser Simulator Δ Yields (MAB–LZM) @ Constant
Slurry = 20 wt%

	Delta (MAB–LZM)
Dry gas wt%	+5
LPG wt%	−2
Gasoline (C5-216°C) wt%	−10
LCO (216–343°C) wt%	+5
Coke wt%	+2
Heavy naphtha aromatics wt%	−8

in contrast, the yield with the MAB catalyst could be reduced to less than half of that value.

Looking at the selectivity differences in greater detail (Table 2.5), the MAB catalyst will produce more gas, coke, and LCO and less gasoline and LPG. Gasoline quality is also affected because of the significant decrease in aromatics resulting in a much lower octane number.

The next stage in process development was to adapt a 1 kg/h feed rate FCC pilot unit to extended residence time cracking. The comparison of catalysts LZM and MAB was once more repeated, this time in the new pilot cracker processing VGO-D (Table 2.1). Figure 2.13 and Table 2.6 present the results. Except for the coke yield that increased very little in the pilot unit, contrary to what had happened in the Riser Simulator, the delta yields between the MAB and the LZM catalysts were basically confirmed. With the larger product sample obtained from the pilot unit it was possible to evaluate gasoline and LCO qualities in greater detail. LCO density was reduced by 0.02 points and the cetane improved by six points whereas the gasoline MON was reduced by 12 points. Aromatic reduction in LCO by the MAB catalyst is not uniform and affected the di-aromatics to a larger extent than the mono-aromatics.

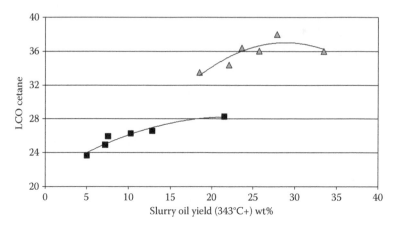

FIGURE 2.13 Pilot unit results: LCO cetane as a function of slurry oil yield (343°C+): (■) LZM catalyst, (▲) MAB catalyst.

TABLE 2.6
Pilot Unit Δ Yields (MAB–LZM) @ Constant
Slurry = 18.5 wt%

	Delta (MAB – LZM)
Reactor temperature °C	+20
Dry gas wt%	+5.1
LPG wt%	−2.6
Gasoline (C5-216°C) wt%	−7.2
LCO (216–343°C) wt%	+4.5
Coke wt%	+0.2
LCO cetane	+6.2
LCO density g/cm³	−0.02
LCO mono-aromatics wt%	−1.3
LCO di-aromatics wt%	−14.0
Gasoline MON	−12

2.4 CONCLUSIONS

By avoiding the acid catalysis mechanism of the conventional FCC zeolite catalyst (optimized over the years for high octane gasoline), the novel MAB catalyst will produce substantially lower aromatics in the liquid products than is possible by less extreme FCC catalyst adaptations. By changing the FCC reaction system, it is possible to overcome the MAB catalyst low activity drawback and achieve slurry yields compatible with those observed in maximum distillate operation in today's FCC units.

Introducing the MAB–FCC process in the refinery will involve a major restructuring of process interconnections and product blending. The standard FCC main product, gasoline, requires little adjustment to meet product specifications, which can be accomplished basically with a relatively mild hydro-desulfurization; however, an unstable, high sulfur mid-distillate will be produced that will require extra room in the distillate hydrotreater. The MAB-FCC light naphtha amounting to less than 15 wt% of the FCC feedstock, because of its low octane and high olefin content, may require either reprocessing, preferably in another FCC, or an octane booster for the gasoline pool. Decreased aromaticity will allow incorporating a larger fraction of LCO into the unstable current of mid-distillate to be hydrotreated and incorporated to the diesel pool and also affect the slurry fraction increasing viscosity somewhat and lowering density. Substituting standard FCC with MAB will involve some choices; it will no longer be possible to accomplish all of the production objectives that are taken for granted in today's FCC units.

Similarities between MAB–FCC and established thermal conversion processes like delayed coking go even further. The higher metal tolerance of the MAB catalyst creates an opportunity for running heavier and more contaminated feedstocks. Choosing a heavier feedstock will actually benefit the process by requiring lower reaction temperatures to achieve the same conversions. Processing atmospheric residue in the FCC in itself is beneficial for low LCO aromatics. Advantageously, the MAB–FCC process compared to thermal conversion will offer higher conversions of the VGO fraction hydrocarbons and greater flexibility.

ACKNOWLEDGMENTS

The authors would like to thank Petrobras for permission to publish this paper and the technicians from the catalyst preparation, catalyst evaluation, and gas chromatography laboratories at CENPES and Professor Sedran's group at INCAPE/CONICET–FIQ/UNL, Santa Fe, Argentina for their contributions.

REFERENCES

1. Anderson, P. C., Sharkey, J. M., and Walsh, R. P. *J. Inst. of Pet.* 58 (1972): 83.
2. ANP. Anuario Estatistico do Petroleo, Gas Natural e Biocombustiveis—2008, http://www.anp.gov.br
3. Boock, L. T., and Zhao, X. *American Chemical Society. Division of Petroleum Chemistry Preprints.* 211th ACS, New Orleans, March 1996.
4. De Lasa, H. I. U.S. Patent 5.102.628, 1992.
5. Dornelles, R. G. *1st International Symposium on Fuels, Bio-Fuels and Emissions.* AEA, São Paulo, May 2008.
6. Guibet, J. C. *Fuels and Engines: Technology, Energy, Environment*, Vol. 1, 220–345. Technip, Paris, 1999.
7. Kayser, J. U.S. Patent 6.069.012, 2000.
8. Mitchell, B. R. *Ind. Eng. Chem. Res. Dev.* 19 (1980): 209.
9. Passamonti, F. J., de la Puente, G., and Sedran, U. doi:10.1016/ j_cattod 2007.12.123, 2008.
10. *Technology Updates. Advanced Hydrotreating and Hydrocracking Technologies to Produce Ultra-Clean Diesel Fuel*, VII–24. Southeastern, PA: Hydrocarbon Publishing Company, 2004.
11. Venuto, P. V., and Habib, Jr., E. T. *Fluid Catalytic Cracking with Zeolite Catalyst.* New York: Marcel Dekker, 1990.

3 Catalyst Evaluation Using an ARCO Pilot Unit on North Sea Atmospheric Residue

Sven-Ingvar Andersson and Trond Myrstad

CONTENTS

3.1 INTRODUCTION

At the end of the 1970s Statoil cracked a North Sea atmospheric residue for the first time in M. W. Kellogg's circulating pilot unit in Texas [1]. This pilot unit was quite large, with a capacity of one barrel a day. The test in this pilot unit was very successful and showed that North Sea atmospheric residues were very suitable feedstocks for a residue fluid catalytic cracker, and that North Sea atmospheric residues gave very promising product yields.

Some years later Statoil decided to start a project within catalytic cracking in order to learn more about residue fluid catalytic cracking in general, and particularly about catalysts suitable for this process. The project started as a prestudy for the residue fluid catalytic cracker unit (FCCU) that Statoil was planning to build at the Mongstad refinery in Norway. The intention was to crack North Sea atmospheric residue directly, without first using a vacuum gas distillation tower followed by cracking of the vacuum gas oil. Therefore all tests in both MAT and pilot unit should be performed with North Sea atmospheric residues and not with vacuum gas oils. At this time Statoil asked Chalmers if it was possible to use the circulating ARCO pilot unit at the university for this purpose.

The ARCO pilot unit is a well-established pilot unit originally designed for vacuum gas oil as feed. This small circulating pilot unit is working at atmospheric pressure [2]. The common way to investigate residue feedstocks at that time, in the mid-1980s, was to mix them into vacuum gas oil and calculate their yields by comparison with the clean vacuum gas oil [3]. Statoil, however, wanted to study the atmospheric residue directly and this was not possible without some modifications of the ARCO pilot unit. The successful result of this was published some years later [4].

3.2 EXPERIMENTAL

The feeds used in all experiments presented in this paper are North Sea atmospheric residues originating from the atmospheric distillation tower at the Statoil Mongstad refinery in Norway. After the start-up of the residue fluid catalytic cracker at this refinery in 1989, the same feed has been used both in the commercial FCCU and in the ARCO pilot unit at Chalmers. Typical data for some North Sea atmospheric residue feeds used in the ARCO pilot unit are shown in Table 3.1.

TABLE 3.1
Typical North Sea Atmospheric Residues Used in the ARCO Pilot Unit at Chalmers

Feed Simulated Distillation	Feed (°C) Standard	Feed B (°C)	Feed C (°C)
0%	256	248	230
10%	373	342	338
30%	438	407	404
50%	481	459	457
70%	544	521	520
90%		619	618
100%		712	712
Density, kg/dm^3	0.922	0.9226	0.9275
Conradson carbon, wt%	2.8	2.6	3.0
Sulfur, wt%	0.69	0.44	0.40
Nickel, ppm	1.7	1.8	1.6
Vanadium, ppm	2.7	2.2	2.0
Sodium, ppm	—	0.8	0.6
Nitrogen (basic), ppm	—	450	420

The catalysts used in the pilot unit are both equilibrium catalysts from the FCCU at the Statoil Mongstad refinery, and impregnated and deactivated fresh catalysts from different vendors. The catalysts have been impregnated with nickel and vanadium naphthenates. The amount of metals has varied over the years, but the nickel to vanadium ratio has usually been 2:3. The deactivation procedure has also changed over the years, as new deactivation methods have been developed and existing deactivation methods have been improved.

The flue gas and the product gas are analyzed with a refinery gas analyzer from Varian. The liquid products are analyzed by simulated distillation on a Varian gas chromatograph.

3.3 RESULTS AND DISCUSSION

3.3.1 ARCO PILOT UNIT COMPARED WITH OTHER UNITS

Already during the first years a lot of important new information was gained. It was demonstrated that the decision to use the same feed for testing as was used in the commercial unit was correct [4,5] because the ranking of the catalysts was found to be dependent on the feed used in the tests. Despite the fact that it is much more troublesome and much more time-consuming to perform the tests with an atmospheric residue than with a vacuum gas oil, the tests should be performed with the same North Sea atmospheric residue as used in the commercial FCCU. The importance of this is always under discussion, but we are very firm at this point.

One important question we have asked ourselves many times since we managed to crack the North Sea atmospheric residue in the ARCO pilot unit for the first time was if the results are reliable. Does the pilot unit show the same trend and ranking as expected for the commercial FCCU, and can the yields from the pilot unit be used for modeling? Initially there was no answer to these questions since no commercial data were available for comparison. The only possibility was to compare the tests done in the ARCO pilot unit with the tests done in the pilot unit at M. W. Kellogg's some years earlier.

The test at M. W. Kellogg's and the test in the ARCO pilot unit were done with different feeds, with different catalysts and in different pilot units, so it was not expected that the yields should be identical. The feed to the M. W. Kellogg's pilot unit was a synthetic Statfjord atmospheric residue and the catalyst used was a Filtrol 900 catalyst containing nickel and vanadium contaminants [1]. This pilot unit was also pressurized. In the ARCO pilot unit at Chalmers the feed was a laboratory distilled Statfjord atmospheric residue and the catalyst was an almost metal-free EKZ equilibrium catalyst from Katalistiks. The ARCO unit is working at atmospheric pressure.

As can be seen from Table 3.2 there are some differences between the results of the two pilot units. The gas yields are close to each other, but the liquid product yields and the coke yields showed differences. But as was explained above, some differences between the results from the two units were expected. The observed differences were defined as acceptable, and we were satisfied with the results from cracking of atmospheric residue in the ARCO pilot unit for this time. Later

TABLE 3.2

Comparison of Pilot Unit Yields at Naphtha Maximum

Pilot Unit	ARCO	M. W. Kellogg
Dry gas (C_2-), wt%	1.7	2.2
LPG, wt%	13.7	15.6
Naphtha, wt%	53.7	47.2
LCO, wt%	12.5	20.0
HCO, wt%	13.4	7.0
Coke, wt%	4.7	8.0
H_2S, wt%	0.3	—

TABLE 3.3

Comparison between Commercial Data and ARCO Pilot Unit Data

Adjusted Yields (wt%)	ARCO (Max Naphtha)	Mongstad
Dry gas (C_2-), wt%	2.4	2.2
LPG, wt%	18.0	16.4
Naphtha, wt%	50.7	50.5
LCO, wt%	13.2	11.7
HCO, wt%	10.5	12.0
Coke, wt%	5.4	7.2

experiments in the ARCO pilot unit with catalysts similar to Filtrol 900 have also shown that this type of catalyst is not optimal for the North Sea atmospheric residue feeds, and give more coke, gas, and lower naphtha yields than a more optimal catalyst.

After the start-up of the residue FCCU at the Statoil Mongstad refinery, a lot of comparisons between results from the pilot unit and commercial data have been made. One such comparison [6] is shown in Table 3.3, and shows good agreement between results from the pilot unit and commercial data. It must be kept in mind that the pilot unit works at atmospheric pressure, while the commercial unit is slightly pressurized. This especially has an effect on the coke yield. The coke yield from the pilot unit is lower than the coke yield from the commercial unit due to the difference in pressure. The table also shows that the yield of dry gas is of the same magnitude in both units. However, the yield of LPG is higher in the pilot unit than in the commercial unit. The naphtha yield is also at the same level in the two units. The light cycle oil (LCO) yield is higher in the pilot unit than in the commercial unit, while the opposite is the case for the HCO yield. But if we add the yields of LCO and HCO in both units, the sums are close to each other for the ARCO pilot unit and the commercial unit.

It must also be remembered that the yields from the pilot unit are dependent on the deactivation conditions for the catalyst tested, such as the deactivation temperature and the metals level. Thus it is not always possible to achieve such a good accuracy as was demonstrated in this comparison.

Unknown new catalysts have to be treated in the same way as the reference catalyst and this is always an uncertainty factor in the test procedure, but this must be accepted if it should be possible to investigate new catalysts.

3.3.2 REPEATABILITY

Another important question that had to be answered was if the results in the ARCO pilot unit were repeatable when North Sea atmospheric residue is used as feed. The repeatability is well demonstrated in ARCO pilot units with vacuum gas oils, but this had to be confirmed for atmospheric residue feedstocks. To be able to achieve repeatability in the ARCO unit, independent of the feed used, the catalyst circulation has to be calibrated for each single catalyst the first time it is used in the pilot unit. If the same catalyst is used more than once, it is not necessary to calibrate the catalyst circulation again according to ARCO [7]. We have experienced the same, but nevertheless the calibration of the catalyst circulation is always checked by a single point measurement if a catalyst is used more than once.

Figures 3.1 through 3.6 show the repeatability of results in our ARCO unit when North Sea atmospheric residue is used as feed. The time between the first test and the repeatability check is about 2 years. As can be seen in Figures 3.1 through 3.6, the two tests give almost the same results in the ARCO pilot unit even with a North Sea atmospheric residue feed. This shows that the ARCO unit is just as suitable for a residue feed as for a vacuum gas oil feed.

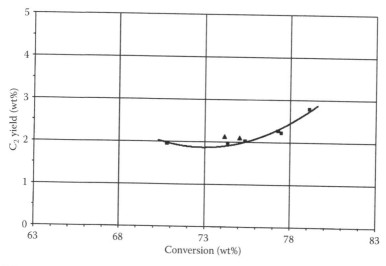

FIGURE 3.1 Yield of dry gas (C_2-) as a function of conversion. Test of repeatability ($\blacksquare$ = 2006, $\blacktriangle$ = 2008).

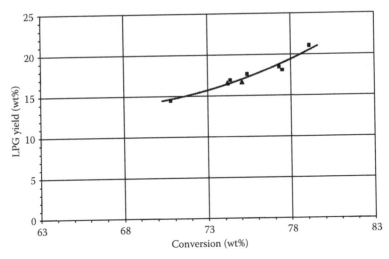

FIGURE 3.2   Yield of LPG as a function of conversion. Test of repeatability (■ = 2006, ▲ = 2008).

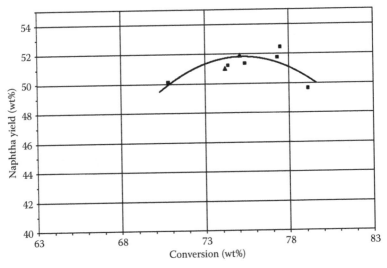

FIGURE 3.3   Yield of naphtha as a function of conversion. Test of repeatability (■ = 2006, ▲ = 2008).

3.3.3   DEACTIVATION OF CATALYSTS

When discussing the suitability of the ARCO pilot unit for cracking atmospheric residues, this cannot be done without touching on the question about how to prepare the catalysts for testing. An equilibrium catalyst used in a commercial residue FCCU contains significant amounts of metal contaminants, especially nickel and vanadium. Fresh catalysts must therefore be impregnated with these metals and deactivated before the catalysts can be used in the pilot unit. We have shown that this

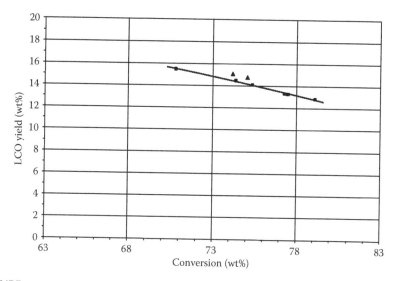

FIGURE 3.4   Yield of LCO as a function of conversion. Test of repeatability (■ = 2006, ▲ = 2008).

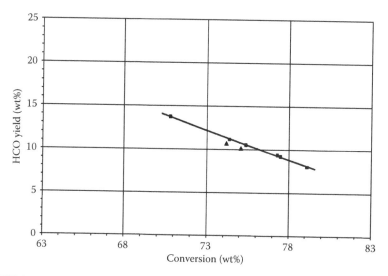

FIGURE 3.5   Yield of HCO as a function of conversion. Test of repeatability (■ = 2006, ▲ = 2008).

is necessary and that the ranking of the catalysts can depend on it [4,5]. When our project was started 25 years ago, the only method available was the Mitchell method [8]. This method consists of a volumetric impregnation of the catalyst with nickel and vanadium metals followed by a steam deactivation of the catalyst. It has been pointed out in the literature that the metals should meet oxygen sometime during the deactivation period before being used as cracking catalyst [9]. This was achieved

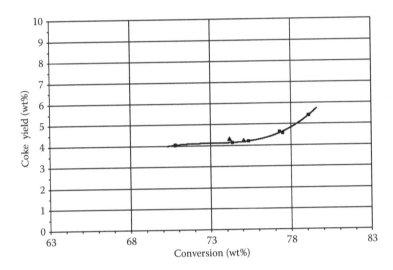

FIGURE 3.6 Yield of coke as a function of conversion. Test of repeatability ($\blacksquare$ = 2006, $\blacktriangle$ = 2008).

in the pilot unit by circulating the catalyst for some days in the pilot unit to reach particle size equilibrium in the catalyst inventory before calibration of the catalyst circulation. Later new methods for impregnation and deactivation of the catalyst were presented such as the cyclic deactivation method (CD) by Albemarle [10], and the cyclic propylene steaming method (CPS) by Grace Davison [11]. Both methods have been improved [12,13], but still neither of them can mimic the equilibrium catalyst completely. The best impregnation method so far seems to be the CD method developed by Shell [14].

We have been using the CPS method since it was published by Grace Davison. The method is simple, and consists of a volumetric impregnation of the catalyst followed by a cyclic ReDox deactivation in 50% steam at constant temperature.

For application in the ARCO pilot unit, quite a large batch of catalyst, 3 kg, has to be impregnated and deactivated. However, the characteristics of the deactivation procedure are influenced of how the volumetric metal impregnation is performed. The impregnation is not a straight forward procedure, because it is difficult to get a homogeneously impregnated catalyst when impregnating such a large batch. From the beginning of the project we started to impregnate the catalyst in three steps as our standard impregnation method; with impregnation of one-third of the metals each time and with each step divided into three 1 kg batches. Between each step the solvent is evaporated. This is time-consuming, but it gives a homogeneously impregnated catalyst and the impregnation step becomes reproducible. We have tried to impregnate the catalyst directly in one step, but this resulted in a very uneven impregnated catalyst sample. This was demonstrated by measuring the amount of metals at different locations of the catalyst batch after impregnation, see Table 3.4.

Despite the different impregnation methods, the surface areas are close to each other as can be seen in Table 3.5. The direct impregnated catalyst has a somewhat lower zeolite surface area than the standard impregnated catalyst but the matrix surface areas are equal for both catalysts.

When the catalysts were evaluated in the pilot unit it was found that the product yields were influenced by the selected impregnation method. The maximum naphtha yield (50.7 wt%) occurred at 76.5 wt% conversion for the standard impregnated catalyst. For the direct impregnated catalyst the maximum naphtha yield (50.9 wt%) occurred at a somewhat lower conversion (75.8 wt%). The comparison of product yields from the two impregnation methods shown in Table 3.6 is made at constant conversion (76.5 wt%). The hydrogen yield as well as the dry gas yield is slightly

TABLE 3.4
Metal Content of Two Different Samples of Direct Impregnated Catalyst

	Sample 1	Sample 2	Target
Nickel, ppm	1480	680	1440
Vanadium, ppm	1780	760	2160
Total, ppm	3260	1440	3600

TABLE 3.5
Catalyst Surface Areas after Different Impregnation Methods

	Standard Impregnation	Direct Impregnation
Surface area, m²/g	180	175
Zeolite SA, m²/g	143	138
Matrix SA, m²/g	37	37
ZSA/MSA ratio	3.83	3.68

TABLE 3.6
Product Yields of the Impregnated Catalysts at 76.5 wt% Conversion

	Standard Impregnated	Direct Impregnated
Hydrogen, wt%	0.07	0.08
Dry gas, wt%	2.4	2.5
LPG, wt%	18.0	17.9
Naphtha, wt%	50.7	50.7
LCO, wt%	13.2	13.7
HCO, wt%	10.5	9.8
Coke, wt%	5.4	5.6

higher for the direct impregnated catalyst than for the standard impregnated one. The LPG yield, however, was slightly lower for the direct impregnated catalyst compared with the standard impregnated catalyst. At the chosen conversion the naphtha yields are equal for both impregnation methods. The coke yield was also higher for the direct impregnated catalyst compared with the standard impregnated catalyst. The differences in the yields between the two impregnation methods can only be explained by the uneven impregnation of the direct impregnated catalyst. Therefore it is recommended to do the impregnation in the best possible way because the results depend on how the impregnation is done.

We have proved that the ARCO pilot unit is repetitive when it is used with an atmospheric residue feed. But it is necessary to impregnate and deactivate the catalysts in a repetitive way if this should be the case.

3.3.3.1 Ranking of Catalysts

One of the objectives when our project started was to use North Sea atmospheric residues as feeds in the ARCO pilot unit. It was comprehensively shown during the very first years of the project that the ranking of the catalysts were dependant on the feed used [4,5]. The results of the tests done at this time showed that all the catalysts tested could be divided into three groups, all with their own characteristics, depending on the matrix surface area.

1. Catalysts with low matrix surface areas
2. Catalysts with medium matrix surface areas
3. Catalysts with large matrix surface areas

Catalysts belonging to the first group performed very poorly in the pilot unit. Most of them coked the injection part of the pilot unit completely and the catalyst circulation stopped after a few minutes. At that time this was very confusing, but it could be explained at a later time when the optimization studies were completed. The explanation could be twofold. One reason could be that the matrix surface area was smaller than what was necessary for the severity used in the pilot unit. Another reason could be that these catalysts also had a low RE content and North Sea atmospheric residues need a high RE content in the catalyst to perform well [4,5].

The second group was characterized by well-performing catalysts with high naphtha yields combined with low yields of coke and gas. At that time this was rather unexpected, since it was commonly accepted in those days that a residue catalyst should have a medium zeolite content and a high matrix surface area [15]. Obviously more studies were necessary within this field.

The yields from catalysts belonging to the second group should be compared with those from the catalysts in the third group. Catalysts in the third group were all excellent residue catalysts and performed well in the pilot unit. But compared with catalysts from the second group they all gave lower naphtha yield and higher yields of coke and gas. It is clear that there was a need for an optimization of the catalyst performance in order to achieve the highest possible naphtha yield when North Sea atmospheric residues are used as feed to a FCCU.

The differences between catalysts from Group 2 and Group 3 are illustrated in the following figures, using test results from two typical catalysts from Group 2 and one catalyst from Group 3. The surface areas for these catalysts are shown in Table 3.7.

As can be seen in Figures 3.7 through 3.13, the two catalysts from Group 2, Medium-A and Medium-B, with medium matrix surface areas, were more active than the catalyst from Group 3, Large, with large matrix surface area. This was in line with the previous results, and confirmed that our observations were valid for all tested Group 2 and Group 3 catalysts. The naphtha yields were also higher for catalysts with medium matrix surface areas compared with catalysts with large matrix surface areas; a fact that was much unexpected and indicated that the old guidelines had to be modified when North Sea atmospheric residue was used as feed. The yield

TABLE 3.7

Surface Areas of Two Typical Catalysts from Group 2 and One Catalyst from Group 3 (3000 ppm Metals)

	Group 2 Medium-A	Group 2 Medium-B	Group 3 Large
Surface area, m²/g	125	142	117
Zeolite SA, m²/g	83	97	42
Matrix SA, m²/g	42	46	75
ZSA/MSA ratio	1.96	2.12	0.56

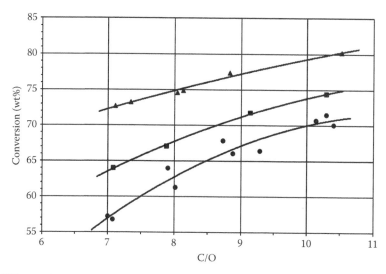

FIGURE 3.7 Conversion as a function of C/O (■ = Medium-A, ▲ = Medium-B, ● = Large).

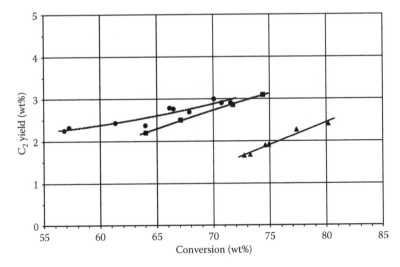

FIGURE 3.8 Yield of dry gas (C_2-) as a function of conversion (■ = Medium-A, ▲ = Medium-B, ● = Large).

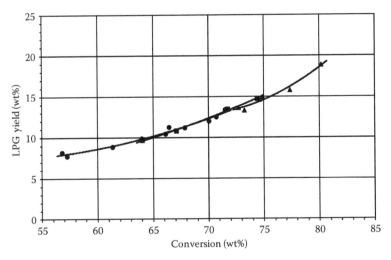

FIGURE 3.9 Yield of LPG as a function of conversion (■ = Medium-A, ▲ = Medium-B, ● = Large).

differences between the two catalysts, Medium-A and Medium-B, from Group 2 also indicated that there was a need for optimization of the residue catalysts with respect to zeolite and matrix surface areas.

3.3.4 INFLUENCE OF RE CONTENT

It has been observed that catalysts aimed for cracking of North Sea atmospheric residues need a high RE content. A fully RE exchanged catalyst was able to crack

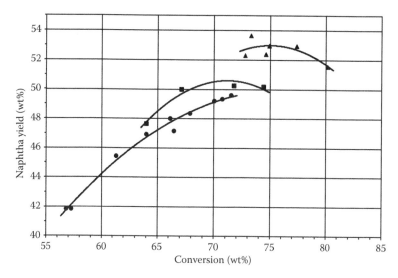

FIGURE 3.10   Yield of naphtha as a function of conversion ($\blacksquare$ = Medium-A, $\blacktriangle$ = Medium-B, $\bullet$ = Large).

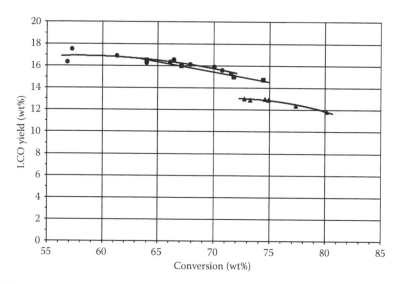

FIGURE 3.11   Yield of LCO as a function of conversion ($\blacksquare$ = Medium-A, $\blacktriangle$ = Medium-B, $\bullet$ = Large).

the North Sea atmospheric residue without any problem in the pilot unit, while a half RE exchanged catalyst did not [4]. It has also been observed during the years that catalysts with low RE content have given much higher coke yield; causing coke plugging of the injection zone shortly after start-up, and making it impossible to maintain catalyst circulation in the pilot unit. A test of two similar catalysts, but with different RE levels, 3.0 wt% and 3.7 wt%, showed that the pilot unit is sensitive to changes

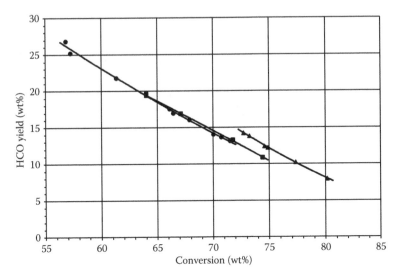

FIGURE 3.12   Yield of HCO as a function of conversion (■ = Medium-A, ▲ = Medium-B, ● = Large).

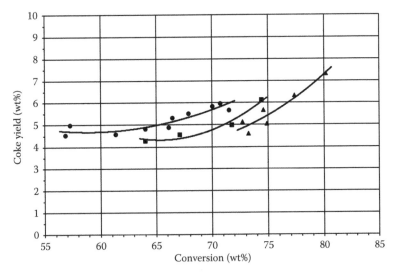

FIGURE 3.13   Yield of coke as a function of conversion (■ = Medium-A, ▲ = Medium-B, ● = Large).

in the RE level and that the pilot unit responds as expected to these changes. The surface areas for the two catalysts are shown in Table 3.8.

The catalyst activity increased when the RE level increased as expected [16]. Product yields compared at the same conversion (76.2 wt%) are shown in Table 3.9. When the RE level increased, the decreased LPG yield was as expected, as well as the increased naphtha yield and the decreased LCO yield. The decreased coke yield was also expected because of the behavior of the catalysts with low RE content in the pilot unit.

TABLE 3.8
Surface Areas for Catalysts with
Different RE Content (3600 ppm Metals)

	RE 3.0	RE 3.7
Surface area, m²/g	197	180
Zeolite SA, m²/g	159	143
Matrix SA, m²/g	38	37
ZSA/MSA ratio	4.2	3.9

TABLE 3.9
Yields from Two Different RE Levels at
Constant Conversion (76.2 wt%)

	RE 3.0	RE 3.7
Coke, wt%	5.0	4.4
Dry gas (C_2-), wt%	2.0	2.2
LPG, wt%	18.4	17.8
Naphtha, wt%	50.9	51.6
LCO, wt%	14.3	14.0
HCO, wt%	9.4	10.0
Hydrogen, wt%	0.08	0.09

All the observed yield responses to changes in the RE content of the catalysts were in the same direction as those observed for vacuum gas oil reported in the literature. This indicated that these changes are true also for North Sea atmospheric residue.

3.4   NEED FOR OPTIMIZATION

It was obvious that the catalysts had to be optimized for North Sea atmospheric residues. In order to find a more useful catalyst than the reference catalyst, two new catalysts were tested. The first one, Catalyst B, was selected based on results from the previous tests; the matrix surface area was reduced to an optimal size. The zeolite surface area was however kept constant. The second new catalyst, Catalyst C, was selected according to the old general recommendation for residue catalysts, and both the zeolite and matrix surface areas were increased compared to the reference catalyst. The surface areas for the three catalysts are shown in Table 3.10.

The reference catalyst A and Catalyst B had both a low activity compared with catalyst C as shown in Figure 3.14. One explanation for this might be the low zeolite surface area for both the reference catalyst and for Catalyst B. Catalyst C had the highest activity of the three catalysts because of its high zeolite surface area and despite its high matrix surface area.

TABLE 3.10

Catalyst Surface Areas after Deactivation (3000 ppm Metals)

	A-Reference	Catalyst B	Catalyst C
Surface area, m²/g	138	99	193
Zeolite SA, m²/g	63	59	108
Matrix SA, m²/g	75	41	86
ZSA/MSA ratio	0.84	1.44	1.26

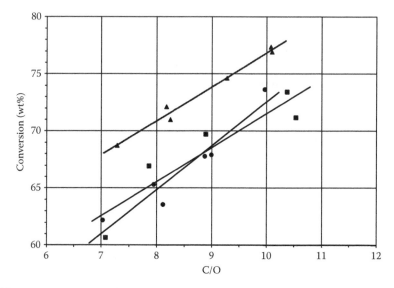

FIGURE 3.14   Influence of catalyst parameters. Conversion as a function of C/O (● = A, ■ = B, ▲ = C).

But as can be seen in Figure 3.15 the large matrix surface area for Catalyst C resulted in the highest dry gas yield of the three catalysts, while the reference catalyst and Catalyst B showed comparable dry gas yields. The hydrogen yield is often of special interest because of its large volume, and in this test it was obvious that Catalyst B with the low matrix surface area gave the lowest hydrogen yield of the three catalysts. The hydrogen yield of Catalyst C was slightly lower than that of the reference.

The LPG yield was lowest for the reference catalyst and highest for Catalyst B, see Figure 3.16.

The naphtha yield was lower for Catalyst B than for the reference and this illustrates the necessity to have enough zeolite surface area in the catalyst to be able to crack all the components in the feed, both those that can be cracked directly and those that must be precracked on the matrix before they can be cracked by the zeolite. Catalyst C had a slightly higher naphtha maximum than the reference catalyst, despite its high matrix surface area. The high matrix surface area of Catalyst C,

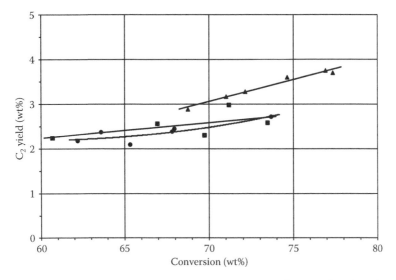

FIGURE 3.15   Influence of catalyst parameters. Dry gas yield (C_2-) as a function of conversion ($\bullet$ = A, $\blacksquare$ = B, $\blacktriangle$ = C).

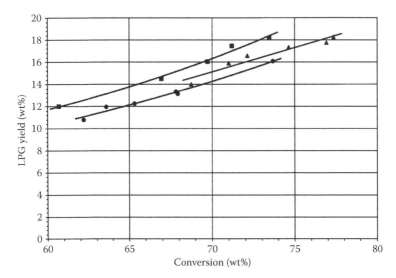

FIGURE 3.16   Influence of catalyst parameters. LPG yield as a function of conversion ($\bullet$ = A, $\blacksquare$ = B, $\blacktriangle$ = C).

however, generated too much dry gas and this indicated that the matrix surface area should not be too high when cracking North Sea atmospheric residue feeds, see Figure 3.17.

However, the high matrix surface area of catalyst C made it possible to crack more heavy components than the other two catalysts, but the matrix cracking was too intense for this catalyst. Catalyst B showed the highest HCO yield of the

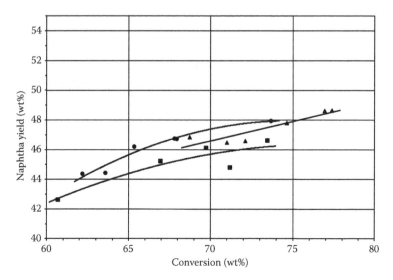

FIGURE 3.17 Influence of catalyst parameters. Naphtha yield as a function of conversion (● = A, ■ = B, ▲ = C).

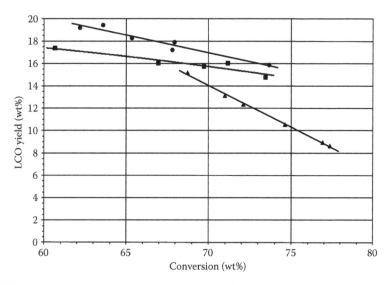

FIGURE 3.18 Influence of catalyst parameters. LCO yield as a function of conversion (● = A, ■ = B, ▲ = C).

tested catalysts and a lower LCO yield than catalyst A, one reason for this might be the low zeolite content of this catalyst be. There is not enough zeolite in this catalyst to crack all the precracked components and consequently some of them are found in the HCO fraction, see Figures 3.18 and 3.19.

The coke yield was highest for the reference catalyst and this might be explained by the fact that the matrix surface area was high but the zeolite surface area was too low to crack all the precracked molecules. These precracked molecules could then

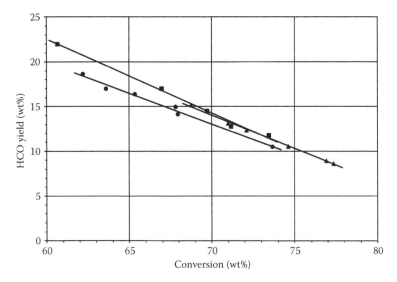

FIGURE 3.19 Influence of catalyst parameters. HCO yield as a function of conversion
(● = A, ■ = B, ▲ = C).

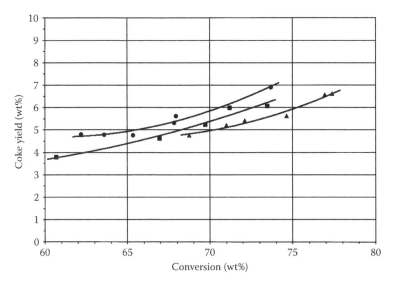

FIGURE 3.20 Influence of catalyst parameters. Coke yield as a function of conversion
(● = A, ■ = B, ▲ = C).

generate coke on the matrix if the zeolite could not crack them to transportation
fuels. For catalyst B the matrix surface area was low enough, but the zeolite content
was too small to crack all the precracked products and consequently the coke yield
was only slightly lower than for the reference catalyst. Catalyst C had the lowest coke
yield of these three catalysts, see Figure 3.20.

This investigation showed that there was a large potential for optimizing the catalyst for the North Sea atmospheric residue feed in order to produce as much valuable products as possible from the feed. It was also obvious that this optimization work for the North Sea atmospheric residue had to be done in the pilot unit [5].

3.5 FEED TESTS

The ARCO pilot unit is also used for evaluation of different atmospheric residue feeds. Reference catalysts are used in these investigations. The results from one such test is presented here, where two residue feeds B and C (see Table 3.1) are compared with each other. The catalyst used in this study was an equilibrium catalyst, see Table 3.11.

The two residue feeds B and C have almost identical boiling point distribution, but the density and the Conradson carbon content value are somewhat higher for feed C than for feed B. This indicates that feed C should be a little bit more difficult to crack than the B feed and this was also notified when the two feeds were to be tested in the pilot unit.

As can be seen in Figure 3.21 the conversion was lower for feed C than for feed B at a given C/O ratio. This means that the test in the pilot unit confirmed that feed C was more difficult to crack than feed B just as the feed analysis data indicated.

Both the dry gas yield and the LPG yield showed very little difference between the two feeds and therefore only the total gas yields are shown in Figure 3.22.

However, the naphtha yield for the somewhat heavier feed C was lower than the naphtha yield of feed B, especially at the naphtha maximum around 75 wt% conversion, see Figure 3.23.

Both feeds had similar LCO and HCO yields at conversions around their naphtha maxima but at lower conversions the heavier feed C gave lower LCO yield and higher HCO yield than the lighter feed B, see Figures 3.24 and 3.25.

As can be seen in Figure 3.26, the coke yield was higher for the heaviest feed.

TABLE 3.11
Characterization of Equilibrium
Catalyst Used for Comparison of
Feed-B and Feed-C

	Equilibrium Catalyst
Surface area, m²/g	136
Zeolite SA, m²/g	112
Matrix SA, m²/g	25
ZSA/MSA ratio	4.48
Nickel, ppm	2870
Vanadium, ppm	3830

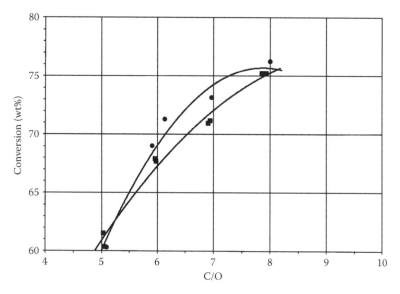

FIGURE 3.21 Comparison of Feed-B and Feed-C. Conversion as a function of C/O (● = Feed-B, ■ = Feed-C).

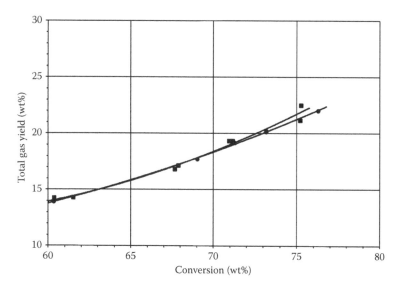

FIGURE 3.22 Comparison of Feed-B and Feed-C. Total Gas yield as a function of conversion (● = Feed-B, ■ = Feed-C).

If the individual hydrocarbon gases from methane to butanes are investigated, it can be observed that they all behaved as expected; they increased with increasing conversion. However, the butylenes go through a maximum and this might be confusing at first sight, but results from the Mongstad FCCU also show the same behavior. The isobutylene yield from this unit goes through the same maximum as

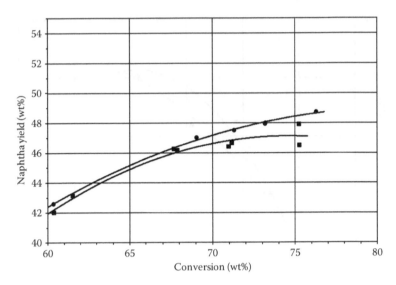

FIGURE 3.23 Comparison of Feed-B and Feed-C. Naphtha yield as a function of conversion (● = Feed-B, ■ = Feed-C).

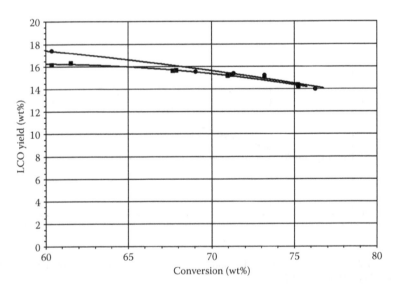

FIGURE 3.24 Comparison of Feed-B and Feed-C. LCO yield as a function of conversion (● = Feed-B, ■ = Feed-C).

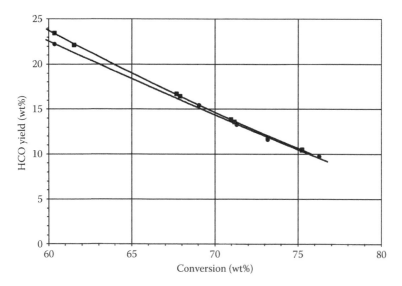

FIGURE 3.25   Comparison of Feed-B and Feed-C. HCO yield as a function of conversion (● = Feed-B, ■ = Feed-C).

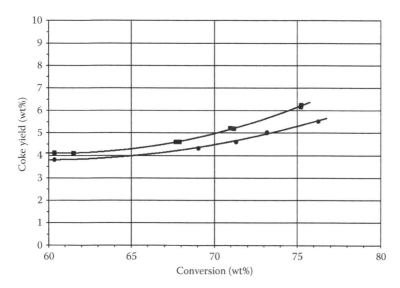

FIGURE 3.26   Comparison of Feed-B and Feed-C. Coke yield as a function of conversion (● = Feed-B, ■ = Feed-C).

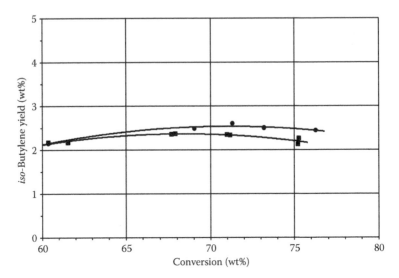

FIGURE 3.27 Comparison of Feed-B and Feed-C. Isobutylene yield as a function of conversion (● = Feed-B, ■ = Feed-C).

the isobutylene yield from the ARCO pilot unit, see Figure 3.27. This behavior of butylenes are typical for catalysts with a high RE content [17].

3.6 CONCLUSIONS

The ARCO pilot unit has shown to be a versatile tool for studying cracking of North Sea atmospheric residues. The yields from this unit were comparable with the yields from the FCCU at the Statoil Mongstad refinery on one occasion.

The pilot unit responds as expected to changes in catalyst parameters and on differences in the feed composition. The ARCO pilot unit shows the correct behavior of all independent gas component yields. The yield of butylenes goes through the same maximum as has been observed in the FCCU at the Mongstad refinery.

The zeolite and matrix surface areas of the catalyst can be optimized using the ARCO pilot unit with North Sea atmospheric residue feeds.

ACKNOWLEDGMENT

The authors are grateful to Statoil for permission to publish this paper.

REFERENCES

1. Torgaard, H. *Oil Gas J.* 81, no. 2 (1983): 100.
2. Wachtel, S. J., Baillie, L. A. Foster, R. L., and Jacobs, H. E. Laboratory Circulating Fluid Bed Unit for Evaluating Carbon Effects on Cracking Catalyst Selectivity, Preprint, Div. of Petroleum Chemistry. *ACS* 16, no. 3 (1971).
3. Otterstedt, J.-E., Mattsson, M., and Röj, A. *Erdöhl und Kohle* 39, no. 2 (1986): 69.

4. Andersson, S.-I., and Otterstedt, J.-E. *Catalytic Cracking of North Sea Resid.* Katalistik's 8th Annual Fluid Cat Cracking Symposium, Budapest, Hungary, June 1–4, 1987, Paper 21.

5. Andersson, S.-I., and Myrstad, T. *Stud. Surf. Sci. Catal.* 166 (2007): 13–29.

6. Gledhill, D., and Pedersen, J. *Operating Experience with the New Riser Termination Technology.* Grace Davison FCC Technology Conference, Lisbon, Portugal, September 1–4, 1998, Paper 14.

7. Humes, W. H. *CEP* February 1983, 51.

8. Mitchell, B. R. *Ind. Eng. Chem. Prod. Res. Dev.* 19 (1980): 209.

9. Hettinger, Jr., W. P., Beck, H. W., Cornelius, E. B., Doolin, P. K., Kmecak, R. A., and Kovach, S. M. *Oil Gas J.* 82, no. 14 (1984): 102.

10. Gerritsen, L. A., Wijngaards, N. J., Verwoert, J., and O'Connor, P. *Akzo Catalyst Symposium*, 1991, 109–124.

11. Cheng, W. C., Juskelis, M. V., and Suarez, W. AICHE Annual Meeting, Miami Beach, FL. November 1–6, 1995.

12. Pimenta, R., Quiñones, A. R., and Imhof, P. *Akzo Nobel Catalyst Symposium,* Paper F-6, 1998.

13. Wallenstein, D., Harding, R. H., Nee, J. R. D., and Boock, L. T. *Appl. Catal. A*, 204 (2000): 89.

14. Mercera, P. D. L., Dirkx, J. M. H., and Hadjigeorge, G. *Akzo Nobel Catalyst Symposium,* Paper F-7, 1998.

15. *Oil Gas J.* 94, no. 42 (1996): 50.

16. *Oil Gas J.* 84, no. 42 (1986): 55.

17. Borodzinski, A., Corma, A., and Wojciechovski, B. W. *Can. J. Chem. Eng.* 58 (1980): 219.

4 Pilot Unit Test of Residue Type Catalysts on North Sea Atmospheric Residue

Sven-Ingvar Andersson and Trond Myrstad

CONTENTS

4.1   INTRODUCTION

The fluid catalytic cracker unit (FCCU) is a very important unit in a refinery. This unit converts low value heavy oil to transportation fuels of a substantial higher value. In order to optimize the catalytic process, it is important to select the catalysts with great care. To maximize the profit, the selected catalyst should fit both the feedstock to the FCC unit as well as the refinery yield demands. However, the feedstocks and the yield demands change continuously and it is therefore necessary to pay particular attention to these issues. For a residue catalytic cracker, like the Statoil unit at the Mongstad refinery in Norway, it is even more important. This FCC unit came on stream in 1989 with a design capacity of 250 t/h [1]. Some years later the rated capacity was increased to 325 t/h [2]. The feed to the catalytic cracker at the Mongstad refinery is mainly a North Sea atmospheric residue [3] and the objective has been to select the most optimal catalyst for this unit; most of the time for maximum naphtha production. For this reason Statoil has had an on-going test and research activity together with Chalmers University in Sweden for many years, using the circulating ARCO pilot unit at the university.

Within the field of catalytic cracking, the phrase residue is commonly used for a number of different hydrocarbon fractions. Heavy vacuum gas oils with a high-Conradson carbon residue (CCR) is sometimes called a residue even though its content of compounds boiling above 550°C is very low. Other residues are the atmospheric tower bottoms (long resid) and the vacuum tower bottoms (short resid). These two types of residues differ from the vacuum gas oil by its much higher boiling range. Substantial parts of the residues are boiling above 550°C, contain a lot of naphthenes, poly aromatics, resins, and asphaltenes, and have a higher CCR than vacuum gas oils. The metals content might also be high, as well as the sulfur and nitrogen contents. Also within each type of residue there are large differences. With this insight it is obvious that different types of catalysts are necessary for optimal cracking of each type of feed.

All feeds require an optimized catalyst for the optimal conversion to lighter and more valuable products. This insight has always concerned FCC professionals. Even with vacuum gas oil as feed the optimization problem was evident. Wear and Mott used a MAT reactor to optimize the zeolite to matrix surface area ratio (ZSA/MSA) for a vacuum gas oil catalyst [4]. The naphtha yield increased with increasing ZSA/MSA ratio, while the coke and dry gas yields decreased. This investigation showed that the optimization of the catalyst indeed was necessary and was very profitable even when vacuum gas oil was used as feed to the catalytic cracker.

An atmospheric residue, however, contains a lot of large molecules boiling above 550°C. Such residual molecules have a carbon number of more than 35 and a molecular size in the range of approximately 10–25 Angstrom. The molecular shape, however, depends on the source of the feed. An aromatic feed has, for example, a more voluminous molecular shape than a paraffinic feed [5] and this influences the diffusion into the catalyst pore system. The diffusion into the pore system is limited by the molecular size; the larger the molecule is the smaller is the penetration depth of the molecule [6]. When cracking atmospheric residue, the large molecules has to be precracked on the matrix before the smaller parts then formed can enter the zeolite supercage to be selectively cracked to transportation fuels. It has been shown that an optimum ratio between the zeolite and matrix surface areas must exist for the maximum formation of naphtha (the most desirable product) in a system like this [7]. The pore structure of the catalyst must fit the feed molecules to get access to the active sites of the matrix. Accessibility, however, is not enough; crackability is just as important. The crackability should be optimized by controlling the strength of the acidic sites [8].

The surface area of the catalyst as well as the pore size distribution can easily be measured, and the zeolite and matrix surface areas of the catalyst can be determined by the t-plot method. The different FCC yields can then be plotted as a function of the ZSA/MSA ratio, zeolite surface area or matrix surface area, and valuable information can be obtained [9]. The original recommendation was that a residue catalyst should have a large active matrix surface area and a moderate zeolite surface area [10,11]. This recommendation should be compared with the corresponding recommendation for a VGO catalyst; a VGO catalyst should have a low-matrix surface area in order to improve the coke selectivity and allow efficient stripping of the carbons from the catalyst [12]. Besides precracking the large molecules in the feed, the matrix also must maintain the metal resistance of the catalyst.

It is obvious that a residue catalyst must be optimized in order to get optimum process economy. The contradiction between the recommendations for optimum performance of VGO and residue catalysts, as well as our own information, also indicated that this was necessary. The optimization investigation was performed with two types of commercial catalysts and with a North Sea atmospheric residue as feed. The results presented in this paper are expected to be dependent on both the types of catalyst and of the residue feed used.

This paper gives a review of earlier presented results [9,13]. The main optimization work was performed with commercial catalysts from two different vendors, but with comparable properties and behavior, such that the results could be used together [13]. In addition, another completely different type of catalyst was optimized as well, in order to show that the optimization technique and the guidelines also could be used on this type of catalysts, even though the figures were different. It has also been shown that there is a lower limit for the matrix surface area of a catalyst type; below this lower limit the catalyst will not be able to precrack all the large components in the feed, and the pore system will be filled with coke. These results have been extended in the present investigation with findings that show there also is an upper limit for the ZSA/MSA ratio for a residue catalyst.

4.2   EXPERIMENTAL

4.2.1   FEED

A paraffinic North Sea atmospheric residue was used as feed, see Table 4.1. This feed is representative for the feedstock used in the catalytic cracker at the Mongstad refinery.

4.2.2   METAL IMPREGNATION AND DEACTIVATION OF THE CATALYSTS

The catalysts were calcined at 600°C for 2 hours, and impregnated with nickel and vanadium naphthenates according to Mitchell [14]; with a nickel to vanadium ratio

TABLE 4.1
Characterization of the North Sea Atmospheric Residue Feed

		Distillation	°C
Density, kg/l	0.922	0%	256
CCR, wt%	2.8	5%	341
Aniline point, °C	89.7	10%	373
Sulfur, wt%	0.48	20%	412
Nickel, ppm	1.7	30%	438
Vanadium, ppm	2.7	40%	458
Nitrogen (total), ppm	2000	50%	481
		60%	508
		70%	544
		75%	567

of 2–3. For the catalysts used in the optimization study, the total metal level was 3000 ppm. The catalysts were deactivated with 100% steam at 760°C for 16 hours. The catalysts used for maximization of the ZSA/MSA ratio were impregnated with 3600 ppm Ni + V and deactivated according to the CPS-1 method [15] at 790°C.

4.2.3 CATALYST CHARACTERIZATION

Three types of commercially available FCC catalysts were used in this investigation. In the first group were catalysts A1–A4 supplied by one vendor, and catalysts A5–A8 from a second vendor. In the overlapping region, the catalysts from the two different vendors showed similar physical properties and product yields [13]. The second group contained three catalysts, B1–B3, all from a third vendor. The third group also contained three catalysts, C1–C3, all from one vendor.

The catalysts were characterized by measuring the total surface area, BET area, and the pore size distribution by nitrogen adsorption in a Micromeritics ASAP 2010 unit. The zeolite and matrix surface areas were calculated by the t-plot method [16,17]. For data see Table 4.2.

The zeolite unit cell size was determined by X ray diffraction according to ASTM-D-3942-80 at SINTEF (SINTEF, Oslo, Norway). For data see Table 4.2.

4.2.4 ARCO PILOT PLANT TESTS

The tests were performed in a modified circulating ARCO pilot unit at Chalmers [18,19] with a North Sea atmospheric residue as feed. The reactor temperature was 500°C and the regenerator temperature was 700°C. Each of the catalyst tests were

TABLE 4.2
Catalyst Characterization

Catalyst	Pore Volume cc/g	Total Area m²/g	Zeolite Area m²/g	Matrix Area m²/g	ZSA/MSA Ratio
A-1	0.22	138	63	75	0.84
A-2	0.20	100	59	41	1.44
A-3	0.30	193	108	86	1.26
A-4	0.37	164	124	40	3.10
A-5	0.35	155	119	36	3.31
A-6	0.31	151	122	30	4.07
A-7	0.40	144	109	35	3.10
A-8	0.28	109	87	22	3.95
B-1	0.29	142	97	46	2.11
B-2	0.38	115	42	73	0.58
B-3	0.45	171	79	92	0.86
C-1	n.a.	127	91	36	2.53
C-2	n.a.	157	122	35	3.49
C-3	n.a.	180	143	37	3.86

run at four different catalyst to oil (C/O) ratios. Flue gases and product gases were continuously collected for analysis with a refinery gas analyzer. Liquid products were collected and analyzed by simulated distillation. The following cutpoints were used:

Gasoline	C_5 to 216°C
LCO	216–344°C
HCO	344°C +

Mass balance, yields, and conversion were calculated. The yield to conversion functions were determined using linear regression. Tests with mass balances between 95 and 99 wt% were accepted.

4.3   RESULTS AND DISCUSSION

Even though the catalysts tested had a wide variation in rare earth (RE) content, they showed very little variation in unit cell size (UCS). No apparent relationship between RE content and UCS was found, and as a result, variations of UCS and RE are not further discussed in this paper.

To be able to select an optimal residue catalyst, many parameters have been proposed; such as the pore volume, the total surface area, and the zeolite to matrix surface area ratio (ZSA/MSA). But the only strong correlation we have found between the catalyst performance and physical properties when North Sea long residue has been used as feed, is between the ZSA/MSA ratio and the catalyst performance [13].

According to the literature, an optimal long residue catalyst should have a pore volume higher than 0.30 cc/g [20]. This limit is however very vague, and we have found catalysts with a pore volume as low as 0.20 cc/g that have performed well in the pilot riser, and opposite is a catalyst with a pore volume of 0.34 cc/g that did not [21]. We have found a slight correlation between the pore volume and the matrix surface area, but not between the pore volume and the zeolite surface area. Due to this lack of correlation, it is not possible to use the pore volume for predicting the performance of a long residue catalyst for our application.

The total surface area of a FCC catalyst is the sum of the zeolite and matrix surface areas and is therefore not useful for optimizing of the catalysts. However, the ratio between the zeolite and the matrix surface areas (ZSA/MSA) is a valuable parameter, and has been used for optimization of vacuum gas oil catalysts [4] as well as catalysts for North Sea long residue feeds [9,13]. Additional information about the catalyst is also gained by studying the yields as a function of the zeolite surface area and as a function of the matrix surface area [9]. The regression analysis in this paper is performed at a constant conversion of 75 wt%.

The objective with our test and research activities has always been to find catalysts with a high-optimum yield of naphtha for the commercial FCC unit at the Mongstad refinery when processing North Sea atmospheric residues. Quite early in our work we became aware that one solution to this issue could be to study the yields as a function of the ZSA/MSA ratio, and in addition also look at the yields as a function of both the zeolite and matrix surface areas. As can be seen in Figure 4.1a the naphtha yield increased for catalysts of type A when the ZSA/MSA ratio increased. This was in line with the VGO results published by Wear and Mott [4], but not in line with the common recommendations for residue catalysts [10,11]. Figure 4.1b also indicated that the zeolite surface area should be high, and Figure 4.1c showed that the matrix surface area should be kept low. But the matrix must be able to precrack all the heavy components in the feed to smaller components that can enter the zeolite supercage for selective cracking to naphtha. As shown by testing catalyst A-8 (see Table 4.2) there was a lower limit for the matrix surface area; below this limit the matrix was not able to crack all the heavy components in the feed, and the catalyst function was blocked.

Additional support for our observations was found when catalysts A-1 to A-3 were studied. Catalyst A-1 was developed according to the old recommendations for a residue catalyst; with a moderate zeolite surface area and a large active matrix surface area. The catalyst did not give as good naphtha selectivity as expected when the North Sea long residue feed was cracked. An attempt to improve this was made with catalyst A-2 where the matrix surface was lowered, while the zeolite surface area was kept the same. The naphtha selectivity was however not improved, and it was concluded that the zeolite surface area was too low. So in catalyst A-3 the zeolite surface area instead was increased compared with the base catalyst A-1. Now the naphtha selectivity increased, but the gas yields also increased dramatically. This catalyst did indicate that a possible way to go could be to increase the zeolite surface

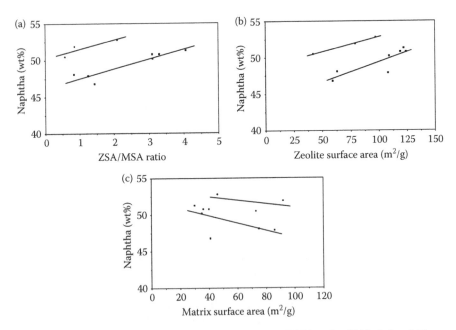

FIGURE 4.1   (a) Naphtha yield as a function of the ZSA/MSA ratio. (b) Naphtha yield as a function of the zeolite surface area. (c) Naphtha yield as a function of the matrix surface area. Conv 75 wt% ■ = A, ● = B.

area, and at the same time decrease the matrix surface area. When this suggestion was followed, when catalysts A-4 to A-7 were tested, it was possible to increase the naphtha yield with more than 3 wt% compared with the base catalyst A-1.

The results above are valid for catalysts of type A when a North Sea long residue is used as feed. As can be seen in Figure 4.1a through c, the conclusions remained the same also for catalysts of the new type B, but the numbers were different. New types of catalysts must therefore be optimized individually, and so must new feedstocks.

It can be observed in Figure 4.2a that the light cycle oil (LCO) yield decreased when the ZSA/MSA ratio increased. This is a common observation for the LCO yield when the naphtha selectivity increases. For the same reason the LCO yield decreased when the zeolite surface area increased, see Figure 4.2b. However, the LCO yield increased when the matrix surface area increased, see Figure 4.2c.

The HCO yield increased when the ZSA/MSA ratio increased, as can be seen in Figure 4.3a. One reason for this is that the ZSA/MSA ratio was changed by simultaneously increasing the zeolite surface area and decreasing the matrix surface area. The increase in the HCO yield is very small for catalysts of Type A, and more pronounced for catalysts of Type B. In Figure 4.3b it can be seen that the HCO yield increased only slightly when the zeolite surface area increased for both catalyst types. As can be seen in Figure 4.3c the HCO yield decreased when the matrix surface area increased. This was expected since a larger matrix surface area will crack the heavy part of the feed more powerfully than a smaller matrix surface area.

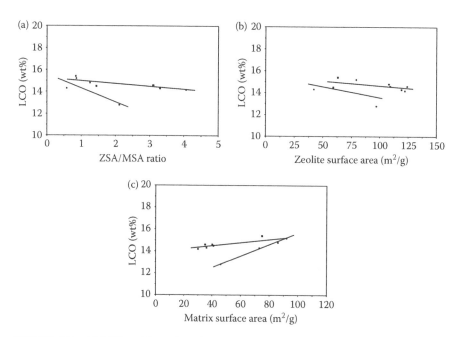

FIGURE 4.2   (a) LCO yield as a function of the ZSA/MSA ratio. (b) LCO yield as a function of the zeolite surface area. (c) LCO yield as a function of the matrix surface area. Conv 75 wt% ■ = A, ● = B.

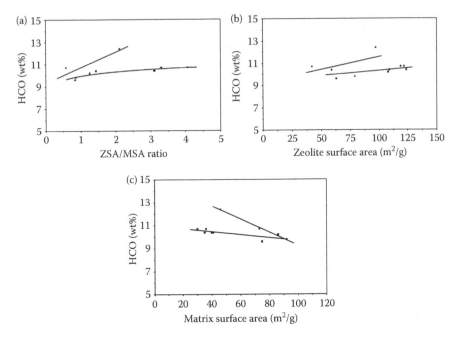

FIGURE 4.3   (a) HCO yield as a function of the ZSA/MSA ratio. (b) HCO yield as a function of the zeolite surface area. (c) HCO yield as a function of the matrix surface area. Conv 75 wt% ■ = A, ● = B.

The LPG yield decreased when the ZSA/MSA ratio and the zeolite surface area increased for both types of catalysts, see Figure 4.4a and b. The reason for this might be that the LPG yield usually decreases when the naphtha selectivity increases. This might also explain the fact that the LPG yield increased for type A catalysts when the matrix surface area increased. However, the LPG yield was almost unaffected of any changes in the matrix surface area for Type B catalysts, see Figure 4.4c.

The dehydrogenation activity of FCC catalysts is highly dependent of the matrix surface area and the feed metals deposited on them. As a result; a large matrix surface area often will give a high-hydrogen yield. This was also the case in this investigation. The hydrogen yields increased with increasing matrix surface area, see Figure 4.5c, and decreased with increasing zeolite surface area; see Figure 4.5b for both types of catalysts. It has also been reported in the literature that the hydrogen yield decreased when the matrix surface area decreased [22].

From Figure 4.6 it can be seen that the coke yields showed different behaviors for the two types of catalysts. For the Type B catalysts the coke yield was almost unaffected by variations in the ZSA/MSA ratio. For the Type A catalysts, however, the coke yield decreased when the ZSA/MSA ratio increased, which means that more naphtha selective cracking gave decreased coke yield. This is also supported by the coke yield as a function of the zeolite surface area, see Figure 4.6b. By comparing catalyst A-1 with catalyst A-3 is it possible to see that the coke

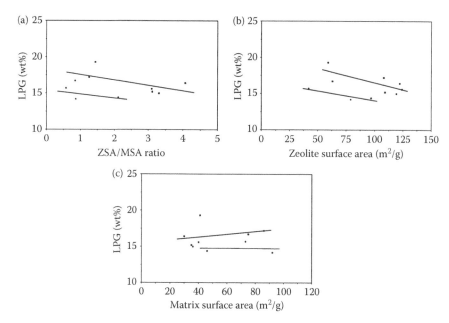

FIGURE 4.4 (a) LPG yield as a function of the ZSA/MSA ratio. (b) LPG yield as a function of the zeolite surface area. (c) LPG yield as a function of the matrix surface area. Conv 75 wt% ■ = A, ● = B.

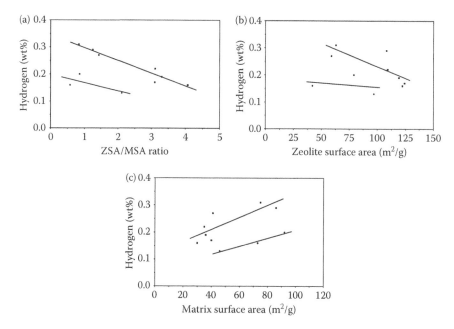

FIGURE 4.5 (a) Hydrogen yield as a function of the ZSA/MSA ratio. (b) Hydrogen yield as a function of the zeolite surface area. (c) Hydrogen yield as a function of the matrix surface area. Conv 75 wt% ■ = A, ● = B.

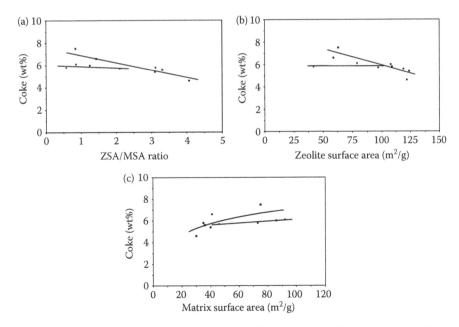

FIGURE 4.6   (a) Coke yield as a function of the ZSA/MSA ratio. (b) Coke yield as a function of the zeolite surface area. (c) Coke yield as a function of the matrix surface area. Conv 75 wt% ■ = A, ● = B.

yield decreases when the zeolite surface areas increases. The difference between these two catalysts is that catalyst A-3 has a much higher zeolite surface area than catalyst A-1, and this caused catalyst A-3 to give a lower coke yield than catalyst A-1 at constant conversion (75 wt%). Figure 4.6c also shows that the coke yield increased by increasing matrix surface area. One explanation for this might be that the matrix dehydrogenation reactions increase when the matrix surface area increases.

To completely optimize the residue catalyst, other parameters than the different surface areas also must be optimized. For a catalyst cracking North Sea atmospheric residues, the pore size distribution also must be optimized. Pores in the mesopore range; that is, pores with diameters between 50 and 500 Angstrom, are most important for precracking of resid molecules [21,23]. The possibility to make nickel and vanadium inactive is also important to optimize.

Pilot unit tests have indicated that there is an upper limit for the zeolite to matrix surface area ratio (ZSA/MSA) for a residue catalyst. This observation was in contrast to the optimization study, which indicated that the ZSA/MSA should be as high as possible for maximum naphtha yield. An increase in the zeolite surface area is, according to the optimization study, expected to increase both the activity of the catalyst and its naphtha yield. But for catalysts with a high ZSA/MSA ratio, close to four or even higher, the observed naphtha yields have been lower than expected in the pilot unit tests, which indicate that there might be an upper limit for the ZSA/MSA ratio in a residue application.

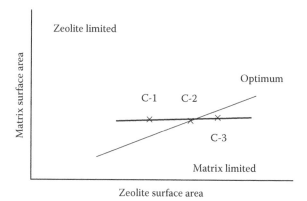

FIGURE 4.7   Simplified figure for test of upper ZSA/MSA limit.

O'Connor has indicated that there should be an optimum ratio between the zeolite and matrix surface areas [7]. This means that if a catalyst has a low-zeolite surface area, and the resulting ZSA/MSA surface area ratio is lower than its optimum, the catalyst is zeolite limited, and both the naphtha yield and the conversion will be lower than their optimal values. By increasing the zeolite content of the catalyst, the ZSA/MSA surface area ratio becomes more optimal, and the naphtha yield will increase toward its maximum value. On the other hand; if the zeolite surface area is too high, the catalyst is matrix limited. These observations can be used in order to investigate if there is an upper limit for the ZSA/MSA ratio for a residue catalyst. A set of three catalysts with the same type of zeolite and matrix was selected for this investigation. All catalysts had the same matrix surface area, while the zeolite surface areas varied. The first catalyst (C-1) has a low-zeolite surface area (zeolite limited), the second catalyst (C-2) had an higher zeolite surface area, with an almost optimal ZSA/MSA ratio, while the third catalyst (C-3) had an even higher zeolite surface area and was limited by its matrix surface area (matrix limited), see Figure 4.7. Characterization data for the catalysts are shown in Table 4.2.

As can be seen in Figure 4.8, the activity of the catalysts increased when the zeolite content of the catalyst increased. Since the matrix surface area was kept the same, the ZSA/MSA surface area ratio also increased. When comparing catalyst C-1 and catalyst C-2, the zeolite surface area was increased with 31 m²/g, and the ZSA/MSA ratio increased from 2.5 to 3.5. As expected from our optimization studies, the activity for catalyst C-2 was significantly improved compared with catalyst C-1. The increase in activity was however not by far so pronounced for catalyst C-3, where the zeolite surface area was further increased with 21 m²/g compared to catalyst C-2, which increased the ZSA/MSA surface area from 3.5 to 3.9.

When the naphtha yields were studied (see Figure 4.9) it was obvious that the results were not as expected. The maximum naphtha yield increased when the zeolite content of the zeolite limited catalyst C-1 was increased, such that the ZSA/MSA ratio became almost optimal in catalyst C-2. This increase was expected, and fully in

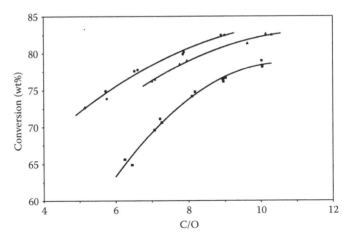

FIGURE 4.8   Conversion as a function of C/O; ■ = C-1, ▲ = C-2, ● = C-3.

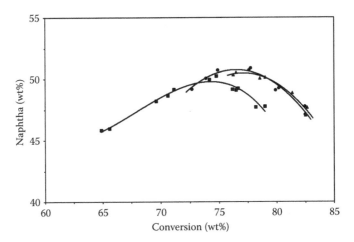

FIGURE 4.9   Naphtha yield as a function of conversion; ■ = C-1, ▲ = C-2, ● = C-3.

line with common FCC knowledge. But when the zeolite content of the catalyst was further increased, such that the catalyst became matrix limited, there was no further increase in the naphtha yield maximum. This was not expected because our optimization studies indicated that there should be an increase in the maximum naphtha yield even if it should be smaller in the matrix limited area than in the zeolite limited area. There was also a small indication that the conversion at the naphtha maximum was somewhat lower for the matrix limited catalyst C-3 than for the optimal catalyst C-2. These observations indicated that there might be an upper practical limit for the ZSA/MSA surface area ratio. For the catalyst type used in this investigation in combination with the North Sea atmospheric residue used, this ZSA/MSA ratio limit is around 3.5. The limit in the ZSA/MSA ratio could only be observed by looking at the naphtha yields. For the other product fractions, no significant yield differences could be observed at constant conversion when the zeolite content of the catalyst increased.

4.4 CONCLUSIONS

The zeolite to matrix surface area ratio can be used for optimization of catalysts for catalytic cracking of atmospheric residues. For North Sea long residues this ratio should be as large as possible, but the ratio should not exceed an upper limit. For the main catalyst type (A) used in this investigation the upper limit of the ZSA/MSA ratio was around 3.5. There is also a lower limit for the matrix surface area. If the matrix surface area is lower than this limit, the catalyst will not be able to crack all the heavy components in the residue feed, and the coke on the matrix will increase dramatically. This will prevent the catalyst from working properly. Different type of catalysts must be optimized individually, as well as different type of long residues.

ACKNOWLEDGMENT

The authors are grateful to Statoil for permission to publish this paper.

REFERENCES

1. Aalund, L. R. *Oil Gas J.* 88, no. 11 (1990): 33.
2. Gledhill, D., and Pedersen, J. *Operating Experience with the New VSS Riser Termination Technology.* GRACE Davison FCC Technology Conference Lisbon Portugal, September 1–4, 1998.
3. Torgaard, H. *Oil Gas J.* 81, no. 2 (1983): 100.
4. Wear, C. C., and Mott, R. W. NPRA, Annual Meeting, March 20–22, 1988, AM-88-73.
5. Young, G. W., Creighton, J., Wear, C. C., and Ritter, R. E. NPRA, San Antonio, TX, March 29–31, 1987, AM-87-51.
6. O'Connor, P., and Humphries, A. P. *ACS Symposium Preprints, Div. of Petrol. Chem.* 38, no. 3 (1993): 598.
7. O'Connor, P., and van Houtert, E. Ketjen Catal. Symposium 1986, Scheveningen, The Netherlands, Paper F-8.
8. Lerner, B. A. *Hydrocarbon Engineering* March 1998, 26.
9. Andersson, S.-I., and Myrstad, T. *Stud. Surf. Sci. Catal.* 134 (2001): 227.
10. Hettinger, W. P., Jr. Chapter 19 in *Fluid Catalytic Cracking, Role in Modern Refining,* ACS Symposium Series 375, Edited by M. L. Occelli, Washington, DC: ACS, 1998.
11. Experts Reveal Catalyst-Selection Methodologies. *Oil Gas J.* 94, no. 42 (1996): 50.
12. Rajagopalan, K., and Habib, Jr., E. T. *Hydrocarbon Processing* 71 (1992): 43.
13. Andersson, S.-I., and Myrstad, T. *Oil Gas Europ. Mag.* 23, no. 4 (1997): 19.
14. Mitchell, B. R. *Ind. Eng. Chem. Prod. Res. Dev.* 19 (1980): 209.
15. Cheng, W. C., Juskelis, M. V., and Suarez, W. AIChE Annual Meeting, Miami Beach, FL, November 1–6, 1995.
16. Johnson, M. F. L. *J. Catal.* 52 (1978): 425.
17. ASTM D 4365-85.
18. Andersson, S.-I., and Otterstedt, J.-E. *Catalytic Cracking of North Sea Resid.* Presented at Katalistiks 8th Annual FCC Symposium, Budapest, 1987, Paper 21.

19. Andersson, S.-I., and Myrstad, T. *Appl. Catal. A.* 159 (1997): 291.
20. Mitchell, Jr., M. M., Hoffman, J. F., and Moore, H. F. *Stud. Surf. Sci. Catal.* 76 (1993): 302.
21. Andersson, S.-I., and Myrstad, T. AIChE, Spring Meeting, New Orleans, March 9–11, 1998, Paper 31d.
22. Bourdillon, G. *Catalagram*, European Edition, May 1993, 8.
23. Magee, J. S., and Letzsch, W. S. *ACS Symposium Series* 571 (1994): 349.

5 Novel FCC Catalysts and Processing Methods for Heavy Oil Conversion and Propylene Production

Long Jun, Da Zhijian, Song Haitao,
Zhu Yuxia, and Tian Huiping

CONTENTS

5.1　INTRODUCTION

Converting more residue (hereafter resid) feed into light products is the key for survival in the refining industry. As one of the greatest refinery moneymakers, the fluid catalytic cracking (FCC) process plays an important role in heavy oil conversion. In China, FCC operations are processing heavier and dirtier feeds. Currently, 80–90% of the FCC Units (FCCU) process resid blended feeds, and the ratio of resid to VGO is increasing. For example, the average vacuum residue (VR) blending ratio reached nearly 35% in FCCUs of SINOPEC and PetroChina in 2005.

Inferior feeds influence FCC operations and product distribution negatively, resulting in the sharp drop of economic benefits. Refineries are seeking technological advances to enhance heavy oil cracking capability and increase the yield of high-value products such as gasoline and propylene. Several examples demonstrate how novel catalysts and technologies can be applied to upgrade catalytic cracking process to convert much heavier feed and yield more propylene.

5.2   EXPERIMENTAL

Chemical composition was measured by a Rigaku 3271E XRF instrument. Crystal intensity of the samples was determined by a Siemens D5005 X ray powder diffraction unit with Ni-filtered Cu Kα radiation. The ^{27}Al NMR measurements were performed on a Varian INOVA300 NMR spectrometer. DTA analysis was performed on TA Instruments SDT Q600 Simultaneous TGA/DSC instrument, and the FT-IR spectra of surface hydroxyl groups of zeolite were collected using a Bruker IFS113V infrared spectrometer.

Nitrogen adsorption and desorption isotherms were performed at 77 K on a Micromeritics ASAP 2400 volumetric adsorption system. The pore size distribution and surface area were deduced from the adsorption isotherms using the BJH method and the BET equation.

The total acidity of different samples was measured by NH_3-TPD with Micromeritics Autochem II (ASAP 2920) chemisorption system. The pyridine FT-IR spectra of the catalysts were recorded on a BIO-RAD FT3000 FT-IR spectrometer after desorption at 250 and 450°C, respectively. Then the total and strong acidity of Lewis and Brönsted acid was obtained from the integrated absorbance of the respective bands.

Catalytic performances of zeolite and catalyst samples were evaluated in an ACE and a fixed-bed MAT unit. Analyses of the products were performed with the following analysis methods: the composition of gaseous products was determined online in an Agilent 6890 gas chromatograph (GC) equipped with a TCD detector; the distillation range of liquid products was analyzed by a packed column using a simulated distillation technique, and the composition of gasoline was analyzed by a capillary column in an Agilent 6890GC equipped with a FID detector. Coke yield was calculated by IR measurement during in-site regeneration.

5.3   CRACKING CATALYSTS FOR RESID PROCESSING

Generally speaking, resid FCC (RFCC) catalysts should be very effective in bottoms cracking, be metals tolerant, and coke and dry gas selective. Based on many years of fundamental research and industrial experiences, a series of RFCC catalysts, such as Orbit, DVR, and MLC, have been developed by the SINOPEC Research Institute of Petroleum Processing (RIPP) and successfully commercialized [1]. These catalysts are very effective in paraffinic residue cracking. However, in recent years more and more intermediate-based residue has been introduced into FCC units, and the performances of conventional RFCC catalysts are now unsatisfactory. Therefore, novel zeolites and matrices have been developed to formulate a new generation of RFCC catalysts with improved bottoms cracking activity and coke selectivity.

5.3.1 STRUCTURE OPTIMIZED Y ZEOLITE

The most widely used zeolite in petroleum refining so far is Y zeolite. Currently, REUSY zeolite is the main active component of RFCC catalysts. However, in the course of hydrothermal preparation of ultrastable Y zeolite, nonframework aluminum debris formed by dealumination could block the channels thus influencing the ion-exchange ratio of rare earth as well as the accessibility of active sites [2].

Recently, RIPP has developed a proprietary method to modify the properties of ultrastable Y zeolite via a treatment for cleaning its pores (CP) [3], which include the selective removal of nonframework aluminum from zeolite pores by a novel acid treatment at optimized pH and temperature conditions.

Tables 5.1 and 5.2 list the main physicochemical properties of the modified zeolite characterized by a series of analyzing methods. XRF, XRD, and ^{27}Al NMR results listed in Table 5.1 showed that with the increasing intensity of CP treatment, nonframework aluminum was removed gradually with little influence on zeolite framework (unit cell size (UCS) changed little), thus the relative crystallinity increased. The removal of nonframework aluminum can also be verified by the FT-IR results shown in Figure 5.1, in which it can be seen that after CP treatment the intensity of the small peak at wave number 3660–3690 cm^{-1} characterizing nonframework hydroxyl groups decreased step by step.

The N$_2$ adsorption (with BET and BJH methods) results listed in Table 5.2 showed that the zeolite surface area and pore volume were apparently increased after CP treatment. For example, compared with DASY(0.0) the specific surface area of SOY0-S3 increased by 53 m^2/g from 618 to 671 m^2/g, and the pore volume increased from 0.352 to 0.393 mL/g. In addition, DTA analysis data listed in Table 5.2 showed that the thermal stability of the zeolite was further improved.

The modified zeolite was then ion exchanged with rare earth to prepare structure optimized Y zeolite (SOY). Due to the removal of nonframework aluminum debris from zeolite pores, SOY can obtain a rare earth content of about 8–10 wt% (by weight RE$_2$O$_3$), which is much higher than that of traditional REUSY zeolite (2–4 wt%). ACE evaluation results showed that SOY zeolite catalysts can perform higher

TABLE 5.1
XRF, XRD, and ^{27}Al NMR Results of Modified Y Zeolites

Samples	XRF Results Chemical Composition, wt%			XRD Results		^{27}Al NMR Results Simulated Peak Area	
	Na$_2$O	Al$_2$O$_3$	SiO$_2$	UCS, nm	ACR, %	~60ppm	~0ppm
DASY(0.0)	1.20	24.4	71.0	2.446	66.3	65.79	34.21
SOY0-S0	0.47	22.0	75.4	2.448	68.9	65.16	34.84
SOY0-S1	0.47	22.1	75.8	2.448	70.1	66.41	33.59
SOY0-S2	0.45	21.2	77.1	2.448	71.4	68.11	31.89
SOY0-S3	0.36	20.0	78.3	2.449	72.4	70.10	29.90

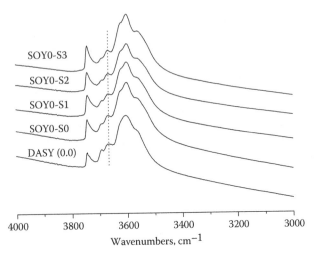

FIGURE 5.1 FT-IR spectra in the OH region of acid treated zeolites before and after CP treatment.

TABLE 5.2
N₂ Adsorption and DTA Results of Modified Y Zeolites

	N₂ Adsorption Results					DTA Results
	Surface Area, m²/g			Pore Volume, mL/g		
Samples	S_{BET}	S_Z	S_M	V_{micro}	V_p	Collapse Temperature, °C
DASY(0.0)	618	561	57	0.257	0.352	1043
SOY0-S0	648	601	48	0.275	0.370	1047
SOY0-S1	652	597	55	0.274	0.375	1052
SOY0-S2	663	605	58	0.277	0.384	1052
SOY0-S3	671	611	60	0.281	0.393	1051

bottoms cracking activity and better coke selectivity compared with traditional REUSY catalysts.

SOY was commercialized at the SINOPEC Catalyst Company in 2006. Owing to its improved ion-exchange properties, SOY zeolite with low sodium content (≤ 0.8 wt% Na₂O) can easily be prepared from NaY by a simplified double ion-exchange/single calcination procedure. As a result, the manufacturing costs of SOY zeolite and its related catalysts decreased significantly. The simplified preparation procedure of SOY has attracted great interest from Chinese Catalyst Companies and has been applied to the preparation of several other Y zeolites. According to the statistical data from November 2006 to October 2008, more than 5000 tons of SOY zeolite has been produced, and a large amount of SOY catalysts have been manufactured and successfully commercialized. For example VRCC-1 catalyst has been applied in the No.2 RFCC unit of SINOPEC Beijing Yanshan Company since December 2007 to process 40–50% VR. Compared with conventional DVR-3 catalyst the light ends yield increased by 1.2 wt%, coke yield decreased by about 1 wt% and slurry yield also decreased by 0.3 wt%.

5.3.2 Silica Modified Matrix

The FCC matrix plays a crucial role in precracking, vaporization, and internal diffusion of heavy feed molecules on catalyst particles. Therefore, many efforts have been made to optimize the acidity and pore size distribution of the matrix to improve reaction performance.

In China, most of the traditional RFCC catalysts (such as Orbit, DVR, and MLC mentioned above) are based on alumina matrix, and the most widely used materials for alumina matrix preparation are alumina sol and modified active alumina [4]. Alumina matrix combines the virtues of alumina-sol (better attrition resistance and coke selectivity) and active alumina (higher cracking activity), thus improving the cracking activity and selectivity of the catalysts. However, the coke selectivity of the alumina matrix is unsatisfactory when processing resid feed due to the insufficient amount of meso/macropores and higher concentration of acid sites.

Compared with alumina matrix, silica matrix has lower strength/concentration of acid sites and better coke selectivity. Accordingly, FCC catalysts based on silica matrix yield lower bottoms at constant coke. However, in the past few decades silica matrix had not been widely used in RFCC catalysts in China for its relatively poor cracking activity and metal tolerance ability [5]. In recent years, more and more VR has been blended into FCC feedstocks, producing higher coke yields. To improve the coke selectivity of traditional catalysts, alumina matrix modification is an important solution. RIPP has developed a proprietary method to prepare silica modified alumina matrix with a unique silica sol [6]. The silicon sol has a particle size range from 5 to 100 nm, and more than 80% of particles have a diameter of about 0.5–1.5 times of the average particle size (APS), which provide the catalyst with a large amount of 5–100 nm meso/macropores. The silica modified matrix further combines the virtues of alumina matrix and silica matrix, thus the acidity and pore size distribution of the matrix can be easily tuned by varying its components and preparation method.

Based on the silica modified matrix, a new catalyst named RSC-2006 was developed for inferior feeds cracking with reduced slurry and coke yields. RSC-2006 was applied in an RFCC unit in SINOPEC Jingmen Company in 2006 for processing intermediate-based VGO blended with ~30% VR. The results are shown in Table 5.3. Compared with the base case (Orbit-3000JM), the feed rate was increased significantly by 10.2%, from 2097 to 2311 t/d after shifting to RSC-2006 due to the reduction of coke making. The slurry yield was decreased from 7.62 to 4.75%, coke yield decreased by 0.52%, and the light ends yield increased by 3.77%. Furthermore, the density of the slurry increased and the paraffin content decreased, which denoted that the bottom conversion achieved by RSC-2006 exceed that of the base catalyst significantly. In conclusion, the RSC-2006 catalyst exhibited excellent bottoms cracking activity and coke selectivity.

82 Advances in Fluid Catalytic Cracking

TABLE 5.3
Commercial Comparison of RSC-2006 and
Orbit-3000JM in SINOPEC Jingmen Company

Items	Orbit-3000JM	RSC-2006
Feed rate, t/d	2097	2311
VR blending ratio, wt%	30.61	31.37
Feedstock properties		
Density (20 °C), g/cm³	0.9266	0.9361
CCR, wt%	4.08	4.54
Product yields, wt%		
Dry gas	4.89	4.52
LPG	17.67	20.16
Gasoline	39.66	36.37
LCO	19.97	24.54
Slurry	7.62	4.75
Coke	9.75	9.23
Conversion, wt%	71.97	70.28
Light ends yield, wt%	77.30	81.07
Slurry properties		
Density, g/cm³	1.0361	1.0636
Paraffin content, wt%	34.38	18.89

5.4. CATALYTIC CRACKING PROCESSES AND CATALYSTS FOR INCREASING PROPYLENE PRODUCTION

The worldwide demand for propylene is expected to increase, primarily driven by the market demand for products made from polypropylene, acrylonitrile, and phenolic resins. Today about 70% of the global propylene is produced by steam cracking using light hydrocarbons as feedstock, and the rest is mostly recovered from the FCC process.

Propylene is a coproduct of steam cracking, the yield of which accounts for nearly half of the ethylene yield. Currently, propylene demand exceeds ethylene demand and steam cracking cannot keep up with the required propylene/ethylene balance. To close the gap, an increase in propylene production from the FCC process is needed.

5.4.1 MAXIMIZING ISO-PARAFFINS WITH CLEANER GASOLINE AND PROPYLENE

In China, the FCC unit is the major gasoline "workhorse"; about 80% of the product streams are from the FCC unit and contribute high olefins and sulfur to the gasoline pool. In recent years, under stringent environmental regulations, refineries have to face another challenge—lowering olefins and sulfur content in FCC gasoline.

Therefore, much attention has been focused on improving the FCC process to simultaneously produce cleaner gasoline and more propylene. Based on the

Maximizing Iso-Paraffins (MIP) process [7,8], RIPP developed the Maximizing Iso-Paraffins with Cleaner Gasoline and Propylene (MIP-CGP) process in 2004 [9].

The MIP and MIP-CGP process (a simple scheme is presented in Figure 5.2) have similar riser configuration featuring two reaction zones that have been described previously [7–12]. Heavy oil vapors mixed with the catalyst are subjected to primary cracking reactions under higher severity (i.e., at higher reaction temperature and catalyst/oil ratio) in the first reaction zone to produce more olefins; after a short residence time the oil vapors and the coked catalysts reach the second reaction zone (which has an enlarged diameter) to perform secondary reactions (cracking, hydrogen transfer, isomerization, etc.) under lower reaction temperature and longer reaction time. However, reactor sizes of MIP and MIP-CGP are different, and the operation conditions of MIP-CGP process can be optimized to promote propylene production.

The specifically formulated CGP-1 catalyst plays a vital role in the MIP-CGP process. Unique catalyst design, such as metal promoted MFI zeolite, phosphorus modified Y zeolite, and a novel matrix with excellent capability to accommodate coke [12] were involved to ensure the primary cracking and secondary reactions to proceed within a defined path. The commercial trial results of the MIP-CGP process in SINOPEC Jiujiang Company showed that, in combination with CGP-1 catalyst, the propylene yield was 8.96 wt%, which increased by more than 2.6% as compared with FCC process. The light ends yield and slurry yield are basically equal. The olefin content of the gasoline produced by MIP-CGP process was 15.0 v%, which was 26.1% lower than that of FCC gasoline. The sulfur content of gasoline was decreased from 400 to 270 μg/g.

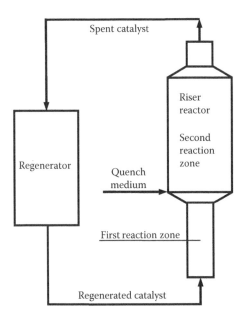

FIGURE 5.2   Scheme of the two reaction zones riser reactor.

Although the olefin and sulfur content of gasoline could be decreased significantly by the application of MIP-CGP technology, further reducing the sulfur content in catalytic gasoline is still a major objective owing to the increased processing of imported sour crude in many refineries located in east-south costal region of China. The custom designed catalyst is a key factor to meet S requirement. Therefore, CGP-2 catalyst was developed on the basis of CGP-1. Some metal-oxides, which could provide Lewis acid–base pairs, were incorporated into the matrix as sulfur-reducing ingredient, and the acidity of Y zeolite was further modified to improve its cracking and hydrogen transfer activity. The acidity of pilot scale CGP-2 samples after adding Lewis acid–base pairs (A) and zeolite modification (B) are listed in Table 5.4. Compared with CGP-1, the total Lewis and Brönsted acidity of CGP-2(A) increased slightly (the strong acidity almost unchanged), however the acidity of CGP-2(B) after zeolite modification increased significantly. Table 5.5 shows the evaluation results of

TABLE 5.4

Comparison of Acidity between CGP-2 Pilot Scale Samples and CGP-1 (Analysis Results of Pyridine FT-IR Spectroscopic)

Catalysts	Lewis Acid–Base Pairs	Zeolite Modification	250°C, µmol/g		450°C, µmol/g	
			Lewis Acid	Brönsted Acid	Lewis Acid	Brönsted Acid
CGP-1	NO	NO	14.8	13.7	11.3	10.2
CGP-2(A)	YES	NO	16.4	15.9	11.4	10.2
CGP-2(B)	YES	YES	22.3	18.1	14.3	10.1

TABLE 5.5

The Evaluation Results of CGP-2 Pilot Scale Samples in an MAT Unit (Reaction Temperature: 480°C; Catalyst/Oil: 4; WHSV: 16 h^{-1})

Items	CGP-1	CGP-2(A)	CGP-2(B)
Product yields, wt%			
Dry gas	2.3	2.2	2.0
LPG	19.2	19.7	19.4
Gasoline	38.7	39.6	40.3
LCO	20.2	19.8	20.3
Heavy oil	16.0	15.2	14.5
Coke	3.6	3.5	3.5
Conversion, wt%	63.8	65.0	65.2
Light ends yield, wt%	78.1	79.1	80.0
Gasoline properties			
Olefins, wt%	27.4	27.8	26.5
Sulfur, µg/g	652.6	320.3	311.8

CGP-2 pilot scale samples (steamed at 800°C for 12 hr) in a MAT unit with VGO feedstock. Compared with CGP-1, the CGP-2 samples with Lewis acid–base pairs could decrease the sulfur content in gasoline by more than 50%. The olefin content of gasoline changed little. Furthermore, the heavy oil conversion ability of CGP-2 samples was enhanced, which is in accordance with the increase of acidity, and the light ends yield of CGP-2(A) and CGP-2(B) were increased by 1.0% and 1.9%, respectively.

Another commercial trail of MIP-CGP for processing intermediate-based sour residual feed has been put on stream in SINOPEC Cangzhou Company in 2005. Table 5.6 shows the commercial comparison of CGP-2 and CGP-1. After shifting to CGP-2 the propylene yield increased by 1.15%, and the light ends yield increased by 0.57%. The sulfur content of gasoline was decreased from 840 to 580 µg/g. The olefin content, RON and MON of gasoline remained essentially constant.

TABLE 5.6
Commercial Comparison of CGP-1 and CGP-2 in SINOPEC Cangzhou Company

Items	CGP-1	CGP-2
Feedstock properties		
Density (20 °C), g/cm³	0.9317	0.9314
CCR, wt%	2.56	3.69
Sulfur, µg/g	6800	6700
Product yields, wt%		
Dry gas	3.53	3.21
LPG	19.44	20.35
Gasoline	35.11	32.70
LCO	27.52	29.59
Slurry	5.61	4.54
Coke	8.62	9.51
Conversion, wt%	66.70	65.77
Light ends yield, wt%	82.07	82.64
Propylene yield, wt%	7.78	8.94
Gasoline properties		
Density, g/cm³	0.7340	0.7273
Induction period, min	300	750
Sulfur, µg/g	840	580
Olefins, v%	33.7	33.9
MON	79.0	81.0
RON	93.5	93.0

The FCC unit can be easily converted to MIP-CGP operation mode with minimal revamping costs. Today, more than 20 FCC units operating in MIP-CGP mode have been put into operation in China, resulting in great economic and social benefits.

5.4.2 DEEP CATALYTIC CRACKING

Deep Catalytic Cracking (DCC) developed by RIPP, SINOPEC is a novel technology derived from FCC process for light olefins production, particularly propylene and isobutylene [13–15]. This technology has opened a new route to produce light olefins from heavy feedstocks. The light olefins produced from DCC units have been used for manufacturing high–quality polypropylene, polyethylene, acrylonitrile, and other petrochemicals.

The DCC process was first demonstrated in 1990 in SINOPEC Jinan Refinery, and has been commercialized since 1994. Shaw is the exclusive licensed provider of DCC technology outside China. The first DCC unit designed and engineered by Shaw was successfully commissioned for Thai Petrochemical Industries in 1997. Up to now, nine units have been put into production worldwide, and several other DCC units (in China and India) are under construction.

Figure 5.3 shows light olefin yields of DCC process in four refineries with different feedstocks at reaction temperatures of 545–565°C. The propylene yield can reach 23 wt% with paraffinic feed, and about 18–19 wt% with intermediate-based feed. The propylene/ethylene ratio is about 3.5–6.2, much higher than that of steam cracking. The DCC operation can be modified to further increase the yield of propylene. For example, recycling a part of DCC cracked naphtha to the reactor resulted in a propylene yield increment of 3.5 wt % in Jinan Refinery [16].

The general characteristics of DCC catalysts include: high-matrix activity for primary cracking of heavy hydrocarbons, consisting of modified mesoporous zeolite for secondary cracking of gasoline range hydrocarbons, good isomerization performance, and lower hydrogen transfer activity. In the past 15–20 years, a series of DCC catalysts have been formulated for various objectives [14]. Recently, to tackle the problem of increasingly heavy DCC feed and ever-rising propylene demand, RIPP has made innovations in the areas of catalytic materials and catalyst preparation technology.

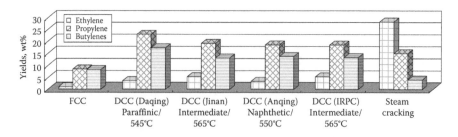

FIGURE 5.3 Light olefin yields of DCC process.

A new zeolite with the MFI structure named ZSP-3, which was modified by new compounds containing metal oxides to enhance its cracking activity and dehydrogenation activity, has been used to replace ZRP as the main active component to promote the formation of propylene. Another new active component, modified Beta zeolite named ZBP was also introduced into catalyst preparation to partially replace USY zeolite for further enhancement of propylene yields. To illustrate the effect of the newly developed zeolites on propylene production, several zeolite samples (USY, ZSP-3, and ZBP mixed at different ratios) were steam deactivated at 800°C for 4 hr and tested in an MAT unit at 500°C with VGO feedstock. The comparative results of propylene production are listed in Table 5.7. When 33.3% ZSP-3 and 33.3% ZBP were separately blended with USY, the propylene yield increased by about 8.3% and 3.4% compared with USY; when USY, ZSP-3 and ZBP were blended together at the same content, the propylene yield increased significantly by more than 9%, which might be attributed to the synergistic effect of different zeolites. It is suggested that ZBP zeolite play a "relay" role between USY and ZSP-3 in the mechanism of catalytic reactions. Large hydrocarbon molecules are selectively cracked on USY zeolite to form middle-sized molecules; then the middle-sized molecules are transformed into C_5–C_8 olefins on the ZBP zeolite prior to being further cracked into propylene on ZSP-3.

Furthermore, a novel catalyst preparation technology has been developed to optimize catalysis kinetics (OCK) based on detailed analysis in DCC reaction network [17]. Through the OCK technology, a new matrix with an increased amount of large pores were prepared to improve residue conversion capability and the accessibility of active sites to promote deep cracking reactions. By adopting the above mentioned innovations, the new generation DCC catalyst DMMC-1 has been developed. The N_2 adsorption analysis results of DMMC-1 and MMC-2 (a previously developed DCC catalyst) are listed in Table 5.8. Compared with MMC-2, the pore volume of DMMC-1 is 13% higher before and after steam deactivation, and the specific surface area of steamed DMMC-1 is 19% higher than that of MMC-2.

DMMC-1 has been applied commercially in the 650 kt/a DCC unit of SINOPEC Anqing Company since July 22, 2006. The commercial results are listed in Table 5.9; DMMC-1 features better bottoms cracking ability and improved product distribution.

TABLE 5.7
Comparison of Propylene Production of Different Zeolite Combination

Samples	S1	S2	S3	S4
Zeolite	USY	USY+ZSP-3	USY+ZBP	USY+ZBP+ZSP-3
Mix ratio	100	66.7:33.3	66.7:33.3	33.3:33.3:33.3
Conversion, wt%	75.6	72.5	76.3	75.8
LPG, wt %	20.1	38.4	30.3	39.6
Propylene, wt %	5.8	14.1	9.2	14.9
Propylene selectivity, %	7.7	19.4	12.1	19.6

TABLE 5.8
N$_2$ Adsorption Analysis Results of DCC Catalysts before and after Steam Deactivation

Items	Fresh		Steam Deactivated	
Catalysts	MMC-2	DMMC-1	MMC-2	DMMC-1
S$_{BET}$, m^2/g	Base	Same	Base	+19%
V$_p$, mL/g	Base	+13%	Base	+13%

TABLE 5.9
Commercial Comparison of DMMC-1 and MMC-2 in SINOPEC Anqing Company

Items	MMC-2	DMMC-1
Product yields, wt%		
Dry gas	7.71	7.89
LPG	34.60	38.90
Gasoline	29.90	26.09
LCO	19.42	19.57
Slurry	0.25	0.00
Coke	7.61	7.05
Conversion, wt%	79.82	79.93
Light ends yield, wt%	83.92	84.56
Light olefin yields, wt%		
Ethylene	2.69	3.11
Propylene	15.37	17.80
Butylenes	12.60	13.01
Gasoline properties		
Density, g/cm^3	0.7458	0.7453
Induction period, min	303	320
Olefins, v%	43.7	39.2
MON	83.8	84.0
RON	97.8	98.0

With the application of DMMC-1 catalyst, the propylene yield is 17.80 wt%, which is higher by 2.43% as compared with the MMC-2 catalyst. The light ends yield increases by 0.64%, and the coke yield decreases by 0.56 wt%. Furthermore, the olefin content of gasoline decreases by 4.5 v%. Thus the worldwide leading position of DCC in propylene production from catalytic cracking has been advanced further.

5.5   CONCLUSIONS

To improve bottoms cracking activity and coke selectivity of RFCC catalysts, novel zeolites and matrices have been developed recently. Commercial results showed that both VRCC-1 catalyst containing SOY zeolite and RSC-2006 based on silica modified alumina matrix have demonstrated excellent bottoms cracking capability and coke selectivity.

To meet the requirements of environmental regulations and increase propylene production, the MIP-CGP process has been developed to simultaneously produce cleaner gasoline and more propylene. Commercial trails have proved that, in combination with custom designed CGP series catalyst, MIP-CGP technology could reduce the gasoline olefins and sulfur content significantly while maintaining high-propylene yield.

DCC is a novel technology derived from the FCC process for light olefins production, particularly propylene and isobutylene. New generation catalyst DMMC-1 can help to convert heavier feedstocks with increased propylene yield.

ACKNOWLEDGMENT

The authors are grateful to the Major State Basic Research Development Program (Grant No. 2006CB202501) and SINOPEC for financial support.

REFERENCES

1. Huiping, T. *Petroleum Refinery Engineering* 36, no. 11 (2006): 6–11.
2. Weilin, Z., Lingping, Z., Shiming, S., et al. *China Petroleum Processing and Petrochemical Technology* 2 (2007): 55–59.
3. Lingping, Z., Mingde, X., Yuxia, Z., et al. China Patent CN 101284243A, 2008.
4. Jiasong, Y., Jun, L., and Huiping, T. *Petroleum Processing and Petrochemicals* 35, no. 12 (2004): 33–36.
5. Lijing, Y., Huiping, T., and Jiasong, Y. *Industrial Catalysis* 14, no. 7 (2006): 18–22.
6. Jiasong, Y., Jun, L., and Huiping, T. China Patent CN 1332765C, 2007.
7. Youhao, X., Jiushun, Z., and Jun, L. *Petroleum Processing and Petrochemicals* 32, no. 8 (2001): 1–5.
8. Jun, L., Youhao, X., Jiushun, Z., et al. *Engineering Sciences* 1, no. 2 (2003): 47–52.
9. Youhao, X., Jiushun, Z., Jianguo, M., et al. *Petroleum Processing and Petrochemicals* 35, no. 9 (2004): 1–4.
10. Youhao, X., Jiushun, Z., Jun, L., et al. *Engineering Sciences* 5, no. 5 (2003): 55–58.
11. Zhonghong, Q., Jun, L., Youbao, L., et al. *Petroleum Processing and Petrochemicals* 37, no. 5 (2006): 1–6.
12. Jun, L., Wei, L., Zhonghong, Q., et al. *Studies in Surface Science and Catalysis, Fluid Catalytic Cracking VII.* Edited by M. L. Occelli, Vol. 166, 55–66. Amsterdam: Elsevier, 2007.
13. Zaiting, L. *Engineering Sciences* 1, no. 2 (1999): 67–71.
14. Zaiting, L., Fukang, J., Chaogang, X., et al. *Petroleum and Petrochemical Today* 9, no. 10 (2001): 31–35.
15. Zaiting, L., Fukang, J., Chaogang, X., et al. *China Petroleum Processing and Petrochemical Technology* 4 (2000): 16–22.
16. Chaogang, X., and Yongcan, G. *China Petroleum Processing and Petrochemical Technology* 4 (2008): 1–5.
17. Jun, L., Huiping, T., Yujian, L., et al. *China Petroleum Processing and Petrochemical Technology* 4 (2007): 1–6.

6 Improving the Profitability of the FCCU

Warren Letzsch, Chris Santner, and Steve Tragesser

CONTENTS

6.1 CATALYST AND ADDITIVES

Important catalyst properties such as gasoline and coke selectivity are listed in Table 6.1 [2]. A catalyst change could be made to improve fluid catalytic cracking (FCC) unit operation for any of the benefits listed in the table. When a change is being sought, however, the trick is to not have another aspect of the FCC operation deteriorate such that the benefit sought is nullified by other changes. All of the catalyst suppliers have a line of products that will enhance each of the benefits shown in the table [3].

Many available catalyst additives that can be used to meet a specific objective are listed in Table 6.2. These have become a necessary adjunct to FCC operations. Originally, catalyst suppliers tried to incorporate the function into the catalyst but a myriad of needs or the catalyst manufacturing scheme employed made a single all-encompassing catalyst formulation impossible.

It was also soon recognized that catalyst additives affected unit activity and conversion. Other objectives, such as CO burning or SO_x control, required a specific amount of control agent and adding more was a waste of money. Therefore a separate additive was desired. Additives could be either liquid or solid but solid additives are the most common.

Specific reasons for adding catalyst or additives are given in Table 6.3. While certainly not all-inclusive, the list gives a few ways that catalysts and additives can be used to improve an FCC operation [1,4]. Only in the catalytic cracking process is the catalyst an operating variable as well as the means of affecting the conversion. Catalyst activity has an independent effect on the unit heat balance as well as the yield selectivities. The dilution effect of additives, the last item listed in Table 6.3,

TABLE 6.1

Important Catalyst Properties

1. Activity	8. CO burning
2. Gasoline selectivity	9. SO_x removal
3. Gas selectivity	10. Metals resistance
4. Coke selectivity	11. Hydrothermal stability
5. Bottoms cracking	12. Octane
6. Density	13. Isobutane/olefin selectivity
7. Attrition	14. Price

TABLE 6.2

Types of Additives

Additive	Function
SO_x removal	Pollution control
Combustion promotion	Regenerator temperature control
Octane enhancement	Increased octane
Metal traps/passivators	Activity maintenance/selectivity control
Bottoms cracking	Feedstock flexibility

TABLE 6.3

Catalyst Modifications

1. Add a catalyst for the desired product(s)
2. Use a catalyst to maximize LPG versus reactor temperatures when an air blower or WGC limit exists
3. Consider lower catalyst activity when at a regenerator temperature limit
4. Use a gasoline desulfurization catalyst system to minimize gasoline sulfur contents
5. Use a catalyst with built-in nickel passivation when nickel is above 1500 ppm or hydrogen production is above 0.13 wt% on fresh feed
6. Use DeSox to reduce or trim SO_x emissions
7. Add CO promoter when operating in a partial CO burn mode
8. Use a vanadium passivator when vanadium level exceeds 3000 ppm on catalyst
9. Add the highest activity additives to minimize dilution of the catalyst inventory

is particularly important when using additives, as illustrated in Figure 6.1, taken from a Chevron publication [5]. At best, additives are inert and dilute the catalyst inventory leading to a loss of conversion and gasoline. At worst, they lead to poorer selectivities (more coke and gas) and can limit the operation; therefore, it is desirable to reduce the amount of additive used.

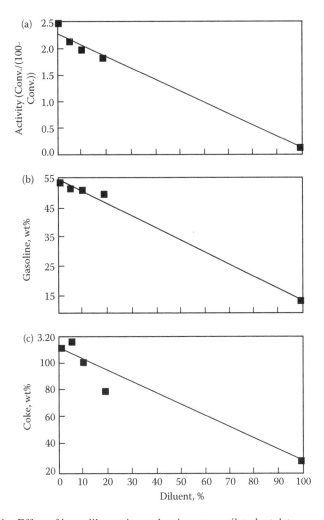

FIGURE 6.1   Effect of inert diluents in catalyst inventory-pilot plant data.

6.2 UNIT MODIFICATIONS

Unit modifications that can improve the operation start with the riser and the feed injection system [6,7]. The riser should be straight and have the desired velocities at both ends, which reduces excessive slip at the bottom and excessive erosion and turbulence at the outlet. A reduced diameter section is normally included at the lower end of the riser, just above the feed nozzles, to accelerate the feed/catalyst mixture. This improves contacting between the catalyst and hydrocarbons and minimizes back mixing of the catalyst.

Feed nozzles can be upgraded or replaced with a modern system. An axial nozzle can be replaced with a radial feed injection system, additional nozzles can be added to existing radial systems and higher performance nozzles can replace the existing feed injectors to improve atomization. Reliability is a key feature of feed nozzles to ensure the nozzles perform well throughout the entire run. If the nozzle exit slots erode during a run, the performance of the feed nozzles will be impacted resulting in unit yield degradation and therefore a reduction in the FCC profitability.

The benefits of an improved feed injection system and riser can be lower delta coke, lower regenerator temperature, higher conversion, higher gasoline yield, and increased liquid yield. More feed can be processed and an air blower or wet gas compressor limitation can be relieved.

An improved riser terminator can lower delta coke, reduce dry gas up to 40%, lower regenerator temperature and allow more feed or higher conversion. Adding post-riser vapor quench can independently reduce dry gas, increase gasoline and reduce diolefins in the alkylation stream and the FCC gasoline.

Post-riser quench can be used if a reactor vessel has a metallurgical limit and a higher riser outlet temperature is desired. Higher octanes and more alkylation feed may be the result. Improved vaporization of the feed could lower delta coke.

Adding recycle streams to improve bottoms cracking is particularly attractive with the high margins for middle distillates. Adding recycle downstream of the feed injectors improves overall selectivity by increasing the cat/oil in the lower section of the riser. If catalyst is in the recycle stream in any significant amount, special nozzles should be used to avoid damaging the primary feed injectors.

Poor stripping robs the benefits of many of the improvements made in the rest of the reaction system [8]. Strippers can be improved by adding additional stages, replacing high maintenance systems with structured packing and using systems capable of running at twice the flux rate of convention disc and donut designs. Koch-Glitsch packing has become very common in stripper upgrade projects because of its performance, reliability, and ease of installation [9].

The benefits are lower regenerator temperatures, less hydrocarbon under-carry to the regenerator, reduced stripping steam usage, and lower dry gas yields. The higher catalyst circulation rate may allow conversion to be raised. The catalyst may have to be reformulated or changed to fully realize the benefits of any of the reactor/stripper modifications.

Improvements in the regenerator [7] can be made to the air distributor to improve air distribution, reduce pressure drop, improve reliability, or add capacity. Distributing the spent catalyst across the entire catalyst-dense bed makes the bed and dilute phase temperatures more uniform and can reduce or eliminate the need for combustion promoters by reducing the tendency to afterburn. A new cyclone system or additional cyclones [10] can improve the FCC reliability, lower catalyst losses, reduce maintenance expense or allow the unit to be run at higher temperatures. Modifying the regenerator's catalyst outlet can eliminate catalyst bypassing and reduce "salt and pepper" catalyst. Afterburn can be reduced since the air distributor can be designed to cover the entire cross-sectional area of the regenerator catalyst bed.

The main fractionator may be the limitation on the unit. Packing can reduce pressure drop, and more efficient trays can give better product separations [11].

Modifications can be made to reduce the amount of middle distillate leaving with the bottoms, or to remove the middle distillate in a separate tower. If more middle distillates are a primary objective, a lower conversion may be desired. A separate heavy cycle oil draw may have to be added or reinstated to give the best overall yields.

Adding more trays or increasing the diameter of the top portion of the tower if it is smaller than the rest of the vessel may improve separations.

Adding a downstream gasoline splitter can help when reducing gasoline sulfur by minimizing octane loss. Other towers may be expanded by using structured packing.

The flue gas system is an area that should not be overlooked. Environmental regulations have tended to make any changes a compliance issue but improvements can add significant value. An expander can improve energy efficiency with a reasonable payout. A more efficient third stage separator can reduce expander fouling and minimize emissions. A downstream filter system can remove essentially all catalyst. A flue gas scrubber can reduce both SO_x and particulate matter to levels that meet the most stringent environmental regulations. Additional capacity can be built-in to meet future unit expansions as well as new regulations such as those designed to limit the amount of fine particles (less than 2.5 microns). One system [12] can also reduce NO_x by the addition of ozone that converts the NO_x to a water soluble form. A metallic filter can be employed to treat a portion of the flue gas when the unit is to be expanded instead of adding another electrostatic precipitator.

Fines removal from the slurry may allow it to be used for anode grade coke rather than fuel grade. Filters have been used successfully in this application and fresh feed or light cycle oil can be used to recycle the fines back to the riser instead of decant oil.

Improved slide valves can lead to a more reliable FCC operation. Better actuators provide a more rapid response to process changes and can prevent unnecessary unit trips. Changing the internals to allow the valves to operate in the desired 35–65% open range gives better control and reduces wear. Expansion joints can be updated with better metallurgy and dual plys with leak detectors can be used. This can help avoid a shutdown and improve both safety and reliability.

Not listed in Table 6.4 of possible unit modifications are improved unit controls. These can be applied in many ways to both the reactor/regenerator and gas plant. Operating closer to limits increases revenue by forcing the operation to several limits rather than one or two. A good FCC model is needed if all the benefits are to be realized.

6.3   OPERATION CHANGES

The tendency when operating an FCC unit is to set the flow rates at a prescribed level and then leave them alone. Over time changes to the unit are made and it pays to revisit these operating conditions. Table 6.5 lists potential operational changes that should be considered for optimization of the FCC unit.

TABLE 6.4
Equipment Modifications

1. Upgrade radial feed nozzles
2. Replace axial nozzle(s) with radial nozzles
3. If resid is added, change feed system so nozzles are capable of vaporizing feed
4. Shaped tips for feed nozzles in horizontal or nonvertical riser
5. Straighten riser and make vertical
6. Swedge riser to minimize velocity
7. Replace riser tee with cyclones
8. Add vapor quench to reactor cyclones
9. Add a close-coupled riser termination system
10. Repair stripper internals
11. Replace stripper internals with packaging
12. Increase the number of stripping stages
13. Add packing to top of stripper to insert oil to increase delta coke when processing very low delta coke feeds
14. Improve spent catalyst distribution to reduce afterburn
15. Revamp spent catalyst distribution to reduce the use of copromoters and to minimize NO_x
16. Replace a plate air distributor with multiple air rings
17. Lower the ΔP of the air distributor to allow more air
18. Add a flue gas scrubber to reduce SO_x, PM, and PM 2.5
19. Consider adding a regenerable scrubber if SO_x is over 1000 ppm
20. Consider adding a vapor line valve in place of a blind
21. Add multiple bed density taps to the reactor
22. Add multiple bed density taps to the regenerator to determine quality of fluidization and identify a damaged or partially plugged air distributor
23. Replace feed system using left gas with one that does not require it to gain WGC capacity
24. Replace old slide valve actuators with new ones to improve response and better control the unit
25. Change slide valve internals if the unit is operating at less than 30% or more than 70% open for better control
26. Add a desalter on front of the FCC if it is processing resid or running with feed taken from tankers
27. Add an elutriator to remove large catalyst particles to improve fluidization and catalyst circulation
28. Replace trays with packing in the main fractionators to reduce pressure drop and increase capacity
29. Install controls to ensure proper flow of wash oil to the main fractionator
30. Add a quench distributor to the bottom of the main fractionators to control coking
32. Convert the unit to cold-wall design
32. Switch to modern, two-ply expansion joints with leak detectors
33. Add a continuous catalyst loading system instead of batching catalyst
34. Add an HCO withdraw tray to the main fractionators to add recycle flexibility
35. Add a flash stage for the decant oil product to remove LCO for recovery

---

TABLE 6.4 (CONTINUED)
Equipment Modifications

36. Add a slurry filter to remove fines and produce a high quality decant oil
37. Install an additive loader capable of adding multiple additives simultaneously
38. Replace a fourth stage cyclone with a filter system to recover fines
39. Add a metallic filter to a unit using an ESP when the ESP is at capacity
40. Install an air distributor that is fully insulated
41. Remove a catalyst circulation limit

Dispersion steam should be adjusted to see if further improvement in the operation is possible. More steam may reduce dry gas and delta coke. If resid is added to the feed or the amount of resid in the feed is increased, the feed dispersion steam is also increased to improve atomization and vaporization.

Stripping steam should be adjusted and its influence on regenerator temperature observed. It should be high enough that no further drop in temperature is observed when the steam rate increases. If resid is added to the unit or coker gas oils are coprocessed, the steam rate to the stripper will usually be higher. When capacity of the unit is increased by raising reactor pressure, the stripper operation will require reoptimization.

Care should be taken not to flood the feed nozzles since this adversely influences atomization. This happens when too little steam is used or too much feed is injected to a nozzle. Restriction orifices used in feed and steam lines may need to be replaced. Strainers on these lines might require periodic cleaning.

Dome steam should be added when any close-connected riser terminator is installed. This will reduce coke formation on the top of the reactor vessel and cyclones. The steam must be superheated so that liquid droplets do not impinge on the cyclones and coke doesn't buildup on the steam ring. Typical dome steam rates are based on a superficial velocity high enough to sweep the hydrocarbons to the reactor catalyst separation system.

In units where a primary regenerator cyclone has failed, it is possible to cut the air back to a superficial velocity of 1.5 ft/sec and limp along until equipment is available for a shutdown. Feed rate would be proportioned to the reduction in air. This technique has been employed in situations where a unit dropped a cyclone and when one of the primary diplegs was plugged with refractory. The pressure drop through the air distributor should not be less than 30% of the bed pressure drop to prevent grid erosion.

Operators should adjust the bed level in the regenerator and observe its effect on catalyst losses, slide valve differentials, afterburn, and regenerator emissions. NO_x may be particularly sensitive to bed level. Recycle should be considered when trying to maximize diesel. It should always be added downstream of the feed nozzles to prevent erosion of the feed injector tips. Mix temperature control can be used when resid is processed to ensure maximum feed vaporization at the lower reactor temperatures used in diesel operations.

TABLE 6.5

Operational Changes

1. Do not flood feed nozzles
2. Add dispersion steam when processing resid
3. Add steam to risers with low superficial velocities
4. Add dome steam with resid feeds
5. Add dome steam with close connected riser terminators
6. Add or delete vapor quench depending on gas prices
7. Lower regenerator superficial velocity with nonoperational primary cyclone until a replacement can be obtained
8. Raise the regenerator bed level to reduce afterburn
9. Add or adjust the aeration to withdraw well inlets to improve catalyst circulation
10. Recycle HCO to increase middle distillate production
11. Put heavy recycle downstream of the feed nozzles to improve dry gas and coke selectivities
12. Adjust the pressure balance to give adequate slide valve pressure differentials

6.4   UNIT TESTING

The pressure balance should be examined to determine the normal pressure readings in the reactor, regenerator, air system, flue gas system, and main fractionator and overhead system. These need to be followed on a time basis and plotted against variables such as feed rate, wet gas rate, and dry gas rate to see if and where problems may occur. Adjustments may be possible if the spent or regenerated catalyst slide valve delta P is at a minimum to provide more operating room.

Reaction mix sampling [13] is a handy technique for determining the extent of secondary cracking in the reactor vessel, the effects of various operating variables such as reactor temperature, recycle rate, variations in stripping, and dispersion steam rates and capacity. Units should be designed with sample ports so that these tests can be run periodically to determine if the unit is running as expected. The advantage of the test method is that changes can be made and it is not necessary to wait for the fractionation system to come to equilibrium.

Radioactive tracers [14] are a useful tool to measure unit parameters such as residence times and distribution of the catalyst and vapors in the reactor, stripper, or regenerator. Bypassing can be detected, slip factors calculated and dilute phase residence times are examples of useful calculations that can point the way to future modifications. This technology is also useful for detecting and analyzing equipment malfunctions. Plugged distributors, erratic standpipes, and main fractionator problems such as salt deposits or flooding can be detected with tracers.

Catalyst performance is always difficult to judge. Many units use too little catalyst due to the pressure to keep costs low. Increase additions by 10–20% and monitor the effect it has on unit performance, provided the unit is not at a regenerator temperature limit.

There are many more ways that an FCC unit operation may be improved than those discussed here and shown in Table 6.6. The importance of the FCC unit to the

TABLE 6.6
Unit Testing

1. Use radioactive tracers to measure both catalyst slip and collection efficiency of reactor separator
2. Use radioactive tracers to monitor stripper performance
3. Use radioactive tracers to determine degree if bypassing of spent catalyst in regenerator
4. Use helium to test heat exchangers for leaks
5. Check the air line from the air blower to the air distributor for excessive pressure drop
6. Adjust stripping stream and observe the regenerator pressure
7. Adjust feed dispersion stream and observe dry gas yield and regenerator pressure
7. Use reaction mix sampling to determine amount dry gas made in the reactor dilute phase
9. Increase catalyst additions 10% or more and observe yield changes

overall refinery performance makes it incumbent upon the refinery staff to continue to find ways that enhance the operation. These include equipment modifications, proper catalyst and additive usage, operation monitoring, and unit testing. Increased computer power and better instrumentation along with more sophisticated FCC models will yield greater safety, reliability, and profitability.

6.5 CONCLUSIONS

The FCC unit is normally the most profitable unit in the refinery due to its large volume and its ability to convert fuel oils into lighter, more valuable products. Unlike most other refining units where the oil is passed over a fixed bed of catalyst, the FCCU is dynamic with tons of catalyst passing between the reactor and regenerator every minute. The complexity of the process presents many ways that improvements can be made. These enhancements can be made by altering the catalyst to meet a specific need or by the addition of an additive. Modifications to the riser, stripper, regenerator, gas plant, and other ancillary equipment can provide fast paybacks. The operating parameters need to be continually monitored to optimize their values. When changes to the unit occur, such as a feedstock switch, these parameters need to be revisited. Useful tools are available to help monitor and troubleshoot the unit. These include Reaction Mix Sampling, the use of tracers, and pressure surveys. The difference in performance between a "tuned" and "untuned" FCC can be as much as $0.30 US/B of feedstock.

REFERENCES

1. Venuto, P. B., and Habib, Jr., E. T. *Fluid Catalytic Cracking with Zeolite Catalysts*. New York: Marcel Dekker, 1979.
2. *Introduction to Refining*. Preston: RPS Training Course, 2005.
3. Catalyst brochures from W.R. Grace, BASF and Albemarle. Cleveland, OH: Freedonia Group, 2008.
4. Hettinger Jr., W. P. Development of a Reduced Crude Cracking Catalyst. Chapter 19 in *ACS Symposium Series #375*, Edited by M. L. Occelli. Washington, DC: ACS, 1988.

5. Krishna, A. S., Hsieh, C. R., Pecoraro T. A., and Kuehler, C. W. *Hydrocarbon Processing*, 59–66. Richmond, CA: Chevron Research and Technology Company, 1991.

6. Letzsch, W. S., and Lawler, D. *Catalyst & FCCU Design for Processing Resid.* 7th International Downstream Technology & Catalyst Conference & Exhibition, London, February 15–16, 2006.

7. Wilson, J. W. *Fluid Catalytic Cracking Technology and Operation.* Tulsa, OK: Pennwell, 1997.

8. Jawad, Z. S. *FCC Fundamentals: Coke Yield/Delta Coke.* PTQ, Q2, 2007.

9. Letzsch, W. S. Revitalize Stripping Operations with Structured Packing. *Hydrocarbon Processing* September 2003, 69–72.

10. Wilcox, J. R. *Troubleshooting Complex FCCU Issues.* PTQ, 3Q, 2009.

11. Dean, C. F., Golden, S. W., and Hanson, D. W. *Manipulating FCC Pressure Balance.* Singapore: ARTC Management, 2004.

12. Weaver, E. H., and Confuorto, N. *An Evaluation of Add-On Technologies for the Reduction of NOx Emissions from the Regenerator.* Singapore: ARTC Management, 2004.

13. Letzsch, W. S., Campagna, R. J., and Kowalczyk, D. *New Optimization Technologies for the FCCU,.* NPRA Paper (AM-93-66).

14. Tracerco Diagnostics FCCU Study. Brochure Tracerco 2007, http://www.tracerco.com.

7 Troubleshooting Complex FCCU Issues*

Jack R. Wilcox

CONTENTS

7.1 TROUBLESHOOTING COMPLEX FCCU UNIT ISSUES

With the increasing lack of refining capacity globally, particularly in North America, to keep pace with the growing demand for transportation fuels including both motor gasoline and on-road diesel fuel, refiners are under constant pressure to continuously improve the performance and reliability of the fluid catalytic cracking unit (FCCU). Despite the recent global economic slowdown contributing to reduced refinery utilization typical FCCU run lengths, or intervals between planned maintenance turnarounds, have been averaging about 4 years with an on-stream factor of about 0.96.

---

* Originally published in *Petroleum Technology Quarterly*, "Troubleshooting Complex FCCU Issues," Jack R. Wilcox, Q3, 2009 (http://www.ePTQ.com). Reprinted with permission.

This would account for a typical 3–4 week turnaround period with the remaining days representing unplanned feed outages. Extending the turnaround interval to 5 years with an improved on-stream factor of about 0.98 is a reasonable target for refiners.

A 50,000 barrels per day (BPD) capacity FCCU, at today's U.S. Gulf Coast crack spread, would generate an incremental $0.5M USD per day in profit. The benefit of the reduced downtime must be weighed against the cost of a potentially longer turnaround, higher operating cost including catalyst usage, and potential revenue from technology upgrades. The lengths of time between turnarounds and causes for unplanned feed outages are obviously very dependent on equipment reliability and consistent operations.

As such, it is incumbent on the FCCU operating personnel to diagnose potential causes for operating problems and return the unit to stable operation as quickly as possible. The following is a discussion of the leading causes for FCCU feed outages and unscheduled shutdowns. Unexplained or increasing catalyst losses, erratic catalyst circulation, transfer line, main fractionator, and slurry circuit coking/fouling, poor spent catalyst stripping, and regenerator afterburn all have a significant impact on unit operating reliability and resulting profitability.

7.2 CATALYST LOSSES

Catalyst is continuously being lost through both the reactor and regenerator. Minimizing these losses is essential to maintaining optimum unit operation as well as environmental compliance and to reduce catalyst costs. The causes for increasing catalyst losses include refractory lining failure, excessive mass flows through the cyclones and diplegs, insufficient dipleg length, mechanical failures with the collection system, and changes with the circulating catalyst quality.

Using the catalyst vendor's equilibrium catalyst report, the physical properties of the circulating catalyst may be monitored for any change. Albemarle routinely analyzes a sample of the circulating catalyst inventory among others for physical characteristics, including surface area (SA), metal content, apparent bulk density (ABD), and particle size distribution (PSD).

For those units utilizing a tertiary separator, regular measurement of the PSD and metal content of the underflow of this separator is also recommended. A shift in catalyst PSD to higher values is an indication of a potential loss problem. A decrease in fines content typically reflects a problem with collection equipment. Depending on the unit configuration, this can have a significant impact on catalyst circulation.

On the other hand, an increase in the fines (particles <40 µ) content of the circulating catalyst usually points to an attrition source or a change in the fresh catalyst PSD. Attrition in the dilute phase will not be reflected in the inventory PSD.

7.2.1 CYCLONE PROBLEMS

Increased catalyst losses will be reflected by an increase in the average particle size of the circulating catalyst inventory. Cyclone separators (see Figure 7.1) are prone to

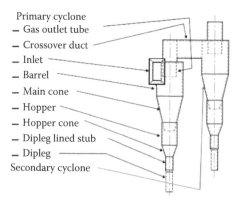

Primary cyclone
— Gas outlet tube
— Crossover duct
— Inlet
— Barrel
— Main cone
— Hopper
— Hopper cone
— Dipleg lined stub
— Dipleg
Secondary cyclone

FIGURE 7.1 Cyclone terminology. (Reprinted from Jack Wilcox, R., Published in Petroleum Technology Quarterly, 'Troubleshooting Complex FCCU Issues,' http://www.ePTQ.com, Q3, 2009. With permission.)

the following problems, all reducing the collection efficiency, leading to increased catalyst losses [1]:

- Damage to the cyclone refractory lining and cyclone itself is a constant concern due to the highly erosive environment experienced in the cyclones. The primary causes for wear leading to holes include excessive inlet vapor and solids loadings, too high gas tube outlet velocities, and gas leakage up the dipleg. The linings used in cyclones utilize refractory materials equal to or harder than the catalyst. As catalyst erodes the lining during normal operation, the refractory is attriting the catalyst to much smaller particles. These particles are eventually re-entrained into the inlet vapor stream to the cyclones, passing through the cyclone system, and adding to the losses.
- High cyclone pressure drop resulting from high gas outlet tube velocities are generally found in the second stage regenerator cyclones. While high outlet tube velocities are usually the result of unit operation at higher than normal design conditions, incorrect cyclone design will frequently generate excess pressure drop. If the cyclone pressure drop is causing a unit constraint, redesign or replacement of the cyclone(s) may be necessary.
- Low cyclone pressure drop resulting from significantly reduced vapor and solids loadings will be reflected in a loss of collection efficiency. In addition to reduced loadings, gas leakage through the cyclone assembly or faulty design will contribute to decreased pressure drop.
- Excessive gas leakage into and up the cyclone diplegs will disrupt the catalyst flow down the dipleg leading to loss of collection efficiency. The probability of this occurring is greater in the second stage diplegs as the mass flux through the primary diplegs is usually high enough to prevent upward gas flow.
- Leaks due to weld cracks resulting from thermal cycling or high stress may occur at any of the weld connections.

- Holes through the cyclone body or diplegs will occur due to refractory failure.
- Blockage or plugging of the dipleg due to an obstruction or catalyst bridging resulting from defluidization or sticky catalyst fines will effectively flood the cyclone, rendering the cyclone inoperative.
- Inoperative dipleg valves, or valves stuck in the closed position will also cause the cyclone to flood, leading to catalyst losses. Deformed or missing valves, particularly on the second stage cyclones, will lead to unsealing the dipleg leaving the cyclone inoperative.

Catalyst losses due to plugged diplegs and/or stuck trickle/counterweight valves may sometimes be reduced by adjusting the dense bed level or with a sudden bump in pressure. During the turnaround, the valve clearances should be verified.

7.2.2 OPERATIONS IMPACT ON LOSSES

The combustion air rate and resulting superficial gas velocity and dense bed catalyst level will have a significant impact on cyclone performance. Increasing air rate will increase both the solids entrainment and pressure drop in the cyclones. This in turn increases the catalyst backup in the diplegs, and eventually the dipleg of catalyst in the dipleg could reach the cyclone vortex. At this point the cyclone floods and catalyst attrition as well as increased erosion of the cyclone cone and hopper will occur. Also, as the catalyst level reaches the dust hopper, some of the descending catalyst will be re-entrained and contribute to increased catalyst losses. As shown in Figure 7.2 below, an excessive regenerator dense bed level can increase losses by increasing dipleg backup and flooding the cyclone.

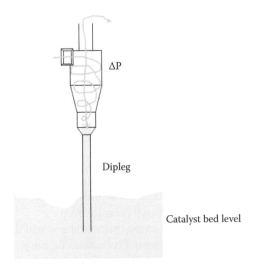

FIGURE 7.2 Catalyst buildup in dipleg. (Reprinted from Jack Wilcox, R., Published in Petroleum Technology Quarterly, 'Troubleshooting Complex FCCU Issues,' http://www. ePTQ.com, Q3, 2009. With permission.)

The backup, or maximum catalyst level in the dipleg, should terminate at least two feet below the dust hopper. The following steps may help relieve a flooded cyclone situation:

- The regenerator dense bed level should be reduced within unit constraints; that is, avoid more of an increase in afterburn (i.e., difference between the temperature of the regenerator dense bed and dilute phase).
- If an air compressor head is available, increase regenerator pressure, reducing cyclone vapor and solids loadings.
- Reduce combustion air rate; this may require a feed rate or operating severity reduction.

Maintaining an optimum regenerator dense phase bed level is essential to stable operation. Increasing the bed level may be required if a "streaming flow" condition develops in the dipleg. Cyclone diplegs typically operate with a relatively dense phase below the dust bowl. However, under certain conditions, the catalyst flow down the dipleg may be in a relatively dilute phase due to an increased quantity of entrained vapors flowing down with the catalyst. This "streaming flow" condition may result from [2]:

- High solid mass flux down the dipleg
- Oversized cyclone diplegs (this is typically more common in the second stage cyclone)
- Insufficient cyclone pressure drop; this may occur during start-up or turn-down operations
- Excessively long diplegs

Excessive vapor entrainment down the dipleg can increase erosion and possibly catalyst attrition. On the reactor side, excessive entrainment will send more cracked product vapors to the stripper.

In order to minimize vapor entrainment down the cyclone dipleg, ensuring the primary cyclone diplegs are sufficiently submerged in the dense bed and maintaining the dipleg valves on the secondary cyclone diplegs are essential.

7.2.3   CATALYST ATTRITION IMPACT

Optimizing the fresh catalyst physical properties including particle density, PSD, and attrition resistance is critical to maintaining acceptable fluidization and resulting circulation of the catalyst inventory. Excessive attrition of the catalyst will lead to nonuniform fluidization and disrupt circulation. Potential sources of attrition include:

- Damaged distributors causing high velocity impingement of catalyst; velocities in excess of 300 feet/s (90 m/s) will break most catalysts
- High velocity impingement of catalyst on the refractory lining in the cyclones
- Improperly installed, designed, eroded, or missing flow/restriction orifices, primarily in steam service

- Wet steam purge or aeration source
- Low fresh catalyst attrition resistance

If the catalyst losses are increasing from the regenerator as measured by increased third stage, precipitator, or scrubber loading, the following options may help to reduce losses and maintain operation until the unit can be shut down for repair:

- Reduce fresh catalyst loading; the fresh catalyst intrinsic activity should be adjusted to maintain a constant equilibrium catalyst performance
- Utilize a denser, coarser grade PSD fresh catalyst while ensuring that the fluidization properties remain acceptable
- Reduce solids and vapor loading to cyclone by raising pressure
- Utilize oxygen enrichment to reduce superficial velocity, reducing solids entrainment to the cyclones
- Reduce FCC feed rate, effectively reducing catalyst circulation and cyclone loading
- Adjust operating conditions to reduce catalyst circulation rate

Note that a higher particle density, not necessarily catalyst ABD, raises cyclone efficiency due to increased centrifugal force as well as reduced solids entrainment to the cyclone.

7.2.4 CATALYST CARRYOVER TO THE MAIN COLUMN

Carryover of catalyst from the reactor disengager vessel to the main fractionator is always a concern and particularly susceptible during start-up operations. The modification of the riser termination as part of a short contact time riser revamp, depicted in Figure 7.3, has increased the concern for catalyst carryover.

Indications that catalyst carryover is occurring include:

- The primary disengager outlet temperature will steadily increase
- The upper (reactor) cyclone outlet temperature will increase
- The cyclone system pressure drop will increase
- The regenerator dense phase level will decrease while the disengager catalyst level remains constant
- The main fractionator inlet and bottoms temperatures will increase
- The regenerator dense bed temperature will increase
- The catalyst content of the fractionator bottoms and slurry circuit will increase

If catalyst carryover starts, the following measures will help avoid a lengthy shutdown:

- Immediately close the regenerated catalyst slide (plug) valve completely; partial closure of this valve usually does not stop the catalyst losses.
- Perform frequent analyses or visual checks of the fractionator bottoms for solids content.

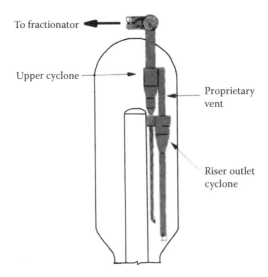

FIGURE 7.3 Closed cyclone system. (Reprinted from Jack Wilcox, R., Published in Petroleum Technology Quarterly, 'Troubleshooting Complex FCCU Issues,' http://www.ePTQ.com, Q3, 2009. With permission.)

- Maintaining slurry circuit circulation is critical; if flow is lost, catalyst will settle and prevent restarting the bottoms circulation pump(s).
- Flush the bottoms circuit with raw oil and pump out the inventory to tankage. Removing settled catalyst from the fractionator if hot, light hydrocarbons are present is extremely difficult.
- Catalyst circulation may be restarted when the solids content in the main column bottoms has been reduced to less than 0.5 wt%.

7.2.5 CYCLONE DESIGN CONSIDERATIONS

Cyclone velocities, vapor and solids loadings, and mass flux rates should be periodically reviewed to ensure acceptable cyclone operation.

Regenerator first stage or primary cyclone inlet vapor velocity should be maintained between 65 and 75 ft/s (~20, respectively, 23 m/s) with second stage inlet velocity at about 75–80 ft/s (~23–25 m/s).

Reactor disengager rough-cut cyclones should be operated at about 50–55 ft/s (~15–17 m/s) and primary cyclones at 60–65 ft/s (~18–20 m/s). Disengager second stage cyclones should operate with an inlet vapor velocity of 60–75 ft/s (~18–23 m/s).

Both the reactor and regenerator first stage cyclones diplegs mass flux rate should be maintained at about 150 lb/ft^3-s (~2400 kg/m^3-s) for optimal performance and reliability. A mass flux rate of about 75 lb/ft^3-s (~1200 kg/m^3-s) is typical for the second stage cyclone diplegs in both reactor and regenerator. Diplegs on all cyclones are typically designed to operate about 90% full at the maximum dense bed catalyst level. Two stage cyclone systems are typically designed to operate with about 2 lb/in^2 (0.14 bar) pressure drop. Operating within these conditions reduces the likelihood of increased catalyst losses due to attrition or abnormal cyclone wear.

7.3   CATALYST CIRCULATION

Diagnosing and rectifying an unstable catalyst standpipe operation can be extremely challenging. Poor standpipe operation leads to erratic catalyst circulation resulting in potential unit upsets, conversion and yield loss, and mechanical damage. Indications of impending catalyst circulation and standpipe problems include:

- Increased vibration or bouncing of the standpipe
- Erratic slide/plug valve differential pressure
- Low slide/plug valve differential pressure
- Erratic riser outlet temperature control
- No change in catalyst circulation when the slide/plug valve changes position

7.3.1   Diagnosing the Problem

As a basis for diagnosing catalyst circulation problems, a reliable standpipe pressure survey must be available, from which a detailed pressure balance may be generated. This pressure balance provides a key tool for diagnosing standpipe operation. The pressure measured at any depth in the standpipe should be approximately proportional to the density of the aerated catalyst and the height of the catalyst above the measurement location. As such, ideally, pressure increases linearly with depth in the standpipe, providing sufficient pressure head at the inlet to the slide/plug valve to maintain adequate valve control. An ideal standpipe pressure profile is illustrated below in Figure 7.4. Nonfluidized solids can support their own weight against the walls of their container. If the circulating catalyst becomes defluidized, it will start supporting a portion of its weight against the standpipe walls, resulting in reduced

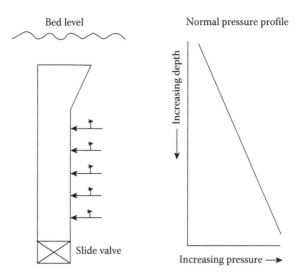

FIGURE 7.4   Pressure in standpipe in case of smooth operation. (Reprinted from Jack Wilcox, R., Published in Petroleum Technology Quarterly, 'Troubleshooting Complex FCCU Issues,' http://www.ePTQ.com, Q3, 2009. With permission.)

pressure generation at the slide (or plug) valve at the bottom of the standpipe. Performing a standpipe pressure survey, and comparing to the reference profile, will usually provide insight into what is occurring inside the standpipe, and may help isolate areas of localized defluidization.

The pressure profile for a standpipe experiencing a moderate or severe catalyst circulation problem will no longer be linear, likely reflecting a loss of pressure buildup below an intermediate aeration tap. The non-linear pressure profiles, illustrated in Figure 7.5, reflect abnormal standpipe operation due to incorrect aeration rates, standpipe obstructions, deteriorating equipment performance (i.e., cyclones), causing the circulating catalyst fluidization properties to change, or changes in the intrinsic catalyst physical properties.

7.3.2 STANDPIPE AERATION

Many FCC units have undergone significant revamp to provide increased feed rate, increased conversion, or increased operating severity. In order to maintain optimum catalyst circulation capability following mechanical modifications and operating condition changes the standpipe aeration system must be evaluated. Excessive aeration gas added to the circulating catalyst will generate gas bubbles, potentially large enough to act as obstacles, impeding uniform catalyst flow as shown in Figure 7.6. Insufficient aeration gas may allow the catalyst to revert to a packed bed regime, again obstructing uniform catalyst flow through the standpipe.

As the fluidized catalyst descends the standpipe, the increasing pressure compresses the fluidizing gas resulting in a decrease in the gas volume. If allowed to continue without adding aeration, the flowing catalyst will defluidize leading to unstable flow and potential loss of catalyst circulation. This is particularly true

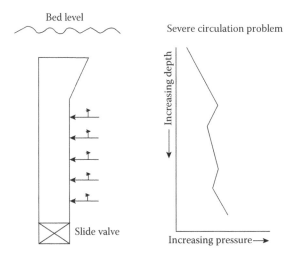

FIGURE 7.5 Pressure in standpipe in case of circulation problems. (Reprinted from Jack Wilcox, R., Published in Petroleum Technology Quarterly, 'Troubleshooting Complex FCCU Issues,' http://www.ePTQ.com, Q3, 2009. With permission.)

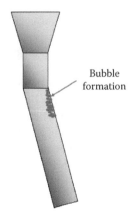

Bubble
formation

FIGURE 7.6 Avoid bubble formation in standpipes. (Reprinted from Jack Wilcox, R., Published in Petroleum Technology Quarterly, 'Troubleshooting Complex FCCU Issues,' http://www.ePTQ.com, Q3, 2009. With permission.)

with units configured with long vertical standpipes. Aeration is required to supply just enough additional gas to the flowing catalyst to maintain a uniformly fluidized system.

Restriction orifices are typically used in each aeration tap to regulate the gas flow. These orifices should be checked for proper size and installation. Pressure drop across the orifices should be sufficient to maintain a constant aeration flow regardless of downstream pressure variations due to normal process changes. The pressure at the orifice inlet should be no less than 10 lb/in^2 (~0.7 bar) greater than the standpipe pressure. These orifices are intended to operate in the critical flow regime along the entire length of the standpipe. Many units utilize flow meters and control valves on each aeration tap for better control of the standpipe aeration. The aeration taps are typically spaced 6–8 feet (1.8–2.4 meters) apart to minimize the chance of any one tap introducing enough gas to allow the formation of large bubbles. Sufficiently large bubbles will impede uniform catalyst flow through the standpipe. Spacing the aeration taps any further apart may allow the flowing catalyst to defluidize due to aeration gas compression resulting from increasing static head. With this spacing, the pressure drop between taps is typically about 1.5 lb/in^2 (0.1 bar).

Assuming a catalyst density at flowing conditions in the standpipe of about 90% of the catalyst bulk density, the amount of excess gas above minimum fluidization that is entrained with the catalyst into the standpipe may be calculated. Sufficient aeration should be added to sustain minimum fluidization along the length of the standpipe.

The choice and properties of the aeration gas are important factors for maintaining stable standpipe operation. The condensate source for steam aeration can cause several problems. If the steam is not kept dry, the condensate can lead to stress cracking of the tap piping, plugging of the tap nozzle with "mud," erratic aeration rates, orifice erosion, and potentially catalyst attrition. Similar problems can occur with wet fuel gas as an aeration source. When possible, dry air and/or nitrogen are preferred rather than steam as aeration media for standpipes. However, in actual operation, air must be limited to the regenerated catalyst standpipe to avoid any possible contact with hydrocarbon. Nitrogen is usually too expensive to use as standpipe aeration.

The circulating catalyst physical properties have a direct impact on fluidization and stable standpipe operation. Mechanical problems may cause a loss of catalyst fines, or a change in catalyst density both of which will impact fluidization and may require adjustment to the standpipe aeration.

Overaeration of an unstable standpipe is a common response to process or catalyst changes not easily recognized. Factors influencing standpipe operation such as catalyst mass flux rate, catalyst density and PSD, standpipe aeration and pressure profile, and unit configuration should be thoroughly evaluated before making large adjustments to the aeration.

7.3.3 PRESSURE BALANCED OPERATION

While not as common, some units rely on the reactor/regenerator pressure balance to control catalyst circulation rather than a slide or plug valve. Since stable reactor temperature control is dependent on uniform catalyst circulation, maintaining smooth, uniform circulation is essential. Assuming the catalyst standpipes are correctly aerated, the key to maintaining smooth circulation is the operation of the withdrawal or overflow well. The catalyst level in the well must be stable for smooth catalyst flow. A fluctuating catalyst level will cause the reactor temperature to cycle. nonideal fluidization in the well or standpipe will cause an unstable level. A low catalyst level in the well constrains the circulation rate, while a high level suggests an obstruction in the well or standpipe.

7.4 COKING/FOULING

The occurrence of coke deposition throughout the reactor system including the riser, reactor disengager vessel, cyclones, overhead transfer line, main column bottoms, and slurry circuit is not unusual [3]. The most probable mechanism of coking is thermal decomposition [4] of heavy feed and unconverted high molecular weight paraffin molecules. Temperatures through the entire system are high enough to initiate the formation of free radicals and subsequent thermal cracking, polymerization, and finally condensation of the heavy polynuclear aromatic compounds.

Indications of a fouling/coking problem include increasing pressure drop through the reactor system, decreased main column bottoms heat removal capability, reduced main column bottoms steam generation, and more frequent cleaning of the coke strainers.

7.4.1 COKE DEPOSIT FORMATION

The precursors for the formation of heavy polymer and coke deposits are initially formed as a result of an ineffective or damaged feed injection system. Loss of pressure drop across the injector nozzle(s) is an indication the flow has been lost; an increase in pressure drop indicates a plugged nozzle. In either case, the catalyst to oil contacting will no longer be optimum leading to increased thermal cracking and production of diolefinic compounds. Diolefins are extremely reactive and rapidly polymerize and condense as coke. Maintaining the feed injector operation within the design operating range is critical to minimizing the formation of diolefins. Unvaporized hydrocarbon droplets resulting from inadequate atomization at the injector readily agglomerate to form coke precursors on any available cold surface. Processing residue feed aggravates this problem, as vaporization of the oil becomes more difficult.

The heavy polymeric hydrocarbons will condense as coke on any heat sink or cold spot present in the unit. As shown in Figure 7.7, coke has deposited directly above the feed injection nozzles on the riser wall, on the disengager vessel walls and dome, in the spent catalyst stripper, on cyclone exterior surfaces and inside cyclone barrels, at instrumentation nozzles, covering the inside surface of the transfer line from the reactor disengager to the main column, at the inlet nozzle to the fraction-ator, and throughout the slurry circuit. The reactor system is particularly vulnerable to stagnant coke deposition with the introduction of feed during start-up operations due to:

- Incomplete feed vaporization
- Nonuniform mixing of feed and catalyst at the feed injection point
- Low catalyst circulation and resulting catalyst to oil
- Low feed temperature
- Low regenerated catalyst temperature
- Low reactor disengager internal temperature
- Low cyclone inlet velocity

During subsequent thermal cycling due to intermittent shutdowns/start-ups, the resulting differential expansion and contraction between the coke deposit and the metal surface will cause the coke to spall in large pieces. These falling coke chunks can cause the following problems:

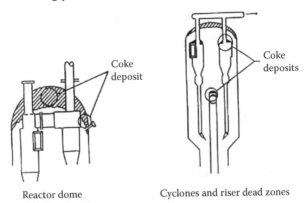

Reactor dome   Cyclones and riser dead zones

FIGURE 7.7 Coking problems—reactor coke deposits. (Reprinted from Jack Wilcox, R., Published in Petroleum Technology Quarterly, 'Troubleshooting Complex FCCU Issues,' http://www.ePTQ.com, Q3, 2009. With permission.)

- Disrupt uniform catalyst flow and catalyst to steam contacting in the spent catalyst stripper zone as reflected by reduced stripping efficiency.
- Increased catalyst losses due to a plugged or severely restricted cyclone(s) dipleg(s) as reflected by increased slurry circuit fouling.
- Partial or complete loss of slide/plug valve control.

Transfer line coke deposition is common, and is normally not disturbed during a turnaround. However, excessive coke formation due to insufficient insulation at bumpers or guides, or at the fractionator inlet nozzle can potentially spall and foul the main column bottoms/slurry circuit. Excessive coke deposition in the vapor line will increase pressure drop, effectively reducing wet gas compressor capacity. Increasing the reactor/regenerator pressure to compensate for the added vapor line pressure drop will reduce air blower capacity, assuming the blower is operating at maximum rate. For both, the operating severity and/or FCC feed rate will be reduced.

In addition, as the pressure balance is obviously affected when the reactor pressure is increased to compensate for increased vapor line pressure drop, catalyst circulation and resulting catalyst/oil ratio may be reduced due to low slide (plug) valve differential pressure.

7.4.2   MAIN COLUMN COKING

A typical FCCU Main Column bottoms circuit is shown in Figure 7.8 below. Thermal gradients due to nonuniform distribution of the recirculated main column bottoms pump-around may be present in the bottom of the main column. Localized bottom pool temperatures combined with relatively high residence time will be high enough to initiate thermal cracking of any unconverted high molecular weight paraffin species present in the bottom of this tower. Many units measure the main column bottoms outlet, or suction line to the bottoms pump. This temperature reflects the tower bottoms pool temperature assuming the bottoms quench is uniformly distributed across the tower cross sectional area. However, the unquenched liquid and quench are typically not thoroughly mixed, causing the localized thermal gradients. The resulting thermal decomposition of the polynuclear aromatic compounds

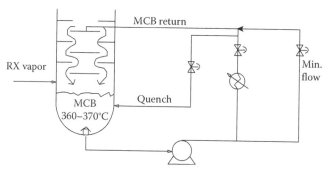

FIGURE 7.8 Slurry circuit. (Reprinted from Jack Wilcox, R., Published in Petroleum Technology Quarterly, 'Troubleshooting Complex FCCU Issues,' http://www.ePTQ.com, Q3, 2009. With permission.)

will leave coke in the bottom of the fractionator and foul the heat exchangers with a varnish-like polymer deposit. This fouling layer reduces heat transfer severely impairing heat removal capability. Since the main column bottoms heat removal can account for as much as 40% of the total heat removal, the impact on FCC feed rate, cracking severity, and product distribution and quality may be significant.

During start-up operations the fractionator and slurry circuit surfaces are relatively cold providing a heat sink for the deposition of these heavy polymers and coke. In order to minimize fouling during initial operations, the quench rate should be increased; reducing bottoms pool temperature, effectively increasing exchanger flow rates to maintain required heat removal. The increased slurry velocity through the tubes will reduce both the polymer-related fouling and coke deposition.

The choice of the appropriate catalyst system will have an impact on the potential formation of the heavy polymers and coke. Cracking the high molecular weight precursors catalytically will significantly reduce the possibility of thermally degrading these components. The zeolite activity should be optimized in combination with an active matrix selective to upgrading the heavy feed components.

State of the art riser termination devices have significantly reduced the coke deposition in the reactor disengager vessel. These modifications have significantly reduced the hydrocarbon residence time and potential thermal cracking in the disengager.

7.5 SPENT CATALYST STRIPPING

The spent catalyst stripper, shown in figure 7.9, is intended to remove entrained and a small portion of absorbed hydrocarbons from the spent catalyst prior to catalyst regeneration by minimizing the carry-under of hydrogen rich hydrocarbons to the

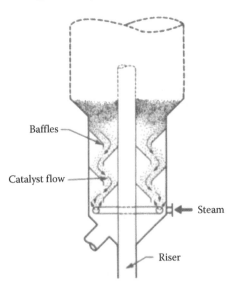

FIGURE 7.9 Catalyst steam stripping—an annual stripper configuration. (Reprinted from Jack Wilcox, R., Published in Petroleum Technology Quarterly, 'Troubleshooting Complex FCCU Issues,' http://www.ePTQ.com, Q3, 2009. With permission.)

regenerator. Stripping is intended to be a mass transfer process with stripping steam moving upward countercurrently to the downward moving spent catalyst. Strippable hydrocarbons not separated in the stripping zone will be entrained with the spent catalyst to the regenerator, burned, and contribute to increased delta coke and catalyst deactivation. The resulting impact on heat balance will adversely affect conversion, yields, and profitability.

7.5.1 DIAGNOSING STRIPPER PROBLEMS

The stripper performance may be evaluated by sampling the spent catalyst standpipe for both hydrogen in the coke on catalyst and the relative amounts of water and hydrocarbon leaving the stripper. If debris disrupting the catalyst and/or steam flow patterns is suspected, tracer scans can be used to identify the location of the blockage [5].

- **Excessive spent catalyst mass flux rate**

    Cold flow tests indicate that a too high superficial mass flux in the stripper causes a dramatic decrease in stripping efficiency. The normal stripping steam flow pattern is disrupted to the point where stripping steam becomes entrained downward [6].

    Catalyst mass flow rates exceeding about 1600 lb/ft²-min (7800 kg/m²-min) results in poor steam/catalyst contacting, flooded trays, insufficient catalyst residence time, and increased steam entrainment to the spent catalyst standpipe. This is reflected by the stripper efficiency and catalyst density shown in Figure 7.10. The primary concern is hydrocarbon entrainment to the regenerator leading to loss of product, increased catalyst deactivation, increased delta coke and potential loss of conversion and total liquid yield, and feed rate limitation. A rapid decrease in stripper bed density is an indication that

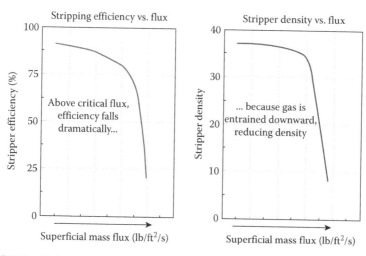

FIGURE 7.10   Stripper mass flow limitation. (Reprinted from Jack Wilcox, R., Published in Petroleum Technology Quarterly, 'Troubleshooting Complex FCCU Issues,' http://www.ePTQ.com, Q3, 2009. With permission.)

a mass flux limit has been exceeded. While the absolute number is suspect, monitoring the hydrogen level in the coke on a weight basis is typically used to indicate changes in the stripper performance. Hydrogen in coke of about 5–6 wt% generally indicates good stripper operation. Increasing hydrogen in coke reflects increasing hydrocarbon carry-under to the regenerator. While not a routine procedure, sampling and analyzing the spent catalyst and vapor exiting the stripper for hydrocarbon types will also provide an indication of stripper performance.

- **Spent catalyst maldistribution**
  Channeling and short circuiting of partially stripped catalyst to the spent catalyst outlet may occur, particularly in asymmetric/annular strippers. As there is typically inadequate temperature monitoring to detect radial maldistribution, this type of problem may be confirmed by gamma scans.

- **Insufficient stripping steam rate**
  The stripping steam rate should be adjusted to maximize stripping; that is, minimize hydrocarbon carry-under to the regenerator. In practice, the stripping steam rate should be increased until there is no visible decrease in regenerator temperatures, and cyclone loadings and sour water handling capability are not exceeded.

- **Stripping steam maldistribution**
  Mechanical damage to the steam distributor or trays is the most common cause for maldistribution. Nonuniform distribution leads to steam bubbles agglomerating to larger bubbles, reducing stripping effectiveness. As mentioned, wet steam will cause significant nozzle erosion, refractory, and tray damage due to the resulting high velocity steam jet. Tracer scanning is a key diagnostic tool to investigate both steam and catalyst distribution.

- **Erratic, nonuniform catalyst flow**
  This could also reflect mechanical damage to the stripping trays or an excessive stripping steam rate. Typically, strippers are designed for about 0.75–1.0 ft/s superficial velocity. This is sufficient to displace the hydrocarbon and allow uniform catalyst flow downward. In efficient stripper, up to 80% of the steam flows upward to the disengager. Tracer scanning is an excellent diagnostic tool to confirm damage to the stripper internals. Nonuniform radial temperature distribution will also provide an indication of possible damage to the internals or a restriction to catalyst flow.

- **Excessive hydrocarbon under-carry**
  Most FCC units are processing more feed and operating at higher severities than originally designed. This leads to high catalyst circulation rates and high stripper mass flux rates, potentially entraining hydrocarbons to the regenerator. The negative impact on the unit operation has previously been described. If increasing the stripping steam rate has little impact, modification or a new stripper design may be required.

- **Plugged steam distributor nozzles**
  Distributor nozzles may become plugged from refractory, slumped or defluidized catalyst, or coke. The steam rate, as reflected by the distributor pressure drop, must be maintained to minimize the possibility of plugging.

- **Corrosion**

  Catalyst will defluidize and pack in stagnant areas and cool, leading to potential corrosion. This is a common problem in annular strippers with the spent catalyst stripper outlet on one side of the bottom of the stripper. Catalyst, if not kept hot by fluffing with steam, will settle, cool, and lead to potential corrosion to the disengager shell and riser.

7.6   REGENERATOR AFTERBURN

The oxidation of carbon monoxide to carbon dioxide is extremely exothermic and must be carefully controlled as much as possible. It is preferable that this reaction be maintained in the regenerator dense phase as the catalyst provides an excellent heat sink for the large heat release. The combustion of carbon monoxide in the dilute phase is referred to as afterburn. Excessive afterburn can result in significant mechanical damage to the regenerator internals and contribute to catalyst deactivation. Many units routinely operate with a limited degree of afterburn, as long as the dilute phase temperature does not exceed the metallurgical limit of the regenerator internals.

7.6.1   INDICATIONS OF AFTERBURN

- Increasing regenerator dense/dilute phase temperature differential
- Increased carbon on regenerated catalyst
- Possible decrease in conversion
- Possible increase in main column bottoms yield

The following steps may be taken to reduce excessive afterburn:

- Marginal use of torch oil may consume some oxygen that contributes to afterburn
- Utilize combustion promoter to catalyze oxidation of carbon monoxide in the dense bed; if already in use, increase addition rate [7]
- Raise feed preheat temperature if possible
- Raise regenerator pressure to increase burning rate in the dense bed
- Repair or revamp the air and/or spent catalyst distributor(s); maintain adequate air distributor pressure drop
- Temporarily stop oxygen enrichment, if in use
- Increase regenerator dense bed level to increase residence time and minimizing channeling of oxygen and/or carbon monoxide through the dense bed

7.7   SUMMARY

Each of the above problems occurs periodically on most FCC units. Recognizing the symptoms and rectifying the upset condition in a timely manner is essential to maintaining a stable and profitable operation while minimizing unwanted feed outages.

REFERENCES

1. Wilcox. J. Diagnosing Catalysts Losses. *Albemarle Catalysts Courier*, Issue 72, 2008.
2. Knowlton, T., and Brevoord, E. PRSI Offers Valuable Insight into Cyclone Dipleg Efficiency. *Albemarle Catalysts Courier*, Issue 65, 2006.
3. McPherson, L. J. *Reactor Coking Problems in Fluid Catalytic Cracking Units*. Akzo Catalysts Symposium, Scheveningen, The Netherlands, 1984.
4. Brevoord, E., and Wilcox, J. Coke Formation in Reactor Vapor Lines of Fluid Catalytic Cracking Units: A Review and New Insights. In *ACS Symposium Series 571* Edited by M. L. Occelli and P. O'Connor. Washington, DC: ACS, 1994.
5. Wilcox J. Optimizing Spent Catalyst Stripping to Enhance FCC Unit Performance. *Albemarle Catalysts Courier*, Issue 71, 2008.
6. Miller, R. B., Yang, Y-L., Gbordzoe, E., Johnson, D. L., Mallo, T. A. New Developments in FCC Feed Injection and Stripping Technologies. Paper AM-00-08, NPRA Annual Meeting, 2000.
7. Yung, K. Y., and Wilcox, J. FCC Operation and Troubleshooting—Part 2. *Akzo Catalyst Courier* 15 (1992).

8 Catalytic Cracking for Integration of Refinery and Steam Crackers

Dilip Dharia, Andy Batachari, Prashant Naik, and Colin Bowen

CONTENTS

8.1 INTRODUCTION

The key refining process that can provide petrochemical advantages is the fluidized catalytic cracking (FCC) process. The related commercial processes are deep catalytic cracking (DCC) and catalytic pyrolysis process (CPP). The DCC process provides high yields propylene (15–25 wt%) depending upon the type of feedstock and operating conditions. Other coproducts include ethylene-rich dry gas, aromatic gasoline, and cycle oil fractions. The CPP process operates at higher severity compared to DCC and uses a bifunctional catalyst producing even higher propylene and ethylene yields.

8.2 DCC PROCESS

Deep catalytic cracking (DCC) is a commercially proven FCC process for selectively cracking a wide variety of feedstocks to light olefins, particularly propylene. Innovations in catalyst development, operational severity, and anticoking conditions, enable the DCC process to produce significantly more olefins. Typical DCC unit feedstock components are listed below:

Feedstock and Abbreviations	
VGO	Vacuum Gas Oil
HVGO	Heavy Vacuum Gas Oil
HTVGO	Hydrotreated Vacuum Gas Oil
AGO	Atmospheric Gas Oil
LSWR	Low Sulfur Waxy Residue
ATB	Atmospheric Tower Bottoms
VTB	Vacuum Tower Bottoms
HCR	Hydrocracker Residue
DAO	Deasphalted Oil

Should it be necessary to reduce sulfur levels and/or increase hydrogen content, then the feedstocks may be hydrotreated.

DCC is a FCC style process developed approximately 15 years ago by Sinopec RIPP [1,2]. Its objective is to convert the feedstock to propylene and gasoline. Shaw (Stone & Webster at that time) licensed several RFCC units to Sinopec. This relationship led to an exclusive partnership for the DCC process and incorporated several of our FCC/RFCC mechanical design aspects.

Propylene yield is increased by processing feedstocks with high hydrogen content. Key features that allow catalytic conversion of fresh feed into propylene yields of 15–25 wt% in the DCC process are

- Modified high severity operating conditions (high reaction temperature of 570–580°C and cat/oil ratio of 10–15)
- Bed cracking to achieve higher conversion of naphtha due to increased residence time
- Low hydrocarbon partial pressure with higher steam rate (total pressure of about 12–15 psig and steam usage of 25 wt% of the feed)
- Light naphtha recycle to maximize conversion of naphtha material to propylene
- Proprietary DCC catalyst

The DCC typical process operating conditions are presented below:

Process Variable	DCC
Reactor temperature °C	530–580
Reactor Pressure Atm. (g)	1–2
Residence time, seconds	2–10
Catalyst to oil ratio, wt/wt	8–15
Dispersion steam, wt% feed	10–30
Cracking environment	Riser & Bed

TABLE 8.1
Typical DCC Yields vs. FCC and Steam Cracking

Wt% on Feed	DCC	FCC with ZSM-5	Steam Cracking
Hydrogen	0.2	0.1	0.6
Dry gas (C1–C2)	11.0	3.5	44.0
LPG	42.5	26.5	25.7
(C2 = Olefin)	5.5	1.2	28.2
(C3 = Olefin)	19.5	8.5	15.0
(C4 = Olefin)	13.5	8.3	4.1
Naphtha (C5–205°C)	26.5	41.8	19.3
Light cycle oil (205–330°C)	9.5	14.5	4.7
Heavy cycle oil (330°C+)	4.3	8.4	5.7
Coke	6.0	5.2	—

For comparative purposes the typical weight percentage yields for a DCC unit, an FCC unit and a steam cracker are shown in Table 8.1. Propylene yields from the DCC unit are considerably higher than those from an FCC unit. The DCC mixed C4s stream also contains increased amounts of butylenes and iso-C4s as compared to an FCC. These high olefin yields are achieved by deeper cracking into the aliphatic components of the initially produced naphtha and life cycle oil (LCO).

The dry gas produced from the DCC process contains approximately 50% ethylene. The cracking reactions are endothermic, and compared to FCC, a higher coke make is required to satisfy the heat balance.

A typical FCC operation can yield around 4–5 wt% propylene. The inclusion of ZSM-5 additives (Zeolite structural matrix-5) coupled with high reactor operating temperature (ROT) yields 5–10 wt% propylene with additive concentrations up to 10 wt% as per the industry standard as shown in the table below. This is based on data from commercially operating units around the world. A comparison of propylene yield from various catalytic cracking options is summarized below:

Option	Propylene Yield, wt%
High severity FCC	3–5
High severity FCC plus ZSM-5 catalyst additive up to 10 wt%	5–10
DCC	15–25

8.2.1 COMMERCIAL STATUS OF DCC PROCESS

Table 8.2 provides a list of licensed DCC units. This table provides start-up date and feedstock information. In mid-1997, the first totally designed and engineered DCC complex was successfully commissioned in Thailand for Thai Petrochemical

TABLE 8.2
List of Licensed DCC Units

No	Location	Capacity, kt/a	Startup Date	DCC Type	Feedstock
1	SINOPEC Jinan, China	150	June 1994	I & II	VGO + DAO
2	SINOPEC Anqing, China	650	March 1995	I	VGO + CGO
3	PETROCHINA Daqing, China	120	May 1995	I	VGO + ATB
4	IRPC (TPI), Thailand	900	May 1997	I	VGO + WAX + ATB
5	SINOPEC Jingmen, China	800	September 1998	II	VGO + VTB
6	CHEMCHINA Shengyang, China	400	October 1998	I	VGO + VTB
7	PETROCHINA Jinzhou, China	300	September 1999	I & II	ATB
8	BLUESTAR Daqing, China	500	October 2006	I	ATB
9	Petro-Rabigh, Saudi Arabia	4600	2009	I	HTVGO
10	HMEL (GGSRL), Punjab, India	2200	2011	I	HTVGO
11	BLUESTAR Tianjin, China	1600	2011	I	HTVGO
12	MRPL, India	2200	2011	I	HTVGO
13	CNOOC Hainan Olefins Project, Hainan, China	1200	2011	I	ATB
14	Yangchang Corp, Shaanxi, China	1500	2012	I	ATB
15	Fushin Mineral Corp Ltd, Liaoning, China	500	2013	I	HTVGO
16	JSC TANECO, Nizhnekamsk, Former Soviet Union	1100	2013 (on hold)	I	HTVGO and hydrocracker bottoms

Total licensed capacity of about 20 million ton per annum (Approx. 400,000 bpsd)

Industries [3]. Six other DCC units are in operation with a total operating experience of more than 50 years. Total licensed capacity is about 20 MMTA. Two major grassroots units currently under construction were licensed to the joint venture of Saudi Aramco & Sumitomo, and to JSC Taneco, Nizhnekamsk, to produce polymer grade propylene as shown in Table 8.2. In addition, RIPP Sinopec recently licensed three DCC units in China and Shaw licensed two DCC units in India.

8.2.2   APPLICATION OF DCC IN INDIA

Two DCC projects are underway as listed in Table 8.2.

8.2.2.1   HPCL-Mittal Energy Limited (HMEL)

A joint venture between Hindustan Petroleum Corporation Limited (HPCL), and Mittal Investments SARL is setting up a 9.0 million metric tons per annum (MMTPA) refinery at Bathinda, in the state of Punjab, India for producing transport fuels of EURO IV specifications. Engineers India Limited (EIL) is retained as project management consultants (PMC) for the entire complex. DCC unit is one of the key units in this grassroots complex. One of the key products from the DCC unit is polymer grade propylene for the poly propylene unit that will be integrated with the refinery.

8.2.2.2   Mangalore Refinery and Petrochemicals Limited (MRPL)

The Mangalore refinery is implementing a Phase-III Refinery Project and will have a crude processing capacity of 15 MMTPA after implementation. MRPL has retained EIL as the PMC for this project. The Phase-III refinery project envisages new facilities that will include Petrochemical FCC or DCC. The key objective of the Petrochemical FCC or DCC unit is to provide secondary processing capacity and to generate petrochemical feedstock, polymer grade propylene.

8.3   CATALYTIC PYROLYSIS PROCESS (CPP)

The next generation of heavy feed catalytic cracking employs a higher temperature modified catalyst that produces light olefins via both carbenium scission and free radical initiation. This process, referred to as CPP process has also been developed by Sinopec RIPP [4], and is at the point of commercialization (commercial unit started in 2009, refer to figure 8.1). The ideal feedstock type is also LSWR, much of which is available in China, Southeast Asia, and North Africa. Operating conditions (ROT; Cat:Oil; S:HC) are significantly higher than those of FCC but considerably lower than equivalent steam cracker conditions.

CPP olefin yields are considerably higher than those of DCC. Depending upon the selected conditions, ethylene/propylene yields can range from 21/18 to 10/25 wt%. Butylene yields range from 7 to 13 wt% and the gasoline fraction is highly aromatic (up to 80 wt%). The key distinction of a CPP unit is the much higher levels of contaminants contained in CPP reactor effluent, which determines to a major extent the sequence of the various clean-up and fractionation stages.

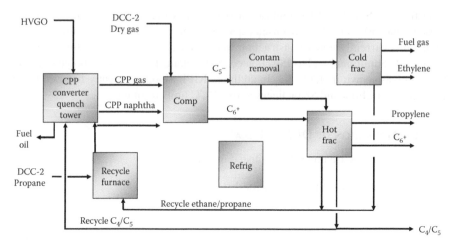

FIGURE 8.1   CPP unit block flow diagram.

8.3.1   COMMERCIAL STATUS OF CPP PROCESS

The commercial prototype CPP unit is at Shenyang Paraffin Wax Co. in China. This is linked with an earlier DCC unit and contains various steam-cracking features. These include recycle ethylene to propylene (E/P) cracking, hot-water belt quench water recovery, cracked gas/acid gas removal, front-end DeC_3 together with various contaminant removal stages, polymer grade ethylene and propylene fractionation, and CPP coproduct recycle cracking.

Since CPP cracking effluent is molecularly similar to that of heavy distillate cracking it will be logical to construct integrated CPP-steam cracker units, or even to add a CPP reactor–regenerator as a revamp side-cracker expansion feature. Various plans are under review for such prospective projects.

8.4   INTEGRATION WITH STEAM CRACKER

Steam crackers provide the traditional cost-effective approach for olefins production from lighter feed stocks such ethane, propane, naphtha, and AGO. However, these options typically provide higher E/P ratio. To meet the increasing demands of ethylene and particularly propylene, refiners and petrochemical producers are planning integrated facilities. The objectives are:

- To attain higher P/E ratio
- To maximize production of olefins from refinery-based heavier feedstocks
- To extend the olefins feedstock flexibility
- To reduce cost of olefins production
- To recover ethylene and propylene from FCC off gases
- To recycle refinery co-product streams such as ethane, propane ... to steam crackers

The petrochemical plant and refinery integration schemes offer lower cost routes to incremental ethylene/propylene production either via revamp modifications or in grassroots application [5,6].

A major current refinery-petrochemical project is under construction for Saudi Aramco–Sumitomo Chemical at their Rabigh, KSA site. In conjunction with our JGC partner we have linked the upstream refinery expansion to a combined mega-DCC unit and mega-ethane steam cracker. Corresponding production rates are 1500 kta ethylene/950 kta propylene. The corresponding integrated layout is shown below in Figure 8.2.

A project of this type achieves major economic advantages. Base petrochemical feedstocks are hydrotreated VGO and ethane. Both are low cost at this site. The combined DCC and ethylene unit reduces ISBL recovery cost significantly due to integration and recycle feedstock access.

Larger scale DCC units (and associated FCC units) offer greater product recovery options. These are particularly attractive when linked to either an existing or new steam cracker. In such schemes several of the individual unit operations can be combined or substituted by a more efficient system.

- Common dilution steam generation system
- Linked wet gas/cracked gas compressor duty
- Dry gas recovery to produce specification ethylene
- Common propylene fractionation system
- Linked cycle oil-quench oil systems
- Common contaminant removal for certain impurities
- Separate C2/C3 fractions without sponge absorber
- Improved overall ethylene/propylene yield by recycle cracking of DCC—derived ethane/propane

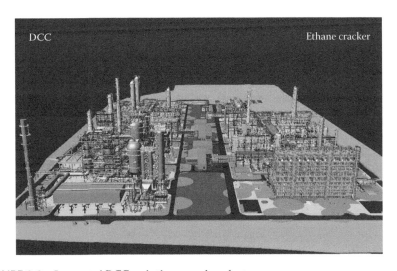

FIGURE 8.2 Integrated DCC and ethane cracker plants.

Another vital part of such projects is contaminant removal. Notwithstanding the hydrogenation of the DCC VGO feedstock, the light hydrocarbon products from such units will contain a range of contaminants at various concentrations depending upon upstream operating conditions:

C_2 and lighter: O_2, NO_x, CO, CO_2, NH_3, AsH_3, NH_3, Hg, acetylenes, dienes, etc.
C_3 fraction: RSH, COS, AsH_3, oxygenates, etc.

This is an area of associated technology that has also been developed and demonstrated in several recent units. We have installed a number of refinery off gas (ROG) units, the purpose of which is to treat and concentrate the combined C_2 and lighter fractions from refinery unsaturated off gas (FCC and delayed cokers).

In our refining and petrochemical industries we need to be more aware of catalytic olefin routes to propylene and ethylene. Several related capacity and cost increases impact this emerging route:

- Increasing propylene demand
- Increasing ethane feedstock, hence less steam cracking propylene production
- Increasing conventional steam cracker feedstock costs
- Increasing energy costs
- Increasing diesel:gasoline ratio

As a consequence of these trends we are increasing various aspects of refining/petrochemical integration. Our DCC and CPP technologies provide the optimum low feedstock/energy cost route to respond to these industry trends.

REFERENCES

1. Zaiting, L., et al, DCC Technology and Its Commercial Experience. *China Petroleum Processing and Petrochemical Technology*, No. 4, 2000.
2. Peiling, Z. Integration of a DCC into a Refinery. *Hydrocarbon Engineering*, October 1997.
3. Fu, A., et al, Deep Catalytic Plant Produces Propylene in Thailand. *Oil & Gas Journal*, January 12, 1998.
4. Xieqing, W., et al, Catalytic Pyrolysis Process (CPP)—An Upswing of RFCC for Ethylene and Propylene Production. AIChE National Spring Meeting, March 2002, (T5a01).
5. Dharia, D., et al, Increase Light Olefins Production. *Hydrocarbon Processing*, April 2004.
6. Chapin, L., Letzsch, W., and Swaty, T. Petrochemical Options from Deep Catalytic Cracking and FCCU. 1998 NPRA Annual Meeting, AM-98-44.

9 Advanced Artificial Deactivation of FCC Catalysts

A. C. Psarras, E. F. Iliopoulou, and A. A. Lappas

CONTENTS

9.1 INTRODUCTION

The worldwide energy management currently demands a more sophisticated utilization of the existing energy sources. The refineries are significantly affected by these energy saving policies requiring improvement/modification of existing or development of new/alternative refining processes. One solution is to upgrade more and more heavier or residual fractions into lighter desirable distillates via catalytic cracking. However, the residual cracking poses numerous problems for the oil companies and the catalyst manufacturers. The increasing portion of larger molecules/compounds containing heteroatoms and metal contaminants in fractions of increasing boiling point mainly accounts for the difficulties in processing heavy oils [1]. In fact, the poisonous metals (mainly Ni and V) are generally in the form of porphyrins that deposit themselves on the catalyst surface. In the case of nickel, the main effect is the favoring of dehydrogenation reactions and the increment in coke selectivity. Similarly to nickel, vanadium is also responsible for the dehydrogenation enhancement, but vanadium is additionally causing a permanent damage of the zeolite structure in the presence of steam at elevated temperatures [2–13]. Thus, problems associated with decreased catalyst stability, activity, and selectivity are becoming more often and more serious during processing residual fluidized catalytic cracking (FCC) feeds, demanding new FCC catalyst technologies. A more detailed knowledge of the deactivation mechanisms could significantly facilitate manufacturing of improved, metal-tolerant FCC catalysts. Moreover, a most realistic and accurate laboratory procedure for the evaluation of such improved FCC catalysts is essential. As a result, one of the biggest challenges in FCC research field is to first simulate how the FCC catalyst is deactivated in a commercial FCC unit and then evaluate its performance in lab-scale testing [14–16]. Several lab-deactivation methods of FCC catalysts are suggested attempting to simulate the E-cat (equilibrium catalyst) properties and characteristics at laboratorial scale. The simplest was proposed by Mitchell and consisted of an incipient wetness impregnation of nickel and vanadium compounds followed by hydrothermal treatment at high temperatures [17]. This approach was abandoned for high metal concentrations. Attempts have been made to develop more realistic bench-scale deactivation tests, focusing on the reproduction of the physicochemical characteristics of the E-cat [16,18]. Unfortunately, most of them overestimate poisonous metal effects leading to incorrect coke and hydrogen selectivity (essential parameters for defining FCC catalyst behavior). This was verified by our earlier work on the conventional methods cyclic propylene steaming (CPS) and cyclic deactivation (CD) [19]. Thus, there is a need for the development of a realistic deactivation technique that would simulate in the laboratory the deactivation of catalysts in a commercial FCC unit, under the combined action of metals, steam, temperature, thermal shock, and so on.

The development and application of most refining processes including catalytic cracking was based mostly on empirical knowledge. The foundation of the technological know-how on a more scientific basis has become essential during the last decades. This need has spurred the research efforts of several groups to describe the chemical and structural properties of the FCC catalysts with sufficient detail and accuracy in order to relate them to the catalytic performance, primarily for a rational basis of further catalyst and process development [20]. Taking for granted that the catalytic cracking of hydrocarbons is realized through the carbocations mechanism, it is easily understood that the catalytic activity of FCC catalysts lies mainly on their acidic properties; that is, the assessment of the concentration, the strength, the density, and the nature of acid sites is an integral part of catalyst development programs. Despite the research efforts on developing a lab-deactivation technique, limited information is available about the changes in catalyst acidity during these deactivations. In our earlier work we investigated the potential benefits of advanced laboratory deactivation methods including the crucial parameter of acidity in our concerns [21].

Vibrational spectroscopy of adsorbed probe molecules is one of the most powerful tools to assess the acidic properties of catalysts. Acidity studies of dealuminated Y zeolites (main active component of FCC catalysts) or other zeolitic catalysts are reported using mostly Fourier Transform Infrared Spectroscopy (FTIR) with CO adsorption at 77 K or FTIR-pyridine/substituted pyridines adsorption at 425 K [22–26]. FTIR acidity studies of commercial FCC catalysts are even more scarce

[23–27]. Among the probe molecules routinely used, pyridine is certainly the most employed. Pyridine is a strong base and easily gives rise to the formation of H-bonded and pyridinium species with weak and strong Brönsted acid sites, respectively, and to coordinated species on Lewis acid sites [28]. The difference of the bonds of pyridine on the two types of acid sites offers the advantage of distinguishing the Brönsted and the Lewis acid sites by creating separate bands for each site within the spectrum. Although pyridine is not able to be adsorbed on the acid sites located inside the sodalite cages of the zeolitic component of the FCC catalyst, this is not a drawback for the method concerning our scope.

We aim to quantify the useful-accessible acid sites. Since the real reactants of the catalytic cracking process are even more bulky than pyridine, they will not be adsorbed on the acid sites located in the hexagonal prisms, but mainly on the acid sites of the supercages. Thus, the acidity measured with pyridine as a probe molecule is more realistic than using a smaller probe molecule like ammonia. Furthermore, the acid site strength distribution can be investigated by monitoring the thermodesorption of pyridine, although this method is just an indication of the acid strength and the correlation between this distribution and the catalyst activity is not yet proven reliable [29].

The scope of the present study is to investigate the time extension of the conventional and the advanced CPS deactivation methods. Moreover, the impact of the deposited metals and their oxidation state during laboratory deactivation on the final properties of the deactivated samples is under research. The correlation of acidity changes during the deactivation with the catalytic performance is inquired for the understanding of the complex phenomenon of deactivation due to interrelation of the several variables in the process.

9.2 EXPERIMENTAL

9.2.1 MATERIALS

A commercial FCC catalyst with high accessibility was supplied by Albemarle Catalysts Company BV and used in the present study (Cat). Definition of accessibility is stated as "the catalyst ability to have active sites accessible to large molecular structures, which are supposed to interact with these sites within a certain time limitation set by the catalytic process" [30,31]. In other words, accessibility is a crucial structural property related with the quick diffusion of the reactants and the products to and from the active sites of the catalytic particle, respectively. Equilibrium sample (E-cat), which is a sample used and deactivated in a commercial FCC unit, was received, characterized and tested. A sample of the corresponding fresh catalyst was also investigated after its deactivation in the laboratory. The CPS deactivation protocol and the improved advanced-cyclic propylene steaming (ADV-CPS) protocol were applied. The samples will be referred as CPS-Cat and ADV-CPS-Cat, respectively. Trials of prolonged protocols of both methods were also realized with 45 and 60 ReDox cycles instead of 30. The samples will be referred as 45CPS-Cat, 60CPS-Cat, 45ADV-CPS-Cat, and 60ADV-CPS-Cat, respectively. Finally, both protocols (CPS and ADV-CPS) were carried

out in the absence of the contaminant metals. The samples will be referred as
NMCPS-Cat and NMADV-CPS-Cat, respectively. The detailed laboratory deactivation procedures are given in a following section. A commercial FCC residual
feedstock was used in all evaluation experiments (S.G.: 0.9169, density (60°C):
0.8856 g/cm^3, density (15.5°C): 0.916 g/cm^3, S:0.6094 wt%, distillation data {wt%,
°C}: {10, 393.5}, {20, 414.5}, {30, 429.3}, {40, 442}, {50, 454.4}, {60, 467.8}, {70,
483.1}, {80, 500.1}, {90, 524.2}, {FBP, 551.6}).

9.2.2 Characterization Techniques

All catalytic samples fresh, commercially (E-cats), or artificially deactivated, were
submitted to a standard series of characterization techniques. More significantly, the
specific surface area and the micropore volume of the catalysts was determined by
nitrogen adsorption (BET method), using an Autosorb-1 Quantachrome flow apparatus. The crystalline structure and especially the unit cell size of the catalysts was
studied recording powder X ray diffraction patterns in a Siemens D500 diffracto-
meter with auto-divergent slit and graphite monochromator, using Cu(Ka) radiation.
The bulk concentrations of deleterious metals and other elements were measured
with ICP/AES analysis, carried out in a Plasma 400 (Perkin Elmer) spectrometer,
equipped with Cetac6000AT+ ultrasonic nebulizer.

The IR spectra were collected using a Nicolet 5700 FTIR spectrometer (resolution 4 cm^{-1}) by means of OMNIC software. Data processing was carried out via the
GRAMS software. All the samples were finely ground in a mortar and pressed in
self-supporting wafers (~15 mg/cm^2). The wafers were placed in a homemade stainless steel, vacuum cell, with CaF$_2$ windows. High vacuum is reached by the means
of a turbomolecular pump and a diaphragm pump placed in series. The infrared cell
was equipped with a sample holder surrounded by a heating wire for the heating
steps and connected to the vacuum line, which is also heated in order to avoid pyridine condensation or its adsorption on the walls. Before IR analysis all samples were
heated at 450°C under high vacuum (10^{-6} mbar) for 1 hour in order to desorb any
possible physisorbed species (activation step). All spectra were collected at 150°C
in order to eliminate the possibility of pyridine condensation. Initially the reference
spectrum of the so-called activated sample is collected. Then adsorption of pyridine
is realized at 1 mbar by equilibrating the catalyst wafer with the probe vapor, added
in pulses for 1 hour. The corresponding bands, used for the quantification of the
Lewis and the Brönsted acid sites are created at 1450 and 1545 cm^{-1}, respectively.
Lambert Beer's law (Equation 9.1) was used for the calculation of the concentration
of the acid sites, normalized however with the mass of the wafers. The spectra are
used in the absorption (A) form:

$$A = -\ln\left(\frac{I}{I_r}\right) = c \times \varepsilon \times d.$$

(9.1)

The symbols I and I_r represent the intensities of the sample and the background (as
a function of energy), respectively, c the concentration, d the sample thickness, and

ε the molar extinction coefficient [32]. There are a number of publications in the literature dealing with the molar extinction coefficients leading to quite different results [26,33–38]. The coefficients chosen in the present study were obtained by the work of Emeis [33].

9.2.3 BENCH-SCALE UNIT FOR THE ARTIFICIAL DEACTIVATION OF FCC CATALYSTS

The CPS and advanced ADV-CPS methods were used for the artificial deactivation of the fresh samples in the laboratory. The description of the standard CPS and the ADV-CPS deactivation procedures is given below.

In the CPS method, V and Ni are initially deposited on the fresh FCC catalytic samples by wet impregnation, followed by deactivation in a series of reduction-oxidation cycles. According to the standard CPS protocol, the samples were at first submitted to a calcination pretreatment in air flow at 205°C for 1 hour and then at 595°C for 3 hours. Then, the wet impregnation was carried out using solutions of nickel and vanadyl naphthenates in toluene [16]. The specific solutions were selected because they mimic metal species in the feed, reported to exist mainly as organic complexes with porphyrins [39]. The metal concentration target in the artificially deactivated samples was the 50% of the metal content of the corresponding E-cats. The solvent was totally removed by heating the mixture at 100–110°C in a rotary evaporator. After that, the sample was submitted to a calcination treatment in air flow at 205°C for 1 hour and then at 595°C for 3 hours in order to decompose the naphthenates. Finally, the catalysts were deactivated via a series of reduction–oxidation cycles at 788°C. It should be underlined that the procedure is starting with a stripping step and finishing with a reducing step. These cycles are repeated up to 30 times to give a total run time of 20 hours.

The ADV-CPS is developed to eliminate the overemphasized effects of vanadium on catalyst performance. Due to burning off of the organic components after impregnation, all the vanadium content is in the +5 oxidation state, which is the most detrimental for FCC catalyst deactivation. So during the first contact of the FCC catalyst with steam, since the first CPS step is not a reduction step, the FCC catalyst is exposed to the attack of the total amount of impregnated vanadium in the + 5 oxidation state. This is not representative of the commercial operation where only the vanadium deposited during the last cracking step is preferably oxidized in the regenerator, and the regenerator is not a perfectly oxidizing environment. Most of the vanadium from previous cracking steps is passivated in the FCC catalysts for instance as rare-earth and alumina vanadates. It is reported in the literature that the ADV-CPS mode showed better agreement as compared with commercially equilibrated catalysts [40].

The main differences from the standard CPS protocol concern:

- The sequence of the steps (stripping–oxidation–stripping–reduction), which was changed to reduction–stripping–oxidation–stripping
- The temperature of the ReDox cycles was increased to 804°C

- The ratio of reduction-time to oxidation-time, which was increased
- The introduction of two prestabilization cycles during the heating-up and a last cycle at the end consisting of only the reductive step [14]

The prolonged trials of both methods were carried out by applying 45 and 60 ReDox cycles, while the experiments in the absence of metals were realized by skipping the wet impregnation step on both methods. The time extended protocols were used in order to perform a deeper investigation of the effect of metals' oxidation state during the laboratory deactivation processes as the metals are being kept in reduced state for a longer time during the ADV-CPS protocol. The application of the two protocols without the presence of the contaminant metals was carried out in order to investigate their contribution to the overall deactivation mechanism.

9.2.4 BENCH-SCALE UNIT FOR THE EVALUATION OF FCC CATALYSTS

All evaluation studies were carried out in a bench-scale, single receiver, short contact time, fixed-bed microactivity test unit (SR-SCT-MAT). The main difference from the conventional SCT-MAT unit [41] is that a single receiver is utilized for the collection of both liquid and gaseous products rendering the gaseous sample that is taken for analysis more representative. There is also a layer of nitrogen introduced above the feed during the injection eliminating the feed losses if backpressure is generated by the cracking conditions. Finally, the reactor is made from stainless steel providing better heat transfer. The reactor is designed in two sections to avoid any hold-up of heavy molecules (LCO, HCO) that could take place because of the design of the SCT-MAT glass reactor. Cracking is realized at 560°C, while reaction time is 12 sec in the SR-SCT-MAT. Details on the design and the experimental conditions can be found elsewhere [42]. Evaluation of catalysts is carried out over a series of cat-to-oil ratios, achieved by altering the mass of catalyst.

9.3 RESULTS AND DISCUSSION

9.3.1 CHARACTERIZATION STUDIES

All catalytic samples were submitted to a series of standard characterization techniques, summarized in Table 9.1. It should be underlined that the E-cat sample proved to be a nonpure equilibrium sample, but a blend of catalysts, as obvious from the standard characterization results. It should also be noticed that the artificial deactivated samples were prepared to contain 50% of the E-cats metals concentration. This is a compromise in order to limit the exaggeration of the metal effects on the lab-deactivated samples. Nevertheless, the undesired effects are still overestimated especially when high metal concentrations are introduced on the catalysts. As obvious from Table 3.1, the losses of the specific surface areas on the ADV-CPS samples are higher than the losses on the corresponding CPS samples. This is an indication that the deactivation is more severe when the ADV-CPS is applied. This was a rather expected observation considering the applied procedures' parameters (increment of temperature in the ADV-CPS protocol). Moreover, it seems that the

TABLE 9.1
Properties of E-Cat and Lab-Deactivated FCC Catalyst

Name	CPS-Cat	45 CPS-Cat	60 CPS-Cat	ADV-CPS-Cat	45 ADV-CPS-Cat	60 ADV-CPS-Cat	NM-CPS-Cat	NM-ADV-CPS-Cat	E-cat
TSA (m²/g)	140.21	129.76	131.19	128.84	125.69	113.18	141.10	131.28	111.54
ZSA. (m²/g)	57.73	51.30	53.19	49.90	46.69	36.32	56.47	48.88	64.88
MSA (m²/g)	82.49	78.45	78.00	78.94	79.04	76.86	84.63	82.40	46.66
UCS. (Å)	24.42	24.41	24.41	24.42	24.41	24.41	24.43	24.43	24.36
Ni (ppm)	3270	3200	3190	3167	3170	3200	—	—	5979
V (ppm)	637	680	660	700	690	725	—	—	1068

absence of metals did not affect the surface areas retention probably because the vanadium concentration was low on the metalated samples and didn't cause signifi- cant zeolite collapse.

Besides standard characterization all the samples were further explored by apply- ing FTIR spectroscopy. The scope was to quantify the acidity of the samples and to discriminate Brönsted from Lewis acid sites. The results of the acidities of the metalated samples are presented in correlation to the corresponding specific surface areas (Figures 9.1 and 9.2). That is the zeolite surface area for the Brönsted acid sites and the total surface area for the Lewis acid sites, obviously because the Brönsted acid sites exist only on the zeolitic component of the catalyst, while the Lewis acid sites are present on both matrix and zeolite.

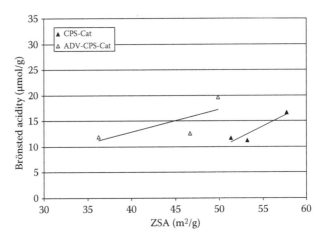

FIGURE 9.1 Correlation between Brönsted acidity and Zeolite surface area for all the trials of CPS and ADV-CPS in the presence of metals.

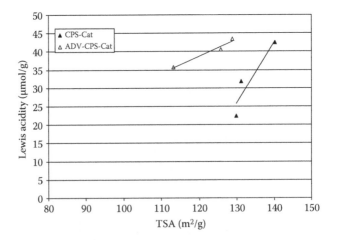

FIGURE 9.2 Correlation between Lewis acidity and total surface area for all the trials of CPS and ADV-CPS in the presence of metals.

It is obvious (Figures 9.1 and 9.2) that keeping the metals reduced for a longer time period during the deactivation process (ADV-CPS protocol) is beneficial for the acidity retention of the sample, despite the extended hydrothermal deactivation. A shift of the acidic properties of the samples toward the desirable direction was observed. It seems that increasing duration of the reduction step results in significant acidity retention at lower specific surface areas. This could be attributed to the diminished destructive activity of vanadium, preventing the acidity loss despite the loss of surface areas due to hydrothermal deactivation as it is well known that oxidized vanadium is far more active than vanadium in reduced oxidation state. We should not pass over the impact of the applied temperature difference between the two protocols. It is possible that the elevated temperature is beneficial for the ability of the metals to change oxidation state. Thus, it could be a matter of the ReDox kinetics of the metals. In any case the ADV-CPS protocol is privileged due to the vanadium effect limitation.

The application of the two protocols without the presence of the contaminant metals was carried out in order to investigate their contribution to the overall deactivation mechanism. The comparison of the acidities deltas between the two deactivation protocols for both cases (with or without metals) is presented in Table 9.2. As is obvious, the deltas in the presence of the metals are positive, while in the absence of the metals the deltas are negative. Thus, the sample deactivated with the ADV-CPS protocol without metals ends up with less acidity than the sample deactivated with the corresponding CPS. This was expected due to more severe hydrothermal deactivation conditions during the ADV-CPS application. The higher loss of specific areas verifies the more intense hydrothermal deactivation.

On the other hand, in the presence of the contaminant metals the acidities deltas are positive. The enhanced reductive part of the ADV-CPS protocol is probably responsible for this alteration. This observation can be attributed to the limitation of the vanadium deleterious effect on the catalyst's structure, as it is less drastic in its reduced oxidation state. Consequently, all the observations are convergent to the fact that keeping the metals reduced for a certain period of time during the laboratory deactivation procedure seems to be beneficial as far as acidity retention is concerned.

Finally, the effect of various steps of the deactivation procedures prior to the ReDox cycles on the acidity of the samples was investigated. Catalytic samples were collected after the first calcination step and after the metal impregnation step just

TABLE 9.2
Acidities Deltas between ADV-CPS and CPS Protocols in the Presence and in the Absence of Metals

Deltas Conditions	ADV/CPS–CPS	
	Presence of Metals	Absence of Metals
Brönsted (μmol/g)	3.02	−1.11
Lewis (μmol/g)	1.09	−5.27

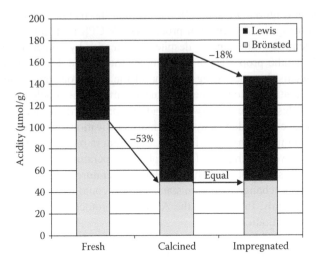

FIGURE 9.3   Impact of various steps prior to ReDox cycles on acidity.

before the ReDox cycles. It must be underlined that a calcination step is also carried out after the impregnation for the removal of the organic precursors of the metals. The acidity of all samples was measured with FTIR-pyridine adsorption revealing that the first calcination step caused significant Brönsted acidity attenuation on the catalyst. As obvious on Figure 9.3, the Brönsted acid sites were reduced by 53% as compared with the fresh sample, while the Lewis acid sites were increased, probably due to the Brönsted sites transformation during the calcination step via dehydroxylation reactions [43]. It is also clear that the impregnation and the successive calcination didn't cause any further Brönsted acid sites elimination. In fact, only the Lewis acid sites were reduced by 18% probably due to neutralization due to metals deposition. The calcination step after the impregnation is not affecting the acidity because it is realized at the same temperature as the first calcination step and the dehydroxylation extent is strongly related to the applied temperature [43].

9.3.2   CATALYTIC PERFORMANCE EVALUATION

The evaluation of the commercially and artificially deactivated samples was realized in the SR-SCT-MAT unit. The activity of the lab-deactivated samples with both protocols is in good correlation to the Brönsted acidity as presented in Figure 9.4. In detail, it is obvious that increment of the Brönsted acidity is reflected in higher activity of the catalytic sample, as the Cat/Oil ratio is decreasing for standard conversion level.

The interpretation of the evaluation results is mostly focused on the undesired metal effects for the purposes of the present study. That is the dehydrogenation effect reflecting in hydrogen and coke extended production. The hydrogen and coke yields are shown in Figures 9.5 and 9.6. As obvious, the hydrogen and coke yields are lower on the catalytic samples deactivated with ADV-CPS than on the samples deactivated with CPS, which means that application of the ADV-CPS protocol is limiting the

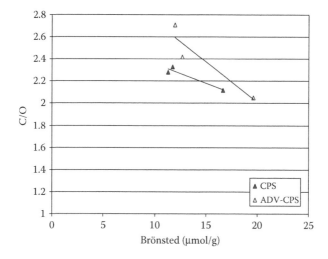

FIGURE 9.4   Correlation between Brönsted acidity concentration and activity (Cat/Oil for 65% conversion) for all the trials of CPS and ADV-CPS in the presence of metals.

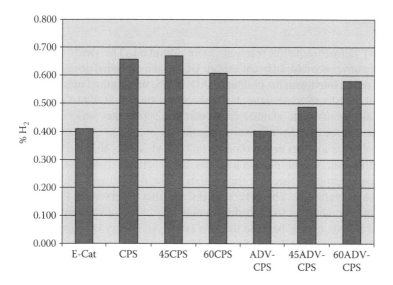

FIGURE 9.5   Hydrogen yields for the deactivated samples with CPS and ADV-CPS in the presence of metals.

overestimation of the metal effects and promoting the simulation of the real deactivation. This observation in combination to the characterization results is attributed to the enhancement of the reducing step of the method, reflecting in limited undesired activity of vanadium. Considering the time extended protocols with 45 and 60 cycles, it seems that the expected better aging of the metals and consequently further limitation of the metal effects was not achieved. In contrast, the hydrogen yields were increased probably due to the extended decay of the catalyst making the

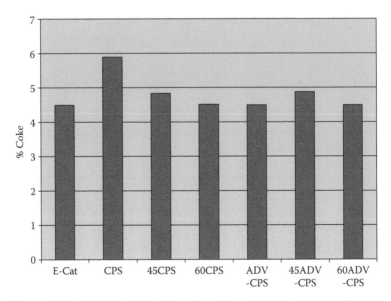

FIGURE 9.6 Coke yields for the deactivated samples with CPS and ADV-CPS in the presence of metals.

secondary dehydrogenation reactions more dominant. The yields are presented at standard conversion but different cat/oil. Due to the extended decay of the catalytic samples deactivated with the prolonged ADV-CPS, the required mass of catalyst was increased in order to achieve the standard conversion (SR-SCT-MAT unit configuration). Thus, the absolute number of the metallic active sites was increased causing the higher degree of dehydrogenation reactions.

This observation was not so obvious on coke yields because the coke production is a contribution of multiple mechanisms and reactions. Thus, the coke yields are quite similar, probably because the catalytic coke is decreased while the contaminant coke is increased. The coke remarks are also observed on the CPS samples taking into account that the dehydrogenation degree is not strongly affected by the extended ReDox cycles, because the lower catalysts' decay is limiting the effect of the required mass of catalyst (C/O ratio). Thus, the small decrement of the coke yield on the CPS samples is possibly related to the descent of the catalyst (less specific area) leaving less available space for coke adsorption and less activity for catalytic coke production. It is clear that prolonging the deactivation procedures is not beneficial as far as the metal effects are concerned.

9.4 CONCLUSIONS

Summarizing our results, the comparison of the effect of both conventional and improved deactivation protocols on the surface areas of the derived catalytic samples suggests that ADV-CPS is more severe than the standard CPS protocol, an observation in total agreement with the varying parameters of the methods. Results from FTIR acidity studies show that keeping the metals reduced during the larger part of the ADV-CPS deactivation protocol is enhancing the acidity retention as compared to the CPS protocol. This observation opposes the more severe hydrothermal deactivation that occurs in ADV-CPS, which leads to lower specific surface areas. The acidity protection can be attributed to the limitation of the undesired effects of vanadium, as it is far more destructive in its oxidative state (+5). The higher temperature of the ADV-CPS protocol could be affecting the ReDox properties of the metals as well, but there is a need of further exploration. Attempted application of the deactivation protocols in the absence of the metals supports the former conclusion, as the hydrothermal deactivation mechanism by itself seems to lead to less acidity on the ADV-CPS deactivated sample. The significance of the duration of the reductive step within the ReDox cycles and consequently the oxidation state of the metals during the laboratory deactivation methods is revealed as mostly beneficial for simulating the real deactivation in a commercial unit.

Combination of evaluation and characterization results reveals that the acidity is an essential parameter for the simulation of the real deactivation as it is strongly related to the activity of the samples. Moreover, it is concluded from the hydrogen and coke yields that the ADV-CPS method limiting the overestimation of the metal effects and promoting the simulation of the real deactivation. Once again this is attributed to the enhanced reduction step of this method. The time extended protocols didn't achieve better aging of the metals but only extended decay of the catalyst making the secondary dehydrogenation reactions more dominant. This observation was not so obvious on coke yields because the coke production is a contribution of multiple mechanisms and reactions.

In general, although proper metal aging is still an open issue, it seems that severe hydrothermal conditions during ReDox cycles with an emphasized reducing step is the direction to optimization of the artificial deactivation methods. The development of such a simulative lab-deactivation protocol will undoubtedly be very essential and a major contribution in the FCC research field.

ACKNOWLEDGMENT

This work was partially funded by the Greek General Secretariat for Research and Technology (GSRT) through the program PENED (Action 8.3.1, Project No.03EΔ859) and Albemarle Catalysts Company.

REFERENCES

1. Otterstedt, J. E., Gevert, S. B., Jaras, S. G., and Menon, P. G. *Appl. Catal.* 22 (1986): 159.
2. Cimbalo, R. N., Foster, R. L., and Wachtel, S. J. *Oil Gas J.* 70, no. 20 (1972): 112.
3. Habib, E. T., Owen, H., Synder, P. W., Streed, C. W., and Venuto P. *Ind. Eng. Chem.,Prod. Res. Dev.* 16, no. 4 (1977): 291.
4. Sporangelo, B. K., and Reagan, W. J. *Oil Gas J.* 30 (1984): 139.
5. Schubert, P. F. In *Proceedings of the ACS Symposium on Advances in FCC*, New Orleans, 1997.
6. Sexton, J., Mays, C., Bartholic, D., Abdillah, R., and Ambarajava K. *AIChE Spring Meeting*, New Orleans, 1998.

7. Vreugdenhil, W., Vogt, E. T. C., Skocpol, R. C., and Yanik, S. J. *Akzo Nobel Catalysts Symposium*, 1998.
8. Trujillo, C. A., Uribe, U. N., Knops-Gerrits, P.-P., Oviedo, L. A., and Jacobs, P. A. *J. Catal.* 168 (1997): 1.
9. Xu, M., Liu, X., and Madon, R. J. *J. Catal.* 207 (2002): 237.
10. Escobar, A. S., Pereira, M. M., Pimenta, R. D. M., Lau, L. Y., and Cerqueira, H. S. *Appl. Catal. A* 286 (2005): 196.
11. Escobar, A. S., Pinto, F. V., Cerqueira, H. S., and Pereira, M. M. *Appl. Catal. A* 315 (2006): 68.
12. Cadet, V., Raatz, F., Lynch, J., and Marcilly, Ch. *Appl. Catal.* 68 (1991): 263.
13. Casali, L. A. S., Rocha, S. D. F., Passos, M. L. A., Bastiani, R., Pimenta, R. D. M., and Cerqueira, H. S. In Studies in Surface Science and Catalysis, Fluid Catalytic Cracking VII: Materials, Methods and Process Innovations. Edited by M. L. Occelli, Vol. 166. Amsterdam: Elsevier, 2007.
14. Wallenstein, D., Roberie, T., and Bruhin, T. *Cat. Today* 127, nos. 1–4 (2007): 54.
15. Vieira, R. C., Pinto, J. C., Biscaia, Jr., E. C., Baptista, C. M. L. A., and Cerqueira, H. C. *Ind. Eng. Chem. Res.* 43 (2004): 6027.
16. Boock, T., Petti, T. F., and Rudesill, J. A. *ACS Div. Petrol. Chem.* 40 (1995): 421.
17. Mitchell, B. R. *Ind. Eng. Chem. Prod. Res. Dev.* 19 (1980): 209.
18. Gerritsen, L. A., Wijngaards, H. N. J., Verwoert, J., and O'Connor, P. *Cat. Today* 11 (1991): 61.
19. Psarras, A. C., Iliopoulou, E. F., Nalbandian, L., Lappas, A. A., and Pouwels, C. *Cat. Today* 127, nos. 1–4 (2007): 44.
20. Zheng, S., Heydenrych, H. R., Roger, H. P., Jentys, A., and Lercher, J. A. *Topics in Catalysis* 22, no. 1/2 (2003): 101.
21. Psarras, A. C., Iliopoulou, E. F., Kostaras, K., Lappas, A. A., and Pouwels, C. *Micropor. Mesopor. Mater.* 120 (2009): 141.
22. Navarro, U., Trujillo, C. A., Oviedo, A., and Lobo, R. *J. Catal.* 211, no. 1 (2002): 64.
23. Tonetto, G., Atias, J., and de Lasa, H. *Appl. Catal. A* 270, nos. 1–2 (2004): 9.
24. Nesterenko, N. S., Thibault-Starzyk, F., Montouillout, V., Yuschenko, V. V., Fernandez, C., Gilson, J.-P., Fajula, F., and Ivanova, I. I. *Micropor. Mesopor. Mater.* 71, nos. 1–3 (2004): 157.
25. Benaliouche, F., Boucheffa, Y., Ayrault, P., Mignard, S., and Magnoux, P. *Micropor. Mesopor. Mater.* 111 (2008): 80.
26. Thibault-Starzyk, F., Gil, B., Aiello, S., Chevreau, T., and Gilson, J.-P. *Micropor. Mesopor. Mater.* 67 (2004): 107.
27. Liu, X., Truitt, R. E., and Hodge, G. D. *J. Catal.* 176 (1998): 52.
28. Travert, A., Vimont, A., Sahibed-Dine, A., Daturi, M., and Lavalley, J.-C. *Appl. Catal. A* 307 (2006): 98.
29. Caeiro, G., Lopes, J. M., Magnoux, P., Ayrault, P., and Ramoa Ribeiro, F. *J. Catal.* 249 (2007): 234.
30. O'Connor, P., Verlaan, J. P. J., and Yanik, S. J. *Cat. Today* 43 (1998): 305.
31. O'Connor, P., and Humphries, A. P. *ACS Preprints* 38, no. 3 (1993): 598.
32. Lercher, J. A., and Jentys, A. *Stud. Surf. Sci. Catal.* 168 (2007): 435.
33. Emeis, C. A. *J. Catal.* 141 (1993): 347.
34. Selli, E., and Forni, L. *Micropor. Mesopor. Mater.* 31 (1999): 129.
35. Hughes T. R., White H. M. *J. Phys. Chem.* 71 (1967): 2192.
36. Datka, J., Tutek, A. M., Jehng, J. H., and Wachs, I. E. *J. Catal.* 135 (1992): 186.
37. Echoufi, N., and Gelin, P. *Catal. Lett.* 40 (1996): 249.
38. Guisnet, M., Ayrault, P., Coutanceau, C., Alvarez, M. F., and Datka J. *J. Chem. Soc., Faraday Trans.* 93 (1997): 1661.
39. Wormsbecher, R. F., Peters, A. W., and Maselli, J. M. *J. Catal.* 100 (1986): 130.

40. Wallenstein, D., Harding, R. H., Nee, J. R. D., and Boock, L. T. *Appl. Catal. A* 204 (2000): 89.
41. Lappas, A. A., Patiaka, D. T., Dimitriadis, B. D., and Vasalos, I. A. *Appl. Catal. A* 152 (1997): 7.
42. Wallenstein, D., Seese, M., and Zhao, X. *Appl. Catal. A* 231 (2002): 227.
43. Magee, A. J. S., and Mitchell, M. M. *Stud. Surf. Sci. Catal.* 76 (1993), Elsevier Science Publishers, 41.

10 Coke Characterization by Temperature-Programmed Oxidation of Spent FCC Catalysts That Process Heavy Feedstock

William Gaona, Diana Duarte, Carlos Medina, and Luis Almanza

CONTENTS

10.1 INTRODUCTION

The mechanism of cracking reactions occurs through carbocations (carbenium ions) formed at the active acid sites of the catalyst. In addition to the formation of light products, during the catalytic cracking reactions deposits of high-molecular weight and high aromaticity coke occurs. This byproduct is subjected to controlled combustion in order to regenerate the catalyst activity. This combustion process is responsible for providing enough energy to carry out endothermic reactions of the feedstock during of the fluid catalytic cracking (FCC) process. The coke produced during cracking reactions can be classified according to the literature in five categories.

The catalytic coke produced by the activity of the catalyst and simultaneous reactions of cracking, isomerization, hydrogen transfer, polymerization, and condensation of complex aromatic structures of high molecular weight. This type of coke is more abundant and constitutes around 35–65% of the total deposited coke on the catalyst surface. This coke determines the shape of temperature programmed oxidation (TPO) spectra. The higher the catalyst activity the higher will be the production of such coke [1].

The second type is the *contaminant or metallic coke*. This coke is produced from the catalytic dehydrogenation reactions of the feedstock caused by the presence of metals in the catalyst such as nickel, vanadium, iron, and copper. This type of coke is of relevant importance in the processing of heavy feedstock. Several authors studied the deactivating effect of these metals and showed that nickel has low mobility, tending to remain on the surface where it was deposited. The amount of nickel is an indication of the catalyst age. On the other hand, vanadium is mobile, showing preference for the exchanged zeolite with rare earths and alumina. Vanadium can deactivate the zeolite by blocking pores temporally and permanently, through the vanadic acid, which reacts with the zeolite structure or with the cations of rare earths. Unlike nickel, the vanadium deactivation does not depend on the severity and characteristics of the catalyst. Nickel promotes a strong dehydrogenation, favoring the reactions of coking and reducing gasoline selectivity, without significantly altering the activity [2]. Vanadium has a hydro-dehydrogenating activity lower than nickel; some researchers have estimated that this capacity is between 25 and 33% of the nickel [2].

The third type is *the additional coke* related with the feedstock quality. FCC feedstock contains a dissolved carbon, polynuclear aromatic compounds, called Conradson carbon residue (CCR; ASTM D-189). It is deposited over the catalyst surface during cracking reactions. In the FCC unit, this material is part of the coke remaining in the catalyst. Some researchers have investigated cracking of heavy feedstock and observed that, in particular cases, the amount of Conradson carbon is linearly related with the carbon–hydrogen ratio of the feedstock [3].

Another coke formed in a FCC unit is *occluded* or *residual coke*. In a commercial unit this coke corresponds to coke formed on catalyst porosity and its content depends on textural properties of the catalyst (pore volume and pore size distribution) and the stripping system capacity in the reaction section. Finally on the FCC catalyst rests some high-molecular weight of nonvaporized hydrocarbons. These molecules do not vaporize or react at the reactor conditions and accumulate in the catalyst pores like a soft carbonaceous residue with high hydrogen content.

The TPO technique has been widely used worldwide with the purpose of studying the evolutions and origins of the different types of coke generated during the cracking reactions in a FCCU. In the case of FCCs, studies have been limited to coke formed during the cracking of light feedstocks. In fact, there are only a few studies of coke characterization formed during the cracking of heavy feedstock. Therefore, for this work the TPO technique was used to characterize coke formed during bottoms feedstocks processing as deasphalted vacuum bottoms or demetallized residual oil (DMO), and to correlate this characterization with the catalyst and feedstock properties.

10.2   EXPERIMENTAL

The catalyst used in this study corresponds to a fresh commercial catalyst used in one FCC unit of ECOPETROL S.A. This solid is hydrothermal deactivated at the laboratory in cycles of oxidation–reduction (air-mixture N_2/Propylene) at different temperatures, different times of deactivation, with and without metals (V and Ni), and different steam partial pressures. Spent catalysts (with coke) are obtained by using microactivity test unit (MAT) with different feedstocks, which are described in Table 10.1.

The catalysts with metals are previously impregnated with solutions of vanadyl and nickel naphtenates based on the Mitchell method [4]. Before hydrothermal deactivation the samples were calcined in air at 600°C. The activity was performed in the conventional MAT test using 5 grams of catalyst, ratio cat/oil 5, stripping time 35 seconds, and reaction temperature 515°C. Elemental analyses to determine the total amount of carbon in the spent catalysts were done by the combustion method using a LECO analyzer.

10.2.1   Temperature-Programmed Oxidation

TPO analyses were performed in a TPD/TPR 2900 (Micromeritics) equipment with a thermal conductivity detector; a trap for sulfur compounds and a Pt/Silica bed for oxidation of CO and hydrocarbons to CO_2. Furthermore, it has a cold trap (isopropyl alcohol/liquid nitrogen) to condense CO_2 and residual moisture. The combustion products are passed through the previous traps connected in series in order to remove other compounds different from O_2 in the carrier gas. This ensures that the conductivity changes observed in the detector are attributed exclusively to changes in oxygen concentration in the carrier gas.

Spent catalysts are pretreated at 110°C during one (1) hour in an inert atmosphere of helium gas in order to remove the moisture retained in the pores of the catalyst. Subsequently the sample is placed in contact with the carrier gas (mixture 5% O_2-He) and heated at a ramp increasing temperature of 10°C/min from 110 to 900°C.

The TPO profiles obtained were analyzed by deconvoluting them using Gaussian peaks and GRAMS 32 software. The peaks obtained were assumed to represent the four different types of coke in the spent catalyst: catalytic coke, contaminant coke, occluded coke, and additional coke (Conradson carbon).

TABLE 10.1
Feedstock Characterization

Feedstock	%CCR	Density (g/cc)	API
VGO	< 0.02	0.8850	28.4
DMO	6.06	0.9663	14.9
VGO 94.5% DMO 5.5%	0.29	0.8921	27.1
VGO 70% DMO 30%	1.86	0.9331	20.1

TABLE 10.2
The Nonvaporized Hydrocarbons as a Function of Feedstock Quality

Feedstock	% Total Coke LECO	% CCR	% Hydrocarbons Extracted	% Total Coke LECO after the Extraction
VGO	0.6969	0.02	0	0.6969
VGO 94.5% DMO 5.5%	0.9074	0.29	0.17	0.9058
VGO 70% DMO 30%	1.5530	1.92	0.92	1.5387

10.2.2 THE NONVAPORIZED HYDROCARBONS

The TPO spectrum of a spent catalyst obtained from a feedstock that contained DMO presents a thermogram with a high level of noise, which makes the identification and quantification of the coke present difficult. However, once the nonvaporized hydrocarbons are extracted with toluene, the noise is removed and a smooth thermogram is obtained. Apparently hydrocarbons of highly molecular weight adsorbed during the reaction desorbed gradually and decompose at the analysis temperatures. For this reason all the samples analyzed by the TPO technique have been pretreated in a Soxhlet system employing as solvent toluene (2 grams of catalyst for 80 ml of solvent), during 24 hours.

Table 10.2 presents the total coke yields and the nonvaporized hydrocarbons produced over a spent catalyst obtained with different feedstocks. The catalyst used was deactivated for 20 hours, 30 ReDox cycles, and 50% steam. When the 100% vacuum gas oil (VGO) is replaced with a mixture of 5%w DMO-VGO and/or 30%w DMO-VGO an increase of 30% and 120% in the coke yields was observed. While the spent catalyst from VGO cracking does not have adsorbed hydrocarbons, the mixture with DMO does, becoming almost 1% for the mixture with 30%w DMO. The SARA (saturates, aromatics, resins, and asphaltenes) analysis of these hydrocarbons showed a high concentration of asphaltenes.

10.3 RESULTS AND DISCUSSION

10.3.1 CATALYTIC COKE

This type of coke depends exclusively on the FCC cracking activity. In order to have samples with different activity and little influence of contaminant coke, the fresh catalyst was deactivated hydrothermally at different severity conditions without metals. MAT test for these deactivated samples was performed with VGO as a feedstock to diminish coke yields.

Table 10.3 shows the characteristics of the catalyst samples after different deactivation temperature, steam partial pressure, and deactivation time. The zeolite and matrix surface area and the unit cell size decrease as the deactivation severity increase. Similarly, VGO conversion and coke yields obtained in the MAT test also decreased with deactivation severity.

The TPO profile of the most deactivated sample is presented in Figure 10.1. After deconvoluting the TPO profile, the four signals that best fit the TPO original

TABLE 10.3

Fresh Catalyst Deactivated at Different Severities Using VGO in MAT Test

Temp. (°C)	Time h + Cycles	Steam %	UCS (Å)	Zeolite Surface Area (m²/g)	Matrix Area (m²/g)	N_{Al} Struct.	Coke Yield %w	Conversion %w
705	6h + 9C	95	24.50	162	64	28	8.63	85.59
788	4h + 6C	80	24.42	150	57	20	6.97	84.00
788	20h + 30C	80	24.35	106	43	12	3.38	78.77

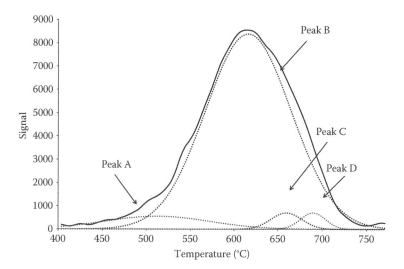

FIGURE 10.1   TPO deconvolution of spent catalyst (sample deactivated during 20 hours, and tested with VGO in MAT analysis).

spectrum can be identified. The Signal A appears with a peak at 500°C, the signal B with a peak at 617°C, the signal C with a peak at 660°C, and the signal D with a peak near to 690°C. The signal B has the highest peak area, which is about 89% of the total coke. Given that this catalyst has no Ni or V, that the CCR of the feedstock is very low (<0.02 wt%) and that literature reports that catalytic coke obtained after gasoil cracking is higher than 65% of the total coke, it can be concluded that the signal B can be attributed to the signal corresponding to catalytic coke.

   Table 10.4 illustrates the TPO data for each of the samples tested, standardized in coke (g of coke associated with each peak)/total coke on the catalyst (g). In this table it is observed how the total coke yield of these samples is inversely proportional to the catalyst deactivation severity. Similarly, it is noted that catalytic coke increases with catalyst activity, but it is not a direct function of conversion.

TABLE 10.4

Temperatures and Quantification of Each TPO Peak

Samples Deactivation Conditions	Signal A T(°C) – (g Coke/g Total Coke)	Signal B T(°C) – (g Coke/g Total Coke)	Signal C T(°C) – (g Coke/g Total Coke)	Signal D T(°C) – (g Coke/g Total Coke)
6 hours, 705°C	525 – (0.042)	617.2 – (1.38)	659.7 – (0.25)	691.9 – (0.039)
4 hours, 788°C	525 – (0.14)	610.1 – (0.95)	652.8 – (0.050)	683.3 – (0.023)
20 hours, 788°C	512 – (0.036)	617.2 – (0.62)	665.6 – (0.020)	690.6 – (0.018)

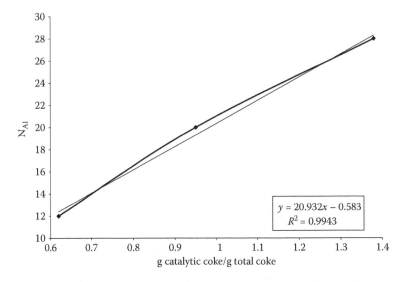

FIGURE 10.2   Catalytic coke vs. N_{Al}.

When the zeolite surface area is plotted as a function of catalytic coke the correlation improves. The best correlation between the physicochemical properties of the catalyst and catalytic coke is the one involving an amount of aluminums in the framework (N_{Al}), estimated from the unit cell size by Equation 10.1 [1], as it is evidenced in Figure 10.2.

$$N_{Al} = 107.1*(UCS–24.238). \qquad (10.1)$$

It is possible to conclude from the TPO spectra that the signal with the peak of highest intensity located between 610°C and 617°C, for the case of light feedstock and low CCR, can be attributed to catalytic coke. This type of coke can reach values up to 89% of the total coke on the catalyst. In addition, it can be also concluded that catalytic coke yields are a direct function of the number framework aluminums in the zeolite structure.

10.3.2 CONTAMINANT COKE

Three catalyst samples were prepared to identify and quantify the coke related with the hydro-dehydrogenation capacity of contaminant metals. The first sample corresponds to the fresh catalyst deactivated 20 hours, at 788°C, with 80% steam, the second and third samples were deactivated at the same conditions but impregnated with 4100 ppm V, and with 4100 ppm V and 4000 ppm Ni, respectively. The MAT test was performed using gas oil.

Table 10.5 summarizes the physicochemical characterization of deactivated solids. The dehydrogenation capacity of Ni and V is represented using the concept of nickel equivalent, which attributes a factor of 1 to the nickel, and a factor of ¼ to vanadium (literature reports vanadium values between 0.25 and 0.33 [2]).

$$Ni_{eq} = \left(\frac{1}{4}\right)[V] + 1[Ni]. \tag{10.2}$$

The Figures 10.1 and 10.3 present the TPO spectra of the samples with and without metals. For the sample impregnated with 4100 ppm vanadium, it was observed the appearance of a shoulder around 680°C that translates in a 10% increase in peak C area, compared to the metal-free catalyst as illustrated in Figure 10.3. Then, the signal C located around 677°C apparently corresponds to the contaminant coke produced by the hydro-dehydrogenation properties of vanadium.

Table 10.6 summarizes data from the TPO profiles for the three samples with and without metals. This table clearly shows how the signal C increase as the concentration of nickel and vanadium increases and supports the hypothesis that this peak corresponds to contaminant coke. It is possible to support the theory that a higher content of vanadium in the catalyst results in a loss of activity because the peak area B, previously attributed to catalytic coke, decreases strongly with vanadium levels.

In Figure 10.4 was plotted the contaminant coke yield as a function of Ni equivalent. In this graph it is observed that the signal C, expressed as grams of contaminant coke, is almost a linear function of Ni equivalent. When the vanadium factor is changed to 0.38 the ratio is completely linear. Then with this technique it is possible to find the real dehydrogenation factor of vanadium with respect to nickel.

TABLE 10.5

Hydrothermal Deactivation of Catalyst Impregnated with Different Levels of Metal

Metals V – Ni (ppm)	%Conversion MAT	Total Coke (g/g Catalyst)	Ni Equivalent (ppm)	Zeolite Area (m²/g)	Matrix Area (m²/g)
0 – 0	78.77	0.6969	0	106	43
4100 – 0	71.15	0.6119	1025	84	39
4100 – 4000	63.64	0.8983	5025	76	38

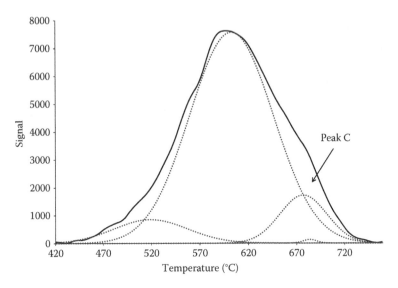

FIGURE 10.3 Spectrum deconvolution obtained using the catalyst deactivated 20 hours with 4100 ppm V.

TABLE 10.6
Signals From Each of the Peaks Found

Metals V – Ni (ppm)	Signal A T(°C) – (g Coke/g Total Coke)	Signal B T(°C) – (g Coke/g Total Coke)	Signal C T(°C) – (g Coke/g Total Coke)	Signal D T(°C) – (g Coke/g Total Coke)
0 – 0	512 – (0.036)	617.2 – (0.62)	665.6 – (0.020)	690.6 – (0.018)
4100 – 0	520 – (0.061)	603.8 – (0.48)	677.9 – (0.067)	684.7 – (0.0013)
4100 – 4000	528.7 – (0.030)	611.9 – (0.40)	665.1 – (0.19)	714.1 – (0.028)

10.3.3 ADDITIONAL COKE

This coke has been related with feedstocks quality and more specifically with their carbon Conradson content. The catalyst used for this analysis was the sample without metals deactivated 20 hours, 30 cycles, and 80% steam. The MAT test was performed with feedstock of different quality: gasoil, DMO, and mixture of gasoil–DMO.

According to the literature, this type of coke is the most refractory of the cokes and is located at the highest temperatures of the TPO spectrum [5]. The variation of the CCR in the feedstock used is presented in Table 10.7.

In Table 10.7 it can be seen that the content of CCR in the mixture DMO–gasoil increase when the amount of DMO increase, resulting in greater total coke yield. Figures 10.5 and 10.6 show the TPO profiles of the mixtures of 5%w DMO and 30%w DMO in gasoil.

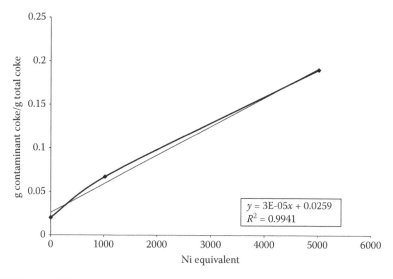

FIGURE 10.4   Ni equivalent vs. contaminant coke.

TABLE 10.7

Catalyst MAT Results With Different Feeds of Sample Deactivated for
20 Hours, 30 Cycles, without Metals

Feedstock	%CCR	Total Coke LECO	Additional Coke (g/g Total Coke)	Zeolite Area (m²/g)	Matrix Area (m²/g)
VGO	0.02	0.6969	0.0176		
VGO 94.5% DMO 5.5%	0.29	0.9074	0.0182	106	43
VGO 70% DMO 30%	1.90	1.5250	0.0758		
DMO	6.06	2.1050	0.1092		

In Figure 10.5 the catalyst sample is tested in a MAT unit using a mixture of 5.5%w DMO in gasoil. As a result, peak D shifts to 703°C. This peak has a larger area compared with the peak D obtained for catalyst tested with gasoil (Figure 10.3) suggesting that peak D corresponds to coke related with the feedstock.

In Figure 10.6 the catalyst sample is tested MAT conditions using a mixture of 30%w DMO in gasoil. In this figure it is possible to observed an increase in peak D (~702°C) intensity, from 700 units of TCD signal with gasoil up to 1000 units for the mixture of 5.5%w DMO and about 4000 units for the 30%w mixture.

Figure 10.7 represents the TPO profile of spent catalyst tested with pure DMO. It can be noted that the area of the peak at 715°C has increased even more owing to this feedstock high-CCR content. It can be concluded that the coke related with the feedstock CCR is located between 700°C and 715°C. As expected, this coke increases with the feedstock Conradson carbon content.

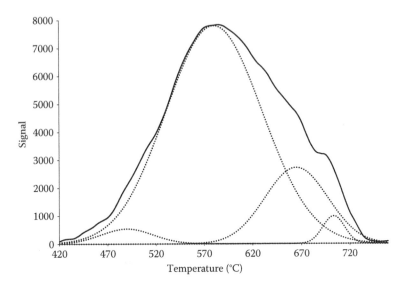

FIGURE 10.5 Spectrum deconvolution corresponding to the mixture 5.5%w DMO in gasoil.

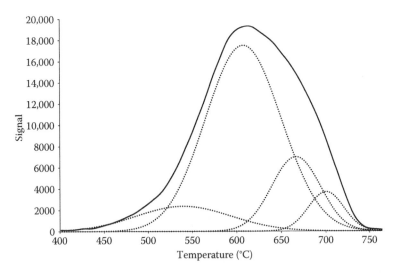

FIGURE 10.6 Spectrum deconvolution corresponding to the mixture 30%w DMO in gasoil.

Figure 10.8 shows that the relative area of signal D (g of additional coke/g of total coke on the catalyst) increases as CCR of the feed increase, confirming the hypothesis that this peak corresponds to the coke related with the feedstock quality or additional coke. The additional coke signal is located between 700 and 716°C and shifts to higher temperatures as the content of CCR in the feedstock increases.

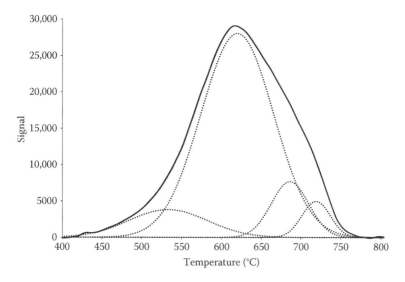

FIGURE 10.7   Spectrum deconvolution corresponding to DMO.

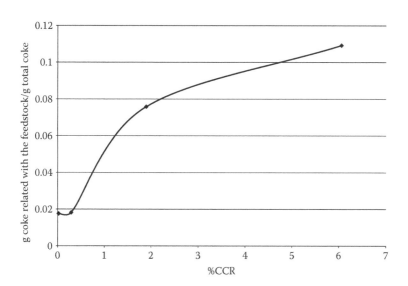

FIGURE 10.8   Conradson carbon versus coke related with the feedstock.

10.3.4   OCCLUDED COKE

Occluded coke is a function of the operating conditions of the FCCU stripper, of catalyst pore volume and pore size distribution. This coke that has the higher content of hydrogen is the lowest refractory between the different kinds of coke produced in the FCC process. Although in this study it was not directly evaluated, it may be associated with the peak A at the lowest temperature (~500°C–550°C). So once the other four types of coke are estimated, the occluded coke can be quantified by the difference with the total coke yields.

10.4 CONCLUSIONS

The technique of TPO is an instrumental analysis tool that allows (knowing the trend of oxidation) identification of the different types of coke produced during the cracking reactions. TPO spectra can distinguish and quantify each type of coke formed by deconvoluting the profiles using the GRAMS software. This technique was used with highly accurate results.

Before TPO analysis it was necessary to Soxhlet extract the FCC samples to remove nonvaporized hydrocarbons and avoid their accumulation in the pores of the catalyst as carbonaceous residue with high hydrogen content. As a result, the interference during TPO analysis caused by the desorption and decomposition of these compounds at high temperatures was eliminated. In this study it was observed that this type of coke is directly related to the Conradson carbon content of the feedstock.

According to the TPO analysis of the spent catalysts, the peak corresponding to catalytic coke is located between 610°C and 617°C for the case of light feedstock and low CCR. This type of coke can reach values up to 89% of the total coke in the catalyst. In addition, as the catalyst activity increases the catalytic coke and the temperature range at which this coke appears increase. There is a direct relationship between the number of aluminums in the framework and catalytic coke production.

The peak related with the contaminant coke is located between 665 and 677°C, and is a direct function of the nickel equivalents in the catalyst. This type of coke decreases the catalytic activity. Activity losses are mainly attributed to the presence of vanadium on the catalyst surface.

The additional coke related with feed quality is a function of their Conradson carbon content. According to the literature, our results show that this coke is the most refractory and consequently is located in the highest temperatures range (~700°C–716°C). Furthermore, it shifts to higher temperatures as the content of CCR of feedstock increases.

REFERENCES

1. Scherzer, J. *Octane-Enhancing Zeolitic FCC Catalyst: Scientific and Technical Aspects.* New York: Marcel Dekker, 1990.
2. Sadeghbeige, R. *Fluid Catalytic Cracking Handbook: Design, Operation and Troubleshooting of FCC Facilities.* Houston: Gulf Publishing Company, 2000.
3. Trasobares, S., Callejas, M., Benito, A., et al. Kinetics of Conradson Carbon Residue Conversion in the Catalytic Hydroprocessing of a Maya Residue. *Ind. Eng. Chem. Res.* 37 (1998): 11–17.
4. Mitchell, B. Metal Contamination of Cracking Catalyst. 1. Synthetic Metals Deposition on Fresh Catalysts. *Ind. Eng. Chem. Res.* 19 (1980): 201–13.
5. Bayraktar, O., and Kugler, E. Characterization of Coke on Equilibrium Fluid Catalytic Cracking Catalysts by Temperature-Programmed Oxidation. *Applied Catalysis A: General* 233 (2002): 197–213.

11 The Effect of Cohesive Forces on Catalyst Entrainment in Fluidized Bed Regenerators

Ray Cocco, Roy Hays, S. B. Reddy Karri, and Ted M. Knowlton

CONTENTS

11.1 INTRODUCTION

Clustering of particles in granular-fluid systems has been known to exist for some time. Wilhelm and Kwauk [1] were among the first to show evidence of particle clustering in fluidized beds with others following suit with similar experiments [2–10]. Kaye and Boardman [10] suggested that particle clustering is significant in many systems where solids concentrations exceed 0.05%. Particle drop experiments by Jayaweera et al. [3] showed that clusters have stable size ranges of two to six particles with clusters greater than six particles having a tendency to split and form stable subgroups of clusters. Fortes et al. [7] observed similar-sized clusters to be stable up to a particle Reynolds number of 1800.

Although often neglected, the role of particle clusters on entrainment can be significant. Yerushalmi et al. [11] first illustrated this with observations of slip velocity in fast fluidized beds that were significantly larger than expected for the particle size used. This was later attributed to particle clusters [12]. Matsen [13] concurred with their findings, and added that cluster size may be dependent on the solids loading. Geldart and Wong [14] observed the reduction of entrainment in a fluidized bed with the addition of Geldart Group C particles to a bed of Group A powder. They

155

suggested that the Group C particles or fines may be adhering to the large Group A particles, adhering to each other or both. Baeyens et al. [15] reported similar findings where the entrainment rate leveled off at some critical particle size between 25 to 40 microns. They proposed that this was the point where cohesive forces exceeded all other forces (such as drag and gravity). Choi et al. [16] also investigated the effect of fines on the elutriation of coarse particles and found that the elutriation of fines was not affected by the particle size distribution in the bed. Li et al. [17] suggested that there may be a bridging effect with fine powders and that the elutriation rate constant of Group C or Group A particles is not only affected by the properties of the elutriated powders or particles and gas velocity, but also by both the weight fraction and size of the Group C powder in the bed.

The underlying questions with these clusters are: What is the mechanism for their existence? Where are the clusters formed? and How do clusters affect entrainment rates? Based on evidence from pilot and commercial scale plants along with high-speed video of a cold-flow fluidized bed, the mechanism of particle clustering in and above fluidized beds and its effect on entrainment were examined.

11.2   ENTRAINMENT CORRELATIONS

Most entrainment rate correlations predicts that smaller particles have a higher entrainment flux or rate than larger particles, as shown in Figure 11.1. This is especially true for the empirical correlations of Stojkovski and Kostic [18], Zenz and Weil [19], and Lin et al. [20]. The first two correlations show orders of magnitude increase in the entrainment flux for particles less than 50 microns. The correlation of Lin et al. [20] shows this same trend, but with the exponential increase in entrainment occurring at 200 microns. The correlations of Colakyan [21], Colakyan and Levenspiel [22],

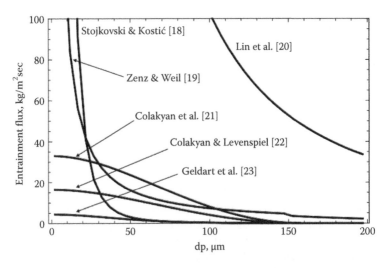

FIGURE 11.1   Empirical correlations for entrainment of equilibrium FCC catalyst with a d_{p50} of 81 microns and a fines level ($d_p < 44$ microns) of 10% in a 3 meter diameter × 12 meter tall bed with a 6 meter bed height at a superficial gas velocity of 1 meter/sec.

and Geldart et al. [23] also show that smaller particles have a higher entrainment rate than larger particles. However, these correlations show a less dramatic increase in the entrainment rate with decreasing particle size. In contrast, some empirical correlations show that entrainment rates decrease for smaller particles sizes. As shown in Figure 11.2, the correlations of Wen and Hashinger [24] and Tanaka et al. [25] suggest that entrainment decreases at below particle sizes of 35 and 70 microns, respectively. All these correlations are presented in Table 11.1.

This maximum in the entrainment flux observed by Wen and Tanaka is in some agreement with the work of Baeyens et al. [15]. Baeyens et al. worked with equilibrium fluidized catalytic cracking (FCC) catalyst powder with varying additions of fines. They found that below a "critical" particle size, entrainment rates leveled off and noted that this reflected the point where Van der Waals forces were in balance with gravitational forces. For their systems, this critical particle size was found to be approximated by the expression

$$d_{crit} = 10325\rho^{-0.725}, \tag{11.1}$$

where d_{crit} is the critical particle diameter in microns. Baeyens also proposed a modification to the Colakyan and Levenspiel correlation for particles smaller than the critical particle size. The Colakyan and Levenspiel correlation, shown in Table 11.1, can be rewritten as

$$K_i = 0.011\rho_p \left(1 - \frac{u_{ti}}{u_o}\right)^2; \quad d_{pi} < d_{crit}$$

$$K_i = 0.0128 u_o^{3.4} \rho_p \left(1 - \frac{u_{ti}}{u_o}\right)^2; \quad d_{pi} < d_{crit}, \tag{11.2}$$

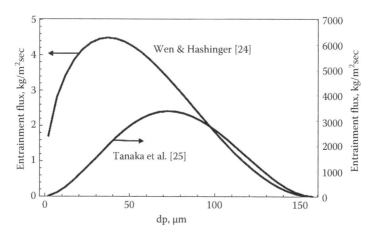

FIGURE 11.2 Empirical correlations for entrainment flux of equilibrium FCC catalyst with a d_{p50} of 81 microns and a fines level (d_p < 44 microns) of 10% in a 3 meter diameter × 12 meter tall bed with a 6 meter bed height at a superficial gas velocity of 1 meter/sec.

TABLE 11.1
Empirical Entrainment Correlations

Reference	Equations (MKS)	Comments
Stojkovski and Kostić [18]	$$K_i = \dfrac{k_1 \left(\dfrac{Fr_i^{1.1225}}{Re_i}\right)^{k_2} u_o \rho_f}{\left(1 - \dfrac{u_{ti}}{u_o}\right)^{1.25}}$$	Based on particles ranging from 150 to 580 microns with particle density of 2660 kg/m³
Colakyan and Levenspiel [22]	$K_i = 0.011 \rho_p \left(1 - \dfrac{u_{ti}}{u_o}\right)^2 ; u_{ti} < u_o$	Based on particles ranging from 300 to 1000 microns with particle density ranging from 920 to 5900 kg/m³ with superficial gas velocities up to 4 m/sec
Lin, Sears, and Wen [20]	$K_i = 0.000943 u_o \rho_f \left(\dfrac{u_o^2}{g d_{pi}}\right)^{1.65}$	Char and sand material ranging from 75 to 150 microns
Colakyan [21]	$K_i = 33\left(1 - \dfrac{u_{ti}}{u_o}\right)^2$	Geldart Group B material
Geldart et al. [23]	$K_i = 23.7 \rho_f u_o e^{-5.4\frac{u_{ti}}{u_o}}$	Over a wide range of particles and conditions including large particles and high velocities
Tanaka, Shinohara, and Tanaka [25]	$K_i = 0.046 \rho_f (u_o - u_{ti})$ $\times Re_{ti}^{0.3} \left(\dfrac{(u_o - u_{ti})^2}{g d_{pi}}\right)^{0.5} \left(\dfrac{\rho_p - \rho_f}{\rho_f}\right)^{1.15}$	Glass beads, steel balls, lead balls ranging from 60 to 2300 microns
Wen and Hashinger [24]	$K_i = 0.000017 \rho_f (u_o - u_{ti})$ $\times Re_{ti}^{0.725} \left(\dfrac{(u_o - u_{ti})^2}{g d_{pi}}\right)^{0.5} \left(\dfrac{\rho_p - \rho_f}{\rho_f}\right)^{1.15}$	Glass beads ranging from 70 to 280 microns and bituminous coal ranging from 100 to 700 microns
Zenz and Weil [19]	$K_i = k_1 \rho_f u_o \left(\dfrac{u_o^2}{g d_{pi} \rho_p^2}\right)^{k_2}$	FCC catalyst powder up to pressure of 14 Barg

$$\text{where} \begin{cases} k_1 = 1.26 \times 10^7, k_2 \\ \quad = 1.88; \text{if } \dfrac{u_o^2}{g d_{pi} \rho_p^2} \le 0.0003, \\ k_1 = 4.31 \times 10^4, k_2 \\ \quad = 1.158; \text{if } \dfrac{u_o^2}{g d_{pi} \rho_p^2} > 0.0003 \end{cases}$$

where a 1.164 $u_o^{3.4}$ prefix was incorporated into the Colakyan and Levenspiel correlation. Baeyens did not promote this correlation as being best for determining the entrainment rates but only because "its form is such that it can be modified." Ideally, such a correction could be applied to entrainment rate correlations that have a form similar to the Colakyan and Levenspiel correlation.

The question these correlations ask is why does the entrainment rate decrease for smaller particles for some systems whereas in other systems, the entrainment rate correlates with the particle terminal velocity or particle drag. Baeyens infers that particles may be clustering due to an interparticle adhesion force that becomes dominant at some critical particle diameter. However, no evidence of particle clusters was reported. Baeyens' assumption was based on fitting their data. Therefore, the role of particle clustering on entrainment rates was difficult to establish from first principles.

11.3   EVIDENCE OF PARTICLE CLUSTERING

Recently, Hays et al. [26] reported on of several cases where particle clustering was inferred in fluidized bed systems. In the first case, they attempted to reproduce why highly variable entrainment rates were observed in a commercial-scale fluidized bed even though steady-state was presumed. Tests were conducted in a 6 inch (15-cm) diameter fluidized column with a static bed height of 52 inches (132 cm) of the same Geldart Group A powder (d_{p50} of 55–60 microns) used in the commercial process. The test unit was operated in batch mode at a superficial gas velocity of 0.66 ft/sec (0.2 m/sec).

Figure 11.3 shows the entrainment flux measured at the outlet of the unit and the fines weight fraction (defined as particle sizes smaller than 44 microns) for the bed and entrained material as a function of time. As the fines concentration began to decrease in the bed, the entrainment rate increased rapidly to a peak approximately 10 times the initial rate. Further depletion of fines from the bed resulted in a subsequent drop in the entrainment rate. For the commercial unit, the highly variable entrainment rates appeared to be due to the extreme sensitivity to the fines concentration. The dipleg was not designed to handle the presumed tenfold increase in the entrainment rate and often flooded as a result.

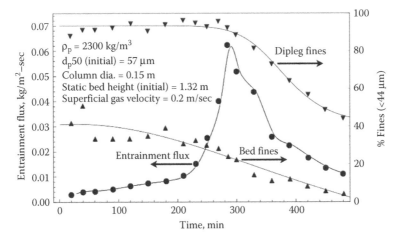

FIGURE 11.3   Entrainment flux from a 6 inch (15 cm) diameter fluidized bed at 0.66 ft/sec (0.2 m/sec) superficial gas velocity. Fines level and solids entrainment rate are tracked with time.

TABLE 11.2

Entrainment Flux for Flexicoke Particles in a 0.15 m Diameter Fluidized Bed for Two Different Particle Size Distributions at Two Different Superficial Gas Velocities

	Entrainment Flux at Superficial Gas Velocities Of	
Particle Size	0.30 m/sec	0.46 m/sec
20 μm	0.002 kg/m²-sec	0.355 kg/m²-sec
76 μm	0.035 kg/m²-sec	0.448 kg/m²-sec

These results are consistent with the earlier studies of Geldart and Wong [14], Choi et al. [16] and Li et al. [17]. The addition or presence of fines in a fluidized bed of Geldart Group A powder results in a reduction of the entrainment rate. Hays study further confirmed the role of fines by showing that the removal of fines from a fluidized bed resulted in a significant increase in the entrainment rate, presumably resulting from a decrease in particle clustering.

For the second case, PSRI observed similar results with coarse and fine coke powders (d_{p50} of the coarse particles was about 76 microns and that of the fines was about 20 microns) in a 15 cm diameter fluidized bed. Table 11.2 shows a comparison of the entrainment fluxes measured in the 15 cm diameter fluid bed at fluidizing gas velocities of 0.3 and 0.46 m/s for both the coarse and finer coke particles. At 0.3 m/s, the entrainment flux for the 76 micron coke particles was about 16 times that of the 20 micron coke particles, suggesting that the 20 micron particles were clustering together. However, increasing the gas velocity from 0.3 to 0.46 m/s caused the entrainment flux of the finer coke particles to increase by a factor of 184. This was similar to the entrainment flux for the 76 micron coke particles and suggests that higher velocities may limit cluster growth or cause them to break up.

In the third case, Hays et al. [26] examined the continuous fluidization of FCC catalyst fines (d_{p50} of 27 microns) in a 6 inch (15 cm) diameter fluidized bed equipped with a cyclone and a dipleg fitted with an automatic L-valve. Experiments were performed at a superficial gas velocity of 1.8 ft/sec (0.56 m/sec). At this velocity, all the material in the fluidized bed was calculated to be entrainable using a terminal velocity correlation. The bed height decreased slowly with time in this system due to the loss of material from the cyclone to the external solids recovery system. However, during this period of gradual bed height reduction, the measured entrainment rate increased with time. In order to determine if it was the decrease in bed height that was causing the increase in entrainment, material was removed from the dense phase region of the bed while it was fluidized after 4500 seconds. As shown in Figure 11.4, the removal of bed material from the column caused a significant increase in the entrainment rate. Hays' data suggest that the entrainment rate was inversely proportional to the bed height, at least for beds with high fines concentrations. This conflicts with the expectation that larger disengaging heights (lower bed heights) result in a steady (if above transport disengagement height) or a reduced entrainment rate.

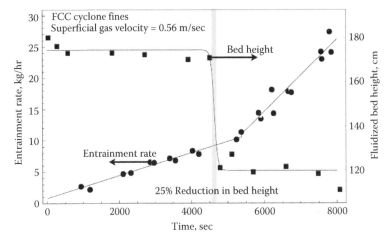

FIGURE 11.4 Entrainment rate from a bed of FCC catalyst fines when bed height was decreased by 25% at 4500 seconds. The superficial gas velocity was at 1.8 ft/sec (0.56 m/sec) at room conditions.

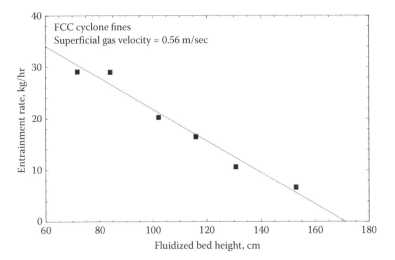

FIGURE 11.5 Entrainment flux from a 6 inch (15 cm) diameter bed of FCC catalyst with and without baffles at various fluidized bed heights.

The effect of bed height was further explored by measuring the entrainment rate of FCC catalyst fines at various bed heights. In separate fluidization experiments, the entrainment rate at six different bed heights was explored. As shown in Figure 11.5, an inverse linear relationship was observed between the fluidized bed height and the corresponding entrainment rate. This dependence of entrainment rate on bed height suggests that clustering is occurring and that the size of the particle clusters may be dependent on bed height.

Cocco et al. [27] used high-speed video (Phantom V7.1 camera) to clearly show that particle clustering was, at least in part, responsible for this behavior in the same

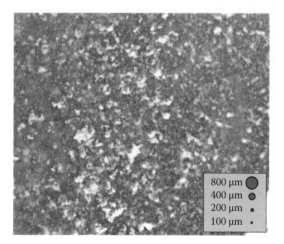

FIGURE 11.6 Frame capture from high-speed video imaging using a Phantom V7.1 camera of FCC catalyst fines (dp50 = 27 microns) in the freeboard region of a 6 inch (15 cm) diameter fluidized bed at a superficial gas velocity of 2 ft/sec (0.6 m/sec). Work was done with H. Jaeger and S. Nagel of the University of Chicago.

test unit. Clusters on the order of several hundred microns were readily observed, as shown in Figure 11.6. The question was not if clusters existed in the freeboard, but to what extent. Figure 11.6 shows the freeboard region at a superficial gas velocity of 2 ft/sec (0.6 m/sec).

The fourth case presented by Hays et al. was with FCC catalyst powder (d_{p50} of 70 microns) with 5% fines in a 6 inch (15 cm) diameter fluidized bed. They modified a rigid achromatic lens boroscope fitted with an optical glass spacer to extend the probe to the necessary focal length. This allowed for the visualization and measurement of particle clusters in and above a fluidized bed. Using the high-speed camera, they showed that the particle size, particle trajectory, cluster size, and cluster trajectory could be obtained, provided a statistically significant amount of frames were analyzed. In addition, varying degrees of magnification could be used to obtain detail of the particles and clusters. Details of this technique can be found in Cocco et al. [27].

Figure 11.7 shows several frames of the video obtained for particles and clusters in the freeboard region. As expected, the FCC catalyst tended to cluster in the freeboard region. A statistical analysis of this video suggested that 30% of the FCC catalyst in the freeboard existed as particle clusters with an average size of 11 ± 5.0 particles.

Figures 11.6 and 11.7 reveal why lower than expected entrainment rates were observed for both these materials. As expected and postulated by Hays et al. [26], Geldart and Wong [14], Baeyens et al. [15] and Choi et al. [16], particle clustering results in particle sizes too large for the drag force to carry the particles out of the unit. Thus, individual fine particles that would have been easily entrained out of the unit now fall back to the fluidized bed after clustering has occurred. Clustering may not only be dependent on the fines level but the material itself. Jayaweera et al. [3] and Fortes et al. [7] also reported of particle clustering with an FCC catalyst

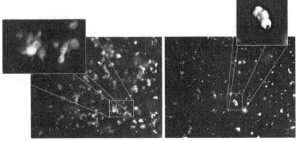

O ← 100 μm diameter

FIGURE 11.7 Selected frames of FCC catalyst in the freeboard region of a 6 inch (15 cm) diameter fluidized bed at a superficial gas velocity of 2 ft/sec (0.61 m/sec). Images were collected at 4000 frames per second with a 20 μs exposure time.

O ← 100 mm diameter

FIGURE 11.8 Several consecutive frames of FCC catalyst in the bed region of a 6 inch (15 cm) diameter fluidized bed at a superficial gas velocity of 2 ft/sec (0.61 m/sec). Images were collected at 4000 frames per second with a 20 μs exposure time.

material, but with smaller-sized clusters than observed in this study. Particle density may also be a factor [15].

Cocco et al. [27] also examined if clusters exist in the fluidized bed as well as in the freeboard. The boroscope, with the high-speed camera arrangement, was inserted into the bottom of a fluidized bed of FCC catalyst powder. The boroscope was positioned at the center of the bed to ensure that wall effects were not an issue. Consecutive images from the collected video in the fluidized bed of FCC catalyst powder are shown in Figure 11.8. Larger and a greater number of clusters were observed in the bed than in the freeboard. A statistical analysis of the video obtained in the fluidized bed suggested that 41% of the FCC catalyst material existed as clusters with an average cluster size of 21 ± 1.7 particles.

Because of the higher density in the bed, cluster imaging was conducted during the period when a gas bubble passed the face of the boroscope. The clusters shown in Figure 11.8 are at a much lower solids concentration than that of the emulsion phase. It is unknown whether the clusters exist in the denser emulsion phase and get ejected into the less dense bubble phase, or if the clusters are solely a product of the bubble phase in the fluidized bed.

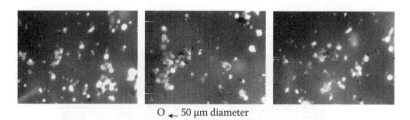

O ← 50 μm diameter

FIGURE 11.9   Several consecutive frames of FCC catalyst in the bed region of a 6 inch (15 cm) diameter fluidized bed at a superficial gas velocity of 2 ft/sec (0.61 m/sec). Images were collected at 4000 frames per second with a 20 μs exposure time. This is a 5 × magnification compared to Figure 11.8.

Figure 11.9 shows three consecutive images where the video was magnified about five times more than images in Figure 11.8. The clusters shown in Figure 11.9 are more definitive, and show that the clusters exist as both large particles with fines or as a large number of fines. Rarely was a cluster of only large particles observed. In some cases, large strands of 10 or more particles were observed consisting of both large and small particles. Similar observations were found with polyethylene in the same test unit [27].

The images in Figures 11.8 and 11.9 clearly show that particle clusters exist in the fluidized bed, even near the distributor. However, it is uncertain if these clusters occur only in the bubble region (cloud phase) or in the emulsion phase as well. Particle concentrations were too high to discern clusters when looking at the emulsion region. Only with the occurrence of a bubble near the probe were particle clusters observed. Thus, for Geldart Group A powders with fines, particle clustering does occur in fluidized beds either in the emulsion and bubble regions or just the bubble regions. Subbarao [28] originally postulated that there was a relationship between bubbles and cluster formation. Perhaps the bubble serves as a concentrator of fines, which promotes clustering. Another explanation could be that the bubbles serve as a concentrator of clusters that were initially formed in the emulsion phase.

To discern the stability of the clusters in the fluidized bed, Hays et al. [26] added baffles to their 6 inch (0.15 meter) diameter fluidized bed. The baffles resembled a simple grating commonly used for floor decking, and were positioned at 1.6 and 2.5 feet (0.5 and 0.76 meters) above the distributor plate. The bed height was 2.5 feet (0.76 meters). As shown in Figure 11.10, the presence of baffles resulted in an increase in the entrainment flux at higher gas velocities. One explanation for this behavior is that the clusters formed in the bed impact the baffles at the high gas velocities and are broken up, which results in smaller clusters and higher entrainment rates.

In a similar experiment with coke (d_{p50} of 150 microns) containing no fines, little difference in the entrainment rate was measured for the test unit with and without baffles, as shown in Figure 11.11. Unlike the fluidization studies using Geldart Group A material, the presence of the baffles with Geldart Group B material did not increase the entrainment flux. Presumably, the larger material did not form significant clusters that were available for breaking up. These data suggest that the particle clusters, possibly responsible for the entrainment rate reduction, may be formed in

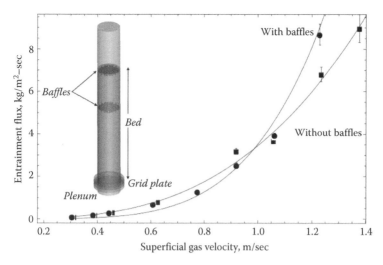

FIGURE 11.10 Entrainment flux from a 6 inch (15 cm) diameter bed of FCC catalyst with and without baffles at various superficial gas velocities.

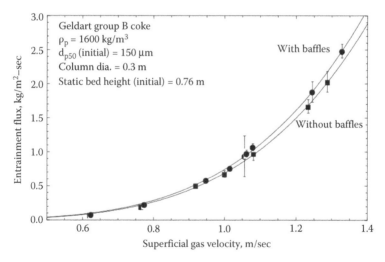

FIGURE 11.11 Entrainment flux from a 12 inch (30 cm) diameter bed of fluid coke powder with and without baffles at various superficial gas velocities.

the bed region. In other words, particle clusters, which form in the fluidized bed and get ejected into the freeboard, have a higher slip velocity than single particles and are entrained at lower rates. The deeper the bed, the larger or more stable the particle cluster.

These results suggest that particle clusters form in the fluidized bed and get ejected into the freeboard instead of clusters forming only in the freeboard region. Particle clustering most likely does happen in the freeboard, but it may not be the dominating contributor to particle cluster concentration in the freeboard. Kaye and

Boardman [10] noted that loadings greater than 0.05% are needed for clustering to occur. Loadings may be too low in the freeboard region, especially at the top of the bed, for sufficient particle concentrations to promote particle clustering. Figure 11.10 also supports the theory that particle clusters occur in the fluidized bed and then get ejected into the freeboard. The addition of baffles in the bed and near the top of the bed resulted in higher entrainment rates for Group A particles, suggesting that baffles can break up the clusters before being ejected into the freeboard. However, this only occurred at high superficial gas velocities. Below superficial gas velocities of 1–1.2 m/sec, the clusters appear to remain intact, suggesting insufficient shear was available to break up the clusters. Conducting the same experiment with larger coke powder showed no difference in entrainment rates with respect to the baffled and unbaffled cases, as shown in Figure 11.11. The coke material had fewer fines available for cluster formation. However, coke clustering with smaller particles is possible as indicated in Table 11.2.

Figure 11.11 suggests that the height of the fluidized bed may control the cluster size or the frequency of cluster formation. As the bed height was increased for the FCC catalyst material, entrainment rates decreased. This result suggests that formation of large clusters may not occur instantaneously, and sufficient time in the emulsion or bubble region is needed for large particle clusters to form. Thus, a particle cluster may form near the bottom of the bed and continue to grow as it migrates to the top of the bed, possibly with the help of bubbles. At the top of the bed, it is either entrained or circulated back down to the bottom of the bed where it may be broken up at the distributor if the distributor jets have sufficient shear. Several cycles of the circulation may be needed to build large clusters. As bed height is increased, this large circulation zone becomes more dominant and the possible residence time of a particle cluster in the bed becomes extended.

The results from placing baffles in a fluidized bed, as shown in Figures 11.10 and 11.11, also support this mechanism. The addition of baffles may not only serve as a mechanism to break up clusters (especially at low superficial gas velocities), but to disrupt the larger recirculation zones in the fluidized bed and reduce the time a cluster spends in this cycle. Instead of one large recirculation zone, several recirculation zones may develop with only the top recirculation zone contributing to the particle clusters that can be entrained.

11.4 PARTICLE CLUSTERING MECHANISMS

For Geldart Group A powders, the forces or possible interactions responsible for particle clustering include: hydrodynamics (drag minimization), inelastic particle collisions (collisional cooling), electrostatic charging (coulombic), capillary, and van der Waals [15,27–29]. However, it is also possible that two or more forces may work in concert to cause particle clustering, depending on the particles and the environment.

Hydrodynamic interactions with particles may certainly play a role in clustering. Horio and Clift [30] noted that particle clusters, "a group of loosely held together particles," are the result of hydrodynamic effects. Squires and Eaton [31] proposed that clustering resulted from turbulence modification from an isotropic turbulent flow field, which is an expansion of the work of Wylie and Koch [32]. Subbarao [28] proposed that a uniform dispersion of particles in a gas is an unstable state and that particle clustering was inherent to the hydrodynamics of bubbles. This study was only able to confirm the presence of such clusters in and near the bubble region, but it did not provide evidence that clusters exist elsewhere in the bed. All of these studies suggest that viscous dissipation of the gas phase promotes the formation of particle clusters whether due to drag or turbulence reduction.

However, viscous dissipation is not a requirement for particle clustering. Royer and coworkers [33] did a series of powder drop experiments using glass beads with an average particle size of 110 microns. Using a high-speed camera falling at the same rate as the powder, Royer showed that particles cluster in a fashion similar to a stream of liquid breaking up into droplets due to surface tension. Their study also showed that particle clusters formed whether the system pressure was 101,325, or 300 Pa. In fact, better-formed clusters were observed at the lower pressure due to fewer particles shedding off from the clusters because of drag. Little change in clustering behavior was observed for the glass beads in an electric field.

Royer and coworkers [33] expanded on this study to include powder drop experiments of copper powder and silver-coated glass beads. As with the standard glass beads, the silver-coated glass beads were also reported to cluster. Yet, copper powder did not show any evidence of particle clustering in the powder drop experiment. Copper has a lower coefficient of restitution than glass (0.9 for copper versus 0.97 for glass [34]). Thus, cluster formation because of collisional cooling does not seem to be significant, at least in this case. Only with the addition of a thin layer of oil did copper powder show clustering similar to that seen with the glass beads. Even more interesting, clustering was prevented by the addition of nanoscale asperities to the glass beads.

Royer and coworkers [33] found that the degree of surface roughness of the particle contributed to cohesive forces. Surfaces, where roughness was significant, such as for the copper powder and the glass beads with nanoscale asperities, seemed to prevent particle clustering. Particles with smooth surfaces such as that of the glass beads and the silver-coated glass beads were prone to particle clustering. One possible explanation for this behavior is that large-scale surface roughness may reduce sliding and lead to different particle rotational and collisional dynamics [27]. In other words, surface roughness does not lower the granular temperature as much as smooth surfaces. If enough energy is dissipated in collisions and not redirected into rotation, the granular temperature (and rotational energy) may decrease enough such that various cohesive forces dominate the hydrodynamics. When the granular temperature is low enough, cohesive forces such as van der Waals, electrostatics or capillary forces can become dominant. For FCC catalyst, it could be capillary forces with a water monolayer (not liquid bridging in the strictest sense) that causes the cohesion. For even smaller particles or lighter particles, van der Waals may have a significant role [15]. It could also be that a rough surface prevents these cohesive forces from interacting whether due to increased radial spacing of the core particles, reduced chemical attractive forces (i.e., hydrogen bonding), or reduced electronic interactions (i.e., coulombic). However, Royer et al. [33] used atomic force microscopy to quantify the force between two glass and two copper particles. The results showed that the cohesive force for the two copper particles was almost twice as high as the smooth glass particles.

Regardless of these short-ranged cohesive forces, the formation and stability of particle clusters in a fluidized bed appears to be a multistep process [27]. Some shear (as in two particles grazing each other) may be needed to promote collisional cooling, but less than that perhaps in the dense emulsion of a fluidized bed. Perhaps the lower particle concentration in a bubble provides the environment where cluster stability is promoted for the smaller particles. Collisional stresses in the emulsion may be too high and the cohesive forces may be too low to have long-lasting particle clusters. Indeed, the only evidence of particle clusters in fluidized beds offered here is that the clusters are located near the bubbles.

11.5   IMPACT ON PARTICLE ENTRAINMENT

Using a modified boroscope and a high-speed video camera, clear evidence of particle clustering was observed in and above a fluidized bed of FCC catalyst powder. The level of clustering appears to be linked to the particle size (and/or density) but also to the surface morphology, as noted by Royer et al. [33]. This may explain why such a large variation of entrainment rate profiles and magnitudes exist throughout the literature. It also highlights that for Geldart Group A particles, it is difficult for a correlation to encompass all aspects of particle elutriation from a fluidized bed. The most accurate predictions of entrainment rate for Group A material are obtained with experimental measurements, preferably at the operating conditions of the process.

The entrainment rate correlations of Geldart et al. [23], Colakyan et al. [21], and Colakyan and Levenspiel [22] may provide a conservative estimate of entrainment rate; but their predictions could lead to an overdesigned cyclone recovery system, which could result in too low of a downward solids flux in the primary cyclone dipleg and result in gas bypassing to the cyclone or plugging of the diplegs. Other correlations where lower particles sizes result in an exponential increase in entrainment rates (such as that by Stojkovski and Kostic' [18], Zenz and Weil [19], and Lin et al. [20]) may not be appropriate for Geldart Group A powders such as FCC catalyst. The entrainment correlations of Wen and Hashinger [24] and Tanaka et al. [25] predicted the lower entrainment rate for smaller particles, but the magnitude of the entrainment rate and the critical particle size predictions need to be validated. Both these correlations predicted different magnitudes and critical particle sizes for the test case illustrated in Figure 11.2.

Entrainment rates measured in a laboratory or pilot plant unit can also underpredict entrainment rates in commercial units if the solids tend to form clusters. This could be the case if the commercial-scale plant contains baffles and the laboratory or pilot plants do not. This was the case illustrated in Figure 11.10. Even the spacing of the baffles can have an effect on the entrainment rate. Adding baffles to a fluidized bed regenerator may actually increase entrainment rates. As discussed above, particle clusters may need sufficient time in the emulsion or bubble region of the bed to form. Thus, a particle cluster may start growing near the bottom of the bed and continue to grow as it travels to the top of the bed. Baffles spaced too close to each other may disrupt the cluster growth process and lead to higher entrainment rates.

Particle clustering can also result in increased cyclone efficiencies. If small particles that would normally not be collected by a cyclone cluster together to form larger particles, the cyclone can capture the clusters. However, the clusters are generally relatively fragile, and an increase in the cyclone inlet velocity could also result in increased solids losses. The higher impact forces produced by the higher velocity can break up the clusters into the smaller particles that the cyclone cannot capture. This is counterintuitive because increasing the gas inlet velocity to a cyclone generally causes higher collection efficiencies.

As mentioned above, particle clustering can also reduce entrainment rates from Geldart Group A systems. The larger clusters will be entrained at lower rates than the individual particles comprising the clusters, as suggested in Figure 11.3.

11.6   SUMMARY

Direct evidence now exists that substantiates early claims of particle clustering in fluidized beds containing FCC catalyst powder. Using a high-speed video camera in conjunction with a modified rigid boroscope allowed images of particle clusters to be visualized. The images of the clusters were obtained beyond the column wall, in and above a fluidized bed of FCC catalyst powder. Cohesive forces such as electrostatics, capillary, and van der Waals forces, may play a significant role in particle cluster formation. A possible clustering mechanism is that particle shear results in collisional cooling that allows the granular temperature to decay to a point where these cohesive forces can dominate. The decrease in the granular temperature appears to be dependent on the particle properties and surface morphology.

Particle clustering has a significant impact on how entrainment is calculated. Currently, no one literature correlation appears to predict entrainment accurately, especially for Geldart Group A powders. Experimental measurements still seem to be the best way to predict entrainment rates for commercial units. However, care needs to be taken when adding baffles to the fluidized bed because it could lead to higher entrainment rates at higher superficial gas velocities. In contrast, adding fines to a Group A fluidized bed could result in reduced solids loss rates because of increased particle clustering. Of course, the opposite is true as well. Reducing fines levels can result in an increase in the entrainment rate. It all appears to be dependent on microscopic properties as well as particle size and concentration.

ACKNOWLEDGMENTS

The authors would like to thank John Roper, Scott Waitukaitis, Sid Nagel, and Heinrich Jaeger of the University of Chicago for their invaluable discussion on particle clustering. The authors would like to thank Alexander Mychkovsky and Steve Ceccio of the University of Michigan for their help and consultation with the development of the boroscope. The authors also thank the National Energy Technology Lab and the Department of Energy for their funding under DE-FC26-07NT43098 as well as their expertise.

REFERENCES

1. Wilhelm, R. H., and Kwauk, M. Fluidization of Solid Particles. *Chem. Eng. Prog.* 44 (1948): 201–18.
2. Happel, J., and Pfeffer, R. The Motion of Two Spheres Following Each Other in a Viscous Fluid. *AIChE J.* 6 (1960): 129–33.
3. Jayaweera, K. O. L. F., Mason, B. J., and Slack, G. W. The Behavior of Clusters of Spheres Falling in a Viscous Fluid. *J. Fluid Mech.* 20 (1964): 121–28.
4. Johne, R. Einfluß der Konzentration einer monodispersen Suspension auf die Sinkgeschwindigkeit ihrer Teilchen. *Chem. Ing. Tech.* 38 (1966): 428–30.
5. Crowley, J. M. Clumping Instability of a Falling Horizontal Lattice. *Phys. Fluids* 19 (1976): 1296–1301.
6. Graham, A. L., and Bird, R. B. Particle Clusters in Concentrated Suspensions. 1. Experimental Observations of Particle Clusters. *Ind. Eng. Chem., Fundam.* 23 (1984): 406–10.
7. Fortes, A. F., Joseph, D. D., and Lundgren, T. S. Nonlinear Mechanics of Fluidization of Beds of Spherical Particles. *J. Fluid Mech.* 177 (1987): 467–83.
8. Wang, M. K., and Li, H. Fluidization of Fine Particles. *Chem. Eng. Sci.* 53 (1998): 377–95.
9. Chen, U.-M., Jang, C.-S., Cai, P., and Fan, L.-S. On the Formation and Disintegration of Particle Clusters in a Liquid—Solid Transport Bed. *Chem. Eng. Sci.* 46 (1991): 2253–68.
10. Kaye, B. M., and Boardman, R. P. Cluster Formation in Dilute Suspensions. In *Proceedings of the Symposium on the Interaction between Fluids and Particles*, 17–21. London: Institute of Chemical Engineering, 1962.
11. Yerushalmi, J., Tuner, D. H., and Squires, A. M. The Fast Fluidized Bed. *Ind. Eng. Chem. Proc. Des. Develop.* 15 (1976): 47–53.
12. Yerushalmi, J., and Cankurt, N. J. Further Studies of the Regimes of Fluidization. *Powder Technol.* 24 (1979): 187–205.
13. Matsen, J. M. Mechanisms of Choking and Entrainment. *Powder Technol.* 32 (1982): 21–33.
14. Geldart, D., and Wong, A. C. Y. Entrainment of Particles from Fluidized Beds of Fine Powder. *AIChE Symp. Ser.* 255 (1987): 1.
15. Baeyens, J., Geldart, D., and Wu, S. Y. Elutriation of Fines from Gas Fluidized Beds of Geldart A-Type Powders—Effect of Adding Superfines. *Powder Technol.* 71 (1992): 71–80.
16. Choi, J.-H., Suh, J.-M., Chang, I.-Y., Shun, D.-W., Yi, C.-K., Son, J.-E., and Kim, S.-D., The Effect of Fine Particles on Elutriation of Coarse Particles in a Gas Fluidized Bed. *Powder Technol.* 121 (2002): 190–94.
17. Li, J., Nakazato, T., and Kato, K. Effect of Cohesive Powders on the Elutriation of Particles from a Fluid Bed. *Chem. Eng. Sci.* 59 (2004): 2777–82.
18. Stojkovski, V., and Kostic', Z. Empirical Correlation for Prediction of the Elutriation Rate Constant. *Thermal Sci.*, 7 (2003): 43–58.
19. Zenz, P. A., and Weil, N. A. A Theoretical-Empirical Approach to the Mechanism of Particle Entrainment from Fluidized Bed. *AIChE J.* 4 (1958): 472–79.
20. Lin, L., Sears, J. T., and Wen, C. Y. Elutriation and Attrition of Char from a Large Fluidized Bed. *Powder Technol.*, 27 (1980): 105–15.
21. Colakyan, M., Catipovic, N., Jovanovic, G., and Fitzgerald, T. J. *AIChE Symp. Ser.* 77 (1981): 66.
22. Colakyan, M., and Levenspiel, O. Elutriation from Fluidized Beds. *Powder Technol.*, 38 (1984): 223–32.

23. Geldart, D., Cullinan, J., Georghiades, S., Gilvray, D., and Pope, D. J. The Effect of Fines on Entrainment from Gas Fluidized Beds. *Trans. Inst. Chem. Eng.* 57 (1979): 269–77.
24. Wen, C. Y., and Hashinger, R. F. Elutriation of Solid Particles from a Dense Phase Fluidized Bed. *AIChE J.* 6 (1960): 220–26.
25. Tanaka, I., Shinohara, H., and Tanaka, Y. Elutriation of Fines from Fluidized Bed. *J. Chem. Eng. Jpn.* 5 (1972): 57–62.
26. Hays, R., Karri, S. B. R., Cocco, R., and Knowlton, T. M. Small Particles Cluster Formation in Fluidized Beds and its Effect on Entrainment. *Circ. Fluid. Bed* 9 (2008): 1–5.
27. Cocco, R., Shaffer, F., Hays, R., Karri, S. B. R., and Knowlton, T. M. Particle Clusters in and above Fluidized Beds *Powder Technol.* 203 (2010): 3–11.
28. Subbarao, D. Clusters and Lean-Phase Behaviour. *Powder Technol.* 46 (1986): 101–7.
29. Visser, J. An Invited Review—Van der Waals and Other Cohesive Forces Affecting Powder Fluidization. *Powder Technol.* 58 (1989): 1.
30. Horio, M., and Clift, R. A Note on Terminology: "Clusters" and "Agglomerates." *Powder Technol.*, 70 (1992): 195–96.
31. Squires, K. D., and Eaton, J. K. Particle Responses to Turbulence Modification in Isotropic Turbulence. *Phys. Fluid A* 2 (1991): 1191.
32. Wylie, J. J., and Koch, D. L. Particle Clustering Due to Hydrodynamic Interactions. *Phys. Fluids* 12 (2000): 964.
33. Royer, J. R., Evans, D. J., Oyarte, L., Guo, Q., Kapit, E., Möbius, M. E., Waitukaitis, S. R., and Jaeger, H. M. High-Speed Tracking of Rupture and Clustering in Freely Falling Granular Streams. *Nature* 459 (2009): 1110–13.
34. Kuwabara, G., and Kono, K. Restitution Coefficient in a Collision between Two Spheres. *Jpn. J. Appl. Phys.* 26 (1987): 1230.

12 Application of ^{1}H-NMR for Fluid Catalytic Cracking Feed Characterization

Dariusz S. Orlicki,* Uriel Navarro,
Michelle Ni, and Larry Langan

CONTENTS

* Author to whom correspondence should be addressed. E-mail:Dariusz.orlicki@grace.com

12.1   INTRODUCTION

Feed quality is considered the most influential single factor affecting the fluidized catalytic cracking (FCC) unit performance. Chemical composition of the FCC feed affects yields and the quality of products. The FCC feed properties strongly depend on two factors: the geographic source of the crude and its initial preprocessing. To make things even more complicated, FCC feeds are usually blends of multiple feed sources and refinery recycle streams, some of which are of very poor quality. Therefore, knowing the molecular composition and concentrations of a complex feed aids in optimizing FCC unit operations, and also supports the process of catalyst selection and product yield predictions. In general, the typical analytical techniques for feed characterization provide helpful information about feed quality, but they are not adequate to provide the identities and concentrations of the molecules in commercial FCC feeds. Over the years, methods have been developed, mostly in the form of heuristic correlations, to predict important FCC feed properties such as hydrogen content, type of carbon (aromatic: Ca, paraffinic: Cp, and naphthenic: Cn), molecular weight, and so on, with a goal to relate these feed quality characteristics to product yields and qualities. However, the range for which these correlations were developed, and the assumptions and mathematical manipulation involved in the derivation of such formulas limit the use of these techniques.

^{1}H-NMR technique alone or in combination with ^{13}C-NMR has already been successfully applied in numerous scientific studies to characterize complex FCC feeds [1]. Although this technique provides an unparalleled level of information about atomic connectivity, it may be less precise in predicting the molecular composition of the feed. There are reports published in literature about the applications of ^{1}H-NMR to predict properties of FCC feeds [2] and FCC products [3,4]. Although these papers proclaimed great potential in using NMR techniques for characterization and prediction of feed properties, these findings should be applied with caution. The analyses were performed on either (i) feeds that originated from the same geographical region (Colombian crudes) or (ii) products of the same nature (gasoline), which narrows the bases of these analyses and imposes some limitations on the broad application of these findings to guide diverse FCC operations.

In this work, a set of 42 FCC feeds representing diverse geographical locations ranging from Asia Pacific through Europe to different regions within the continental United States and Canada were selected from the Grace Davison Feed Database. The selection criterion was based solely on The American Petroleum Institute (°API) and the individual feeds were chosen to cover the widest possible range. In this study the range of °API spanned between 13.8 and 32.3. All the feeds were fully characterized by measuring bulk physical properties, distillation curves, Conradson carbon residues, refractive indices, and metals content (V, Ni, Fe, Na, and others). Table 12.1 lists some of the properties measured along with the average and minimum and maximum values for the set of feeds addressed in this work. Consequently, feeds were characterized by ^{1}H-NMR and chemical shift spectra over the range of 0.5–9.3 ppm were collected. This information was used to find statistical correlations between feed spectra and properties. In addition to careful feed characterization work, an

TABLE 12.1
Properties of the 42-Feed Sample

Property	Minimum	Maximum	Average
°API	13.8	32.3	22.0
Average molecular weight	264	624	375
Ca [%]	13.1	43.6	23.8
Cn [%]	5.6	37.1	18.5
Cp [%]	44.1	67.4	55.3
Basic nitrogen [wt%]	0	0.20	0.05
Total nitrogen [wt%]	0.009	0.36	0.12
Sulfur [wt%]	0.01	3.994	1.264
Nickel [ppm]	0	6.9	1.21
Vanadium [ppm]	0	16.5	2.0
Hydrogen content [wt%]	10.4	13.7	12.0
Conradson carbon [wt%]	0	9.8	1.28
K factor	10.8	12.3	11.6
Refractive index	1.477	1.557	1.517
Initial boiling point (IBP) [°F]	287	709	376
5% [°F]	375	881	529
10% [°F]	458	940	603
20% [°F]	542	1015	674
30% [°F]	576	1064	718
40% [°F]	611	1103	756
50% [°F]	643	1143	792
60% [°F]	675	1185	829
70% [°F]	710	1238	871
80% [°F]	747	1311	922
90% [°F]	810	1383	1002
Final boiling point (FBP) [°F]	1007	1383	1183

Advance Catalyst Evaluation unit (ACE) [5] was used to study catalyst-feed interactions on two commercially available, laboratory deactivated catalyst materials.

12.2 EXPERIMENTAL

The ¹H-NMR spectra of FCC feeds were recorded on a Bruker DRX 400 MHz NMR spectrometer. The concentration of the samples of ~5 wt% in $CDCl_3$ was recommended by Molina, Navarro Uribe, and Murgich [2] to avoid concentration dependence of the chemical shift. A 30° pulse sequence was applied, with 4.089 s acquisition time, 2 s pulse delay [2], 8012.8 Hz spectral width, and 64 scans. Hexamethyldisiloxane (HMDSO) was used as a reference. NMR processing was realized using MestReNova software. The phase and baseline of the resulting spectra were manually adjusted and corrected. The spectra were integrated six times and average values were taken for the purpose of calculations. The spectra were divided

TABLE 12.2
^{1}H-NMR Regions and Their Corresponding Hydrogen Types

Chemical Shift Region (ppm)	Area	Hydrogen Type
0.5–1.0	H1	γCH_3 and some naphthenic CH and CH_2
1.0–1.5	H2	βCH_2 and some βCH
1.5–2.1	H3	CH and CH_2 in β positions
2.1–2.4	H4	αCH_{3s} in olefins
2.4–2.7	H5	αCH_{3s} in aromatic carbons
2.7–3.5	H6	αCH and αCH_2 in aromatic carbons
3.5–4.5	H7	bridging CH_2
4.5–6.0	H8	olefins
6.0–7.2	H9	mono-aromatics
7.2–8.3	H10	di-aromatics and some tetra-aromatics
8.3–8.9	H11	tri- and tetra-aromatics
8.9–9.3	H12	some tetra- aromatics

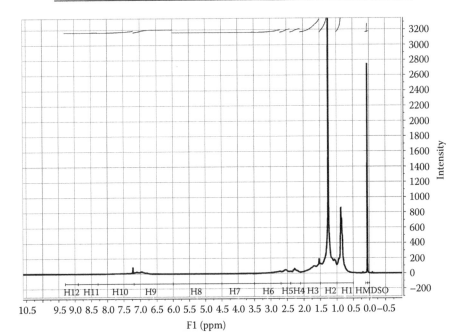

FIGURE 12.1 ^{1}H-NMR spectrum of one feed. The assigned regions (H1–H12) are listed in Table 12.2.

into 12 regions, representing 12 different hydrogen types in feed (see Table 12.2) [2]. Figure 12.1 shows an example of a feed ^{1}H-NMR spectrum. The average integrals of each segment were correlated with the physicochemical properties.

In connection with the study of catalyst-feed interations, the ACE unit temperature was set to 980°F (799.8 K). Every feed was tested on a commercially available

TABLE 12.3
Properties of Two Commercial Catalysts Deactivated in the Laboratory by CPS Method

Catalyst	Total Surface Area [m²/g]	Zeolite Surface Area [m²/g]	Matrix Surface Area [m²/g]	Unit Cell Size [Å]
Catalyst A 5% (Re₂O₃)	187	149	38	24.28
Catalyst B 2% (Re₂O₃)	175	92	83	24.29

FCC catalyst, supplied by Grace Davison, at three different cat-to-oil ratios, 4, 6, and 8. The feed was injected at a constant rate of 3 g/min for 30 seconds. The catalyst to oil ratio was adjusted by varying the amount of catalyst in the reactor. Two catalysts used for this evaluation were laboratory deactivated using the cyclic propylene steaming (CPS) method [6]. Properties of these catalysts after deactivation are listed in Table 12.3.

12.3   RESULTS AND DISCUSSION

12.3.1   PHYSICOCHEMICAL PROPERTIES

The physicochemical properties of the feeds were determined by appropriate ASTM methods. One of the key objectives of this study was to evaluate the possibility of using ¹H-NMR spectra to predict these properties.

12.3.2   API GRAVITY

The American Petroleum Institute °API gravity is a measure of feed density compared to water. The formula used to obtain the °API gravity of petroleum liquids is as follows:

$$\text{API gravity} = \frac{141.5}{\text{SpGr}(60°\text{F})} - 131.5.$$

Specific gravity of petroleum products is measured at 60°F. The values of specific gravity vary over a relatively narrow range from about 0.8 for very light paraffinic crudes to about 1.0 for heavy, highly asphaltic crudes. The °API formula maps that narrow band of specific gravity values onto a stretched range with values between 10 and 40. The °API is one of the most important properties of petroleum products because it can be easily and precisely measured, and it is a good indicator of the feed quality. It correlates well with many feed properties such as:

- $\%C_A$, $\%C_N$, and $\%C_P$, representing the distribution of carbon atoms between the three structures, aromatic, naphthenic, and paraffinic [7]

- Refractive Index that is proportional to a feed's aromatic content [8]
- Watson K-factor, a composite parameter that estimates either paraffinicity or naphthenicity [9]
- Molecular weight [7] and structure of hydrocarbons through the concept of molar volume [10]

SAS Enterprise Guide was used to analyze data. The best linear regression model to predict °API based on ¹H-NMR spectra has the form:

$$API = 31.170 + 0.458 \times H1 - 1.191 \times H3 + 1.779 \times H4 - 3.934 \times H6 + 2.932 \times H8.$$

The coefficient of determination, R^2, for the °API model is 0.98. Figure 12.2 shows °API observed versus that calculated by the model. The two points with the highest error of 6.7% and 6.5% are marked as triangles. Figure 12.3a and b demonstrates the dependence of the °API on ¹H-NMR spectra intensities integrals over the individual regions H1–H12, as explained in Table 12.2.

The two feeds for which the correlation gave the highest error are heavy, low °API, and already preprocessed feed streams. One is DMO from Gulf Coast and the other is Coker Gas Oil from Mid West. The ¹H-NMR spectra features summarized in Figure 12.3a and b don't provide good clues about any special properties or features of the two feeds that resulted in the highest error for °API prediction. One possible explanation for the poor °API prediction for these two feeds is that in the analyzed set of feed samples the low °API population was underrepresented.

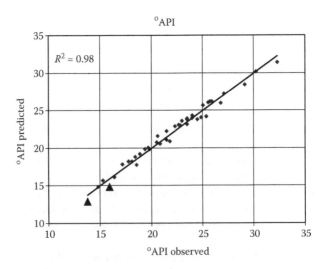

FIGURE 12.2   °API predicted based on ¹H-NMR vs. observed.

12.3.3   Watson Characterization Factor, *K* Factor

The *K* factor is a composite parameter used for feed classification. It was originally defined as a ratio of cube root of average molal boiling point and specific gravity. Instead of the average molal boiling point, other boiling points are sometimes used that include cubic average, mean average, volumetric average, and for simplicity, the

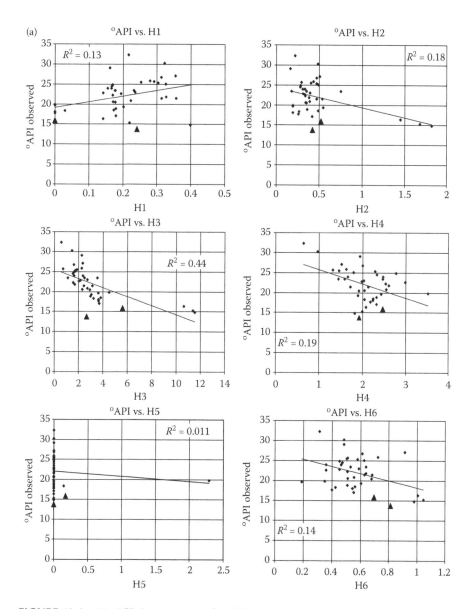

FIGURE 12.3   (a) °API dependence on ¹H-NMR spectra intensity integral over individual regions H1–H6.

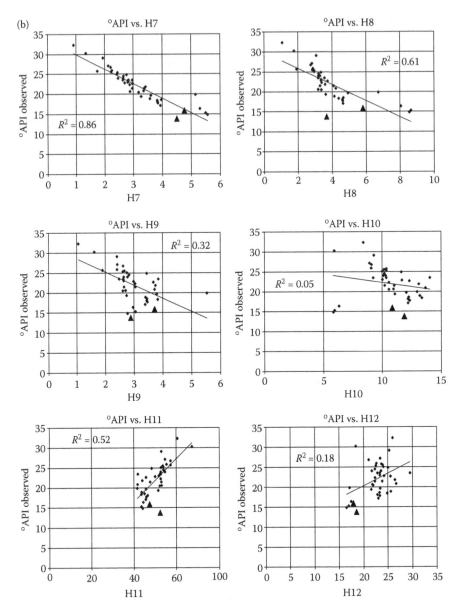

FIGURE 12.3 (CONTINUED) (b) °API dependence on ¹H-NMR spectra intensity integral over individual regions H7-H12.

unaveraged 50% volume boiling point. The *K* factor based on mean average boiling point is now called the Watson *K* factor. Values of 11.8 or higher indicate a product of paraffinic nature, while naphthenic in nature feeds have values between 11.5 and 11.8. Feeds in the range of 11.0–11.5 are aromatic. This composite factor can be calculated according to the equation

$$K = \frac{\sqrt[3]{(\text{MeABP} + 460)}}{\text{SpGr}},$$

where MeABP stands for the average of molal average and cubic average boiling points, and SpGr is specific gravity at 60°F. The best linear regression model to predict Watson K factor based on ¹H-NMR spectra has the form:

$$K = 13.183 - 0.09 \times \text{H3} - 0.295 \times \text{H4} + 0.198 \times \text{H6} \\ + 0.128 \times \text{H9} - 0.203 \times \text{H10}.$$

The coefficient of determination, R^2, for Watson K factor model prediction is 0.96 and the maximum error is less than 1.9%. Note that the predicted values based on ¹H-NMR follow the calculated Watson K factors in the paraffinic and aromatic regions more closely than in the intermediate, naphthenic region. The coefficient of determinations, R^2, for the paraffinic and aromatic regions are 0.98 and 0.96, respectively, while for the naphthenic region it is 0.81. Figure 12.4 shows Watson K factor observed versus that estimated based on ¹H-NMR spectra.

12.3.4 REFRACTIVE INDEX

The refractive index (n) of a feed sample is proportional to its aromatic content; the higher the value the more aromatic compounds are present in the feed. There are many methods to predict composition of petroleum products based on refractive index measurements [11]. The best linear regression model to predict refractive index (n) based on ¹H-NMR spectra has the form:

$$n = 1.2614 + 0.0027 \times \text{H2} + 0.005 \times \text{H3} + 0.0173 \times \text{H6} \\ - 0.0091 \times \text{H8} + 0.0046 \times \text{H10}.$$

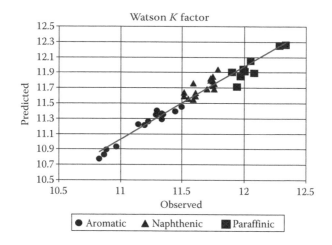

FIGURE 12.4 Watson K factor predicted based on ¹H-NMR vs. observed.

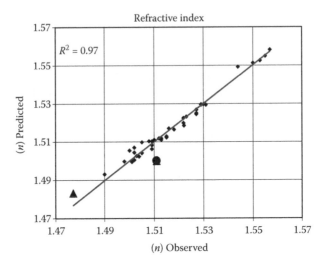

FIGURE 12.5   Refractive index (n) predicted based on ^{1}H-NMR vs. observed.

The coefficient of determination, R^2, for the model prediction is 0.97 and the maximum error is less than 0.7%. Figure 12.5 shows refractive index (n) observed versus that estimated from ^{1}H-NMR spectra. The model shows the largest discrepancies for either low aromatic content feeds (triangles), or high aromatic content feeds (circle).

12.3.5   DISTRIBUTION OF AROMATIC CARBON (CA), NAPHTHENIC CARBON (CN), AND PARAFFINIC CARBON (CP)

Information about feed composition can be inferred from other measured properties of the feed such as refractive index (n), density (d), and molecular weight (M). In the n–d–M method [12], the distribution of carbon atoms among aromatic, naphthenic, and paraffinic groups of the petroleum product is calculated using empirical formulas. The Ca, Cn, and Cp represent the percentage of carbon atoms in aromatic, naphthenic, and paraffinic structures, respectively. In theory, ^{1}H-NMR should be able to distinguish hydrogen distribution between aromatic and aliphatic structures thus yielding information on feed composition. The best linear regression model to predict Ca based on the ^{1}H-NMR spectra has the form:

$$Ca = 24.8728 - 0.455 \times H1 + 4.5356 \times H6 - 2.5847 \times H9 + 6.0229 \times H11 - 10.8938 \times H12.$$

Figure 12.6 shows the percentage of aromatic carbon content estimated from ^{1}H-NMR spectra versus observed Ca (n–d–M method). Although the coefficient of determination, R^2, for the Ca model prediction is 0.97, the relative error for Ca prediction can be substantial approaching 23% for low Ca values. Interestingly, the Ca correlates the most with protons in di-aromatics and tetra-aromatics, H10 region,

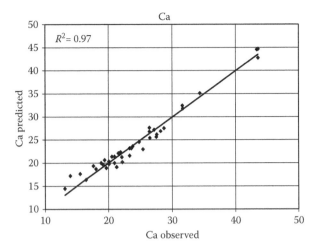

FIGURE 12.6   Ca predicted based on ¹H-NMR spectra vs. observed.

TABLE 12.4
Correlation Coefficients between Ca and Individual Regions, H1–H12, from ¹H-NMR Spectra

Hydrogen Type Region	H1	H2	H3	H4	H5	H6	H7	H8	H9	H10	H11	H12
Correlation (Ca, Hn)	**−0.79**	−0.52	−0.37	0.31	0.92	**0.91**	0.61	0.18	**0.15**	0.93	**0.77**	**−0.19**

and protons in α-CH$_3$ in aromatics, H5 region, which are not a part of the Ca model. The correlation coefficients between Ca and different regions from ¹H-NMR spectra, H1–H12, are collected in Table 12.4. The H-regions printed in italic are associated with aromatic compounds. The bold underlined fonts are used to mark regions contributing to the model. Figure 12.7a and b demonstrates the dependence of the Ca on ¹H-NMR spectra intensities integrals over the individual regions H1–H12, as explained in Table 12.2.

The best linear regression model to predict Cp based on the ¹H-NMR spectra has the form:

$$Cp = -11.8104 + 1.2682 \times H2 + 1.3808 \times H5 + 4.9076 \times H7 \\ - 11.1448 \times H11 + 11.6713 \times H12.$$

The coefficient of determination, R^2, of the Cp model is 0.95. The maximum error for Cp prediction can be as high as 7%. Figure 12.8 shows the percentage of paraffinic carbon content Cp estimated from ¹H-NMR spectra versus observed (n–d–M method).

The percent of naphthenic carbon Cn can be found directly by a regression method or more accurately by difference:

$$Cn = 100 - Ca - Cp.$$

The coefficient of determination, R^2, for Cn model prediction is 0.88 and the maximum error is below 8%.

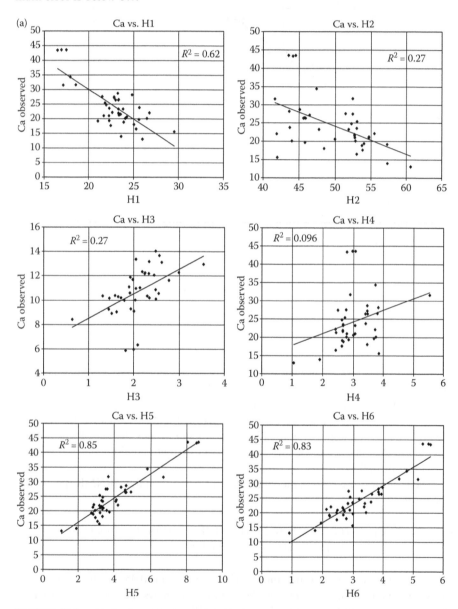

FIGURE 12.7 (a) Ca dependence on ^{1}H-NMR spectra intensity integral over individual regions H1-H6.

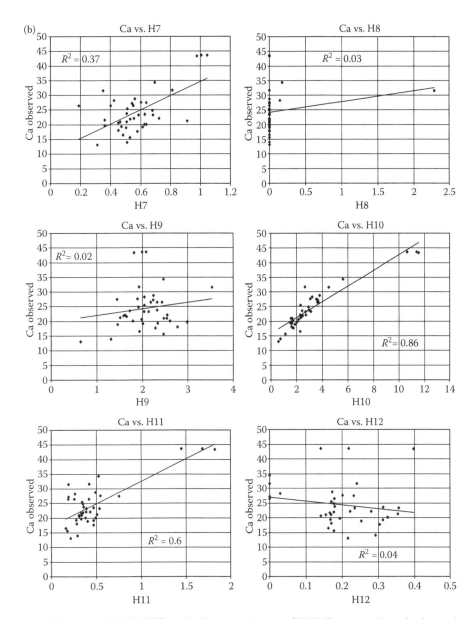

FIGURE 12.7 (CONTINUED)   (b) Ca dependence on ¹H-NMR spectra intensity integral over individual regions H7–H12.

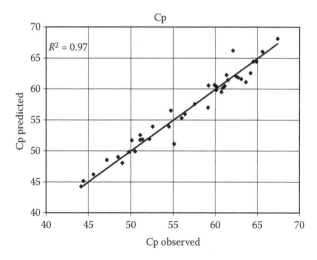

FIGURE 12.8 Cp predicted based on ^{1}H-NMR spectra vs. observed.

12.3.6 CORRELATION INDEX CI

The correlation index, CI, was originally devised by H. M. Smith from the U.S. Bureau of Mines in 1940 and it relates the average boiling point of a feed to its specific gravity. It is defined by the following empirical formula:

$$CI = \frac{87552}{(MeABP + 460)} + \frac{67030}{(°API + 131.5)} - 456.8,$$

where MeABP (in Fahrenheit) is the average boiling point, defined as the arithmetic average of the boiling temperatures taken at 10% intervals from 10 to 90% distillation fractions. The index can be used to classify feeds: values below 29.8 indicate paraffinic feeds while those higher than 57 are naphthenic feeds. The best linear regression model to predict correlation index, CI, has the form:

$$CI = -45.3116 + 0.4727 \times H2 + 4.478 \times H3 + 2.37 \times H6 + 5.3186 \times H10.$$

The coefficient of determination, R^2, of the CI model is 0.98. The maximum error for CI prediction is below 7%. Figure 12.9 shows CI estimated from ^{1}H-NMR spectra versus that calculated from its original formula.

12.4 CATALYST FEED INTERACTIONS: ACE YIELD PREDICTIONS

An advanced cracking evaluation–automatic production (ACE Model AP) fluidized bed microactivity unit was used to study the catalyst and feed interactions. The fluidized bed reactor was operated at 980°F (800 K). Every feed was tested on two different catalysts at three cat-to-oil ratios 4, 6, and 8. Properties of laboratory

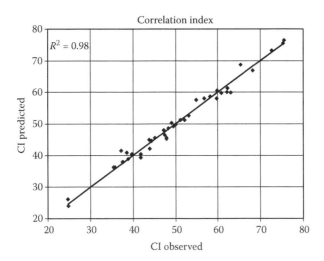

FIGURE 12.9 Correlation index CI predicted from ¹H-NMR spectra versus calculated from original H. M. Smith's formula.

deactivated catalysts A and B are shown in Table 12.3. The feed temperature was kept at 170°F (350 K). The injection rate was 3 g/min and the injection time was 30 s. Products of the cracking reactions were analyzed by GC, PONA, SYMDIS, and LECO.

12.4.1 CONVERSION

By definition, conversion is calculated as:

$$\text{Conversion} = 100 - \text{LCO} - \text{Bottoms}.$$

The best linear regression model to predict conversion for catalyst A has the form:

$$\text{Conversion(A)} = 72.66 + 2.77 \times (\text{C/O}) - 10.125 \times \text{H4} + 7.303 \times \text{H9} - 3.063 \times \text{H10},$$

and for catalyst B

$$\text{Conversion(B)} = 72.722 + 2.998 \times (\text{C/O}) - 9.657 \times \text{H4} + 6.572 \times \text{H9} - 3.016 \times \text{H10}.$$

Although the coefficients in the model for both catalysts are slightly different, the functional dependence on ¹H-NMR spectra is identical. The coefficient of determination, R^2, of the conversion model is 0.94 for both catalysts. Catalyst A gives higher conversion compared to catalyst B at the same cat-to-oil ratio for almost all tested feeds. The exceptions are the heaviest feeds in the set. Figure 12.10a and b shows conversion estimated from ¹H-NMR spectra versus conversion observed for both catalysts A and B. The triangles represent the heaviest feed in the study at three cat-to-oil ratios, 4, 6, and 8. The circles correspond to three different feeds with the

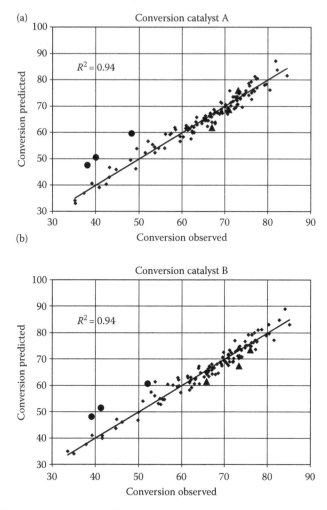

FIGURE 12.10 Conversion predicted from ^{1}H-NMR spectra vs. conversion observed:
(a) catalyst A conversion; (b) catalyst B conversion.

highest level of total nitrogen at cat-to-oil ratio 4. These points suggest that nitrogen
compounds interfere with catalyst activity, lowering the conversion.

12.4.2 GASOLINE

The term of gasoline includes all liquid products with boiling points below 430°F
(494 K). The best linear regression model to predict gasoline yield has the form for
catalyst A:

$$\text{Gasoline(A)} = 56.816 + 1.057 \times (\text{C/O}) - 4.961 \times \text{H4} - 2.105 \times \text{H6}$$
$$+ 5.389 \times \text{H9} - 1.952 \times \text{H10},$$

and for catalyst B

$$Gasoline(B) = 56.922 + 1.009 \times (C/O) - 4.301 \times H4 - 2.692 \times H6 \\ + 5.017 \times H9 - 1.843 \times H10.$$

The coefficient of determination, R^2, of the gasoline model is 0.94 for both catalysts. Under comparable conditions, catalyst A always makes more gasoline than catalyst B for every feed in the study. Figure 12.11a and b shows gasoline yield predicted from ¹H-NMR spectra versus gasoline yield observed by traditional gas chromatography characterization. The labeling of triangles and circles is the same as that for the conversion graphs.

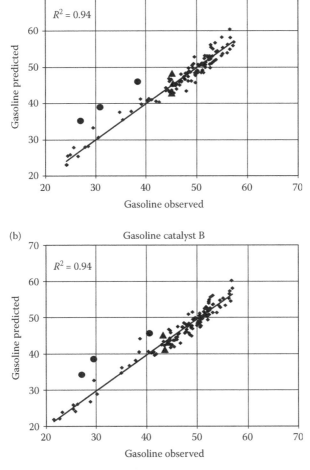

FIGURE 12.11   Gasoline yield predicted from ¹H-NMR spectra vs. gasoline yield observed: (a) gasoline yield for catalyst A; (b) gasoline yield for catalyst B.

12.4.3   RESEARCH OCTANE NUMBER (RON)

Research Octane Number is a measure of the knocking characteristics of gasoline in a laboratory engine and can be used to characterize gasoline quality. The best linear regression model to predict RON for catalyst A has the form:

$$RON(A) = 85.677 + 0.168 \times (C/O) + 0.379 \times H3 - 0.824 \times H4$$
$$+ 0.621 \times H6 - 2.359 \times H7 + 0.737 \times H9,$$

and for catalyst B

$$RON(B) = 86.385 + 0.162 \times (C/O) + 0.362 \times H3 - 0.806 \times H4$$
$$+ 0.672 \times H6 - 2.545 \times H7 + 0.699 \times H9.$$

The coefficient of determination, R^2, of the RON model is 0.84 for both catalysts. Figure 12.12a and b shows RON predicted from ^{1}H-NMR versus RON calculated from gas chromatography measurements for catalysts A and B. The uniform spread of the data points suggests that ^{1}H-NMR does not provide necessary information to accurately predict gasoline quality.

12.4.4   LIGHT CYCLE OIL (LCO) YIELD

Light Cycle Oil (LCO) is composed of liquid products boiling in the range approximately 430°–700°F (494–644 K). The best linear regression model to predict LCO yield for catalyst A has the form:

$$LCO(A) = 76.99 - 1.486 \times (C/O) - 0.732 \times H2 - 1.603 \times H3$$
$$+ 7.179 \times H4 + 2.389 \times H8 - 4.99 \times H9,$$

and for catalyst B

$$LCO(B) = 80.86 - 1.53 \times (C/O) - 0.782 \times H2 - 1.655 \times H3$$
$$+ 7.035 \times H4 + 2.563 \times H8 - 4.985 \times H9.$$

The coefficient of determination, R^2, of the LCO yield model is 0.96 for catalyst A and 0.94 for catalyst B. For the same feed and under the same processing conditions, catalyst B makes more LCO than catalyst A for most of the feeds tested. There are a few cases (the heaviest feeds in the study) with no statistical difference for LCO yield between both catalysts. LCO yield predicted from ^{1}H-NMR spectra versus LCO yield measured by traditional gas chromatography for both catalysts are shown in Figure 12.13a and b.

12.4.5   BOTTOMS

Bottoms is defined as a blend of liquid products boiling above the LCO cut point, 700°F (644 K). The best linear regression model to predict bottoms yield for catalyst A has the form:

$$\text{Bottoms(A)} = 14.418 - 1.275 \times (C/O) + 2.456 \times H6 - 1.494 \times H8$$
$$- 2.424 \times H9 + 6.836 \times H11 - 14.315 \times H12,$$

and for catalyst B

$$\text{Bottoms(B)} = 13.177 - 1.269 \times (C/O) + 2.382 \times H6 - 1.336 \times H8$$
$$- 2.239 \times H9 + 6.843 \times H10 - 13.258 \times H11.$$

The coefficient of determination, R^2, of the bottoms model is 0.84 for catalyst A and 0.82 for catalyst B. Under comparable processing conditions, catalyst A always makes more bottoms than catalyst B for any feed in the study. Figure 12.14a and b shows bottoms yield predicted from ¹H-NMR spectra versus bottoms yield measured

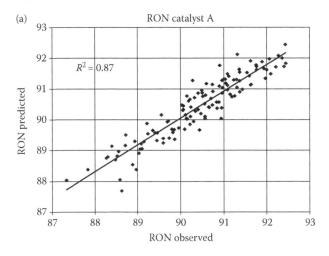

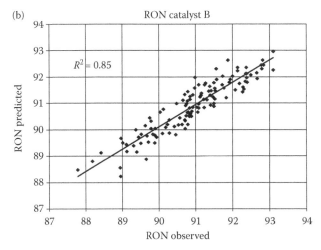

FIGURE 12.12 Research octane number predicted from ¹H-NMR vs. RON observed: (a) catalyst A; (b) catalyst B.

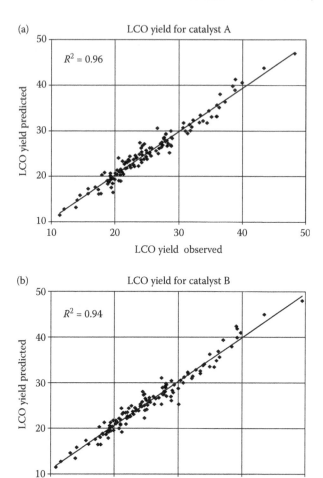

FIGURE 12.13 LCO yield predicted from ^{1}H-NMR spectra vs. LCO yield observed: (a) LCO yield for catalyst A; (b) LCO yield for catalyst B.

by difference from more traditional gas chromatography analyses. For heavy aromatic feeds, triangles in the picture, the model overpredicts bottoms yield. On the other hand, the feeds with high levels of nitrogen compounds, circles, tend to produce more bottoms than the model forecasts. The effect of nitrogen is not captured by the ^{1}H-NMR spectra.

12.4.6 COKE

The best linear regression model to predict coke has the form for catalyst A:

$$\text{Coke(A)} = 6.157 + 0.332 \times (\text{C/O}) - 0.467 \times \text{H1} + 0.658 \times \text{H3} - 0.586 \times \text{H4} + 2.413 \times \text{H7} - 0.537 \times \text{H8} - 0.764 \times \text{H9},$$

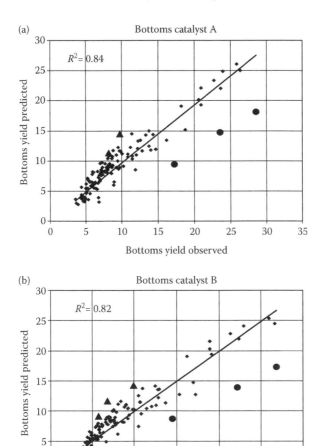

FIGURE 12.14 Bottoms yield predicted from ¹H-NMR spectra vs. bottoms yield observed:
(a) bottoms for catalyst A; (b) bottoms for catalyst B.

and for catalyst B

$$Coke(B) = 6.444 + 0.424 \times (C/O) - 0.308 \times H1 - 1.337 \times H5$$
$$+ 2.179 \times H6 + 1.364 \times H7 - 0.501 \times H8 - 0.575 \times H9.$$

The coke model is relatively poor quality. The coefficient of determination, R^2, of the coke model for catalyst A is 0.75, and it is 0.82 for catalyst B. Both coke models for catalyst A and B have different dependencies on ¹H-NMR spectra ranges. It is interesting to notice that even though both catalysts were deactivated using a metals-free protocol, the feeds with very high levels of vanadium and nickel gave substantially higher coke yield than predicted from the model. The feeds with high levels of vanadium and nickel are marked as magenta squares in Figure 12.15. Catalyst B produces more coke than catalyst A except a few cases for feeds with no vanadium at all.

12.4.7 Dry Gas Yield

Dry gas consists of hydrogen, methane, and ethane products. The best linear regression model to predict dry gas yield for catalyst A has the form:

$$\text{Dry Gas(A)} = -2.18 + 0.158 \times (C/O) + 0.034 \times H2 - 0.093 \times H4 \\ - 0.399 \times H5 + 0.908 \times H6,$$

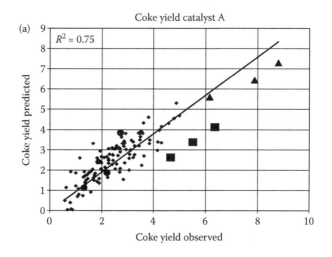

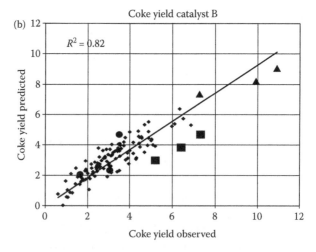

FIGURE 12.15 Coke yield predicted from ¹H-NMR spectra vs. coke yield observed: (a) coke yield for catalyst A; (b) coke yield for catalyst B.

and for catalyst B

$$\text{Dry Gas(B)} = -2.376 + 0.165 \times (\text{C/O}) + 0.037 \times \text{H2} - 0.107 \times \text{H4} - 0.474 \\ \times \text{H5} + 1.059 \times \text{H6}.$$

The coefficient of determination, R^2, of the dry gas yield model is 0.86 for catalyst A and 0.82 for the catalyst B. Based on the model, under the same operating conditions and for the same feed, catalyst B always makes more dry gas than catalyst A. Figure 12.16a and b shows dry gas yield predicted from ¹H-NMR spectra versus dry gas yield measured by gas chromatography. For heavy aromatic feeds, shown by triangles, and feeds with high level of nickel and vanadium, indicated by squares, the model underpredicts dry gas yield.

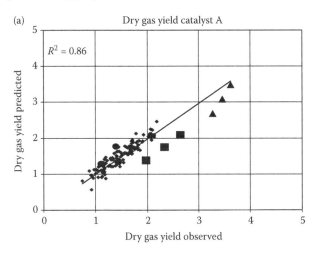

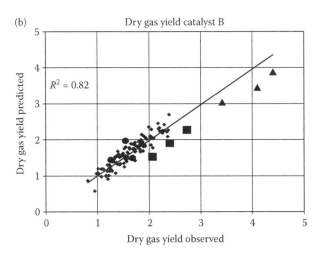

FIGURE 12.16 Dry gas yield predicted from ¹H-NMR spectra vs. dry gas yield observed: (a) dry gas yield for catalyst A; (b) dry gas yield for catalyst B.

12.4.8 WET GAS YIELD

Wet gas yield includes dry gas and all propane and butane products. The best linear regression model to predict wet gas yield for catalyst A has the form:

$$\text{Wet Gas(A)} = 16.534 + 1.374 \times (\text{C/O}) - 4.574 \times \text{H4} + 1.919 \times \text{H8} + 2.87 \times \text{H9} - 0.593 \times \text{H10},$$

While for catalyst B is:

$$\text{Wet Gas(B)} = 17.33 + 1.365 \times (\text{C/O}) - 4.59 \times \text{H4} + 1.59 \times \text{H8} + 2.865 \times \text{H9} - 0.659 \times \text{H10}.$$

The coefficient of determination, R^2, of the wet gas yield model is 0.92. Figure 12.17a and b shows wet gas yield predicted from ¹H-NMR spectra versus wet gas yield measured by chromatography.

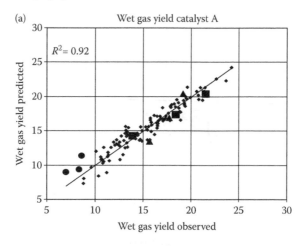

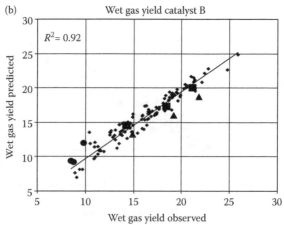

FIGURE 12.17 Wet gas yield predicted from ¹H-NMR spectra vs. wet gas yield observed: (a) wet gas yield for catalyst A; (b) wet gas yield for catalyst B.

12.5   CONCLUSIONS

A feasibility study on the application of ¹H-NMR petroleum product characterization to predict physicochemical properties of feeds and catalyst-feed interactions has been performed. The technique satisfactorily estimates many feed properties as well as catalyst-feed interactions to forecast products yield. There are, however, limitations that have to be understood when using the ¹H-NMR method. The technique, in general, is not capable either to estimate the level of certain contaminants such as nitrogen, sulfur, nickel, and vanadium when evaluating feed properties or the effect of these contaminants on products yields while testing catalyst-feed interactions.

Some of the findings on catalyst-feed interactions have to be properly interpreted taking into account that the catalysts used for the ACE study were deactivated in the laboratory using metals-free protocol.

ACKNOWLEDGMENTS

Some of the feed characterization procedures were conducted in the Resonance Magnetic Nuclear Laboratory at the National University in Colombia. In particular, the authors would like to thank Dr. Ricardo Fierro and Eliseo Abella for performing ¹H-NMR analyses, and Dr. Carlos A. Trujillo for the UV-Vis analyses of the feeds.

REFERENCES

1. Behera, B., Ray, S. S., and Singh, I. D. Structural Characterization of FCC Feeds from Indian Refineries by NMR Spectroscopy. *Fuel* 87 (2008): 2322–33.
2. Molina Velasco, D., Navarro Uribe, U., and Murgich, J. Partial Least-Squares (PLS) Correlation between Refined Product Yields and Physicochemical Properties with the ¹H Nuclear Magnetic Resonance (NMR) Spectra of Colombian Crude Oils. *Energy & Fuels* 21 (2007): 1674–80.
3. Kapur, G. S., Singh, A. P., and Sarpal, A. S. Determination of Aromatics and Naphthenes in Straight Run Gasoline by ¹H NMR Spectroscopy. Part I. *Fuel* 79 (2000): 1023–29.
4. Sarpal, A. S., Kapur, G. S., Mukherjee, S., and Tiwari, A. K. PONA Analyses of Cracked Gasoline by ¹H NMR Spectroscopy. Part II. *Fuel* 80 (2001): 521–528.
5. Kayser, J. C. U.S. Patent 6 069 012—Versatile Fluidized Bed Reactor, assigned to Kayser Technology, 2000.
6. Wallenstein, D., Harding, R. H., Nee, J. R. D., and Boock, L. T. Recent Advances in the Deactivation of FCC Catalysts by Cyclic Propylene Steaming (CPS) in the Presence and Absence of Contaminant Metals. *Appl. Catal. A: General* 204 (2000): 89–106.
7. Van Nes, K., and van Westen, H. A. *Aspects of the Constitution of Mineral Oils.* New York: Elsevier, 1951.
8. Smittenberg, J., and Mulder, D. Relations Between Refraction, Density, and Structure of Series of Homologous Hydrocarbons. I. Empirical Formula for Refraction and Density at 20° of n-Alkenes and n-Alpha-Alkenes. *Rec. Trav. Chim.* 67 (1948): 813–25, 826–38.
9. Watson, K. M., Nelson, E. F., and Murphy, G. B. Characterization of Petroleum Fractions. *Ind. Eng. Chem.* 27 (1935): 1460–64.

10. Hirsch, E. Relations Between Molecular Volume and Structure of Hydrocarbons at 20°C. *Anal. Chem.* 42 (1970): 1326–29.
11. Riazi, M. R., and Daubert, T. E. Prediction of the Composition of Petroleum Fractions. *Ind. Eng. Chem. Proc. Des. Dev.* 19 (1980): 289–94.
12. *Standard Test Method for Calculation of Carbon Distribution and Structural Group Analysis of Petroleum Oils by the n-d-M Method.* ASTM D3238-95 (2005).

13 Surface Acid–Base Characterization of Containing Group IIIA Catalysts by Using Adsorption Microcalorimetry

Georgeta Postole and Aline Auroux

CONTENTS

13.1 INTRODUCTION

The third family of main-group elements is characterized by many interesting structural features and a fascinating period property relationship [1]. Long overshad-
owed by boron, the group IIIA elements are now acknowledged to have rich and
distinctive chemical features of their own, often far removed from boron's image.
This change has come not only due to the commercial importance of aluminum and
alumina-based solids, but due to the explosion of activity within the past two decades involving solid materials or devices chemically and physically tailored to have
specific electronic, optical, thermal, or other properties. The group IIIA elements are
of strategic importance between the groups III–V element compounds, ranging in
properties from refractory, wide-bandgap (>3 eV) semiconductors, through narrow-
bandgap (1–2 eV) semiconductors to intermetallic species [2–4].

Physical properties of these elements (B, Al, Ga, In, Tl) are influenced by the
type of bonding between their atoms, in the solid state. Boron is a network covalent
metalloid—black, hard, and with a very high melting temperature. The other group's
members are metals—shiny, relatively soft, and with a low melting point. The low
density of aluminum and three valence electrons make it an exceptional conductor.
Gallium has the larger liquid temperature range of all elements. Its metallic bonding
is too weak to keep the atoms fixed when the solid is warmed, but strong enough to
keep them from escaping the molten metal until it is very hot [4].

Although the group IIIA elements possess all ns^2np^1 electronic configurations,
they present a wide range of different chemical behaviors. The metallic nature of the
elements in group IIIA increases as one goes down in the group. It is well known
that the metals form basic oxides and the nonmetals form acidic oxides. It is also the
case for group IIIA element oxides: as atomic size increases, the ionization energy
decreases, and oxide basicity increases. The elements of group IIIA present multiple oxidation states: all members exhibit the +3 state, but also the +1 state that first
appears with some compounds of gallium and becomes the only important state of
thallium, because the lower state is more prominent going down the group [5]. The
lower state occurs when the atoms loose their np electrons only, not their two ns
electrons; this effect is often called the *inert-pair* effect. The oxides with the element
in a lower oxidation state are more basic than the oxides with the element in a higher
oxidation state [4].

Despite the parallels, the chemistry of some of these elements is so different
from the others that it deserves particular attention. For example, boron is much

less reactive at room temperature than the other members and forms covalent bonds exclusively. Aluminum acts like a metal physically but its halides exist in the gas phase as covalent dimmers and its oxide is amphoteric rather than basic. Elements in this group, especially boron and aluminum, form three-coordinate Lewis acids capable of accepting an electron pair and increasing their coordination number [4,6].

The group IIIA elements form some common classes of compounds, like halides and oxides and many of them can form molecular sieves networks. Due to these capabilities to form such significant classes of compounds, the group IIIA compounds have useful applications in ceramics [7–12], homogeneous [13–17], and heterogeneous catalysis [18–25]. Besides their use as supports for active metal centers (particularly in Al_2O_3-containing materials) the group IIIA elements play an important role in novel catalytic materials such as promoters [26,27], supported metal oxides and mixed oxides [28,29], supported bimetallic catalysts [30,31], and modified zeolites [32–34]. Used as promoters, these elements enhanced the catalytic performances of metallic oxides based materials by increasing the active phase dispersion and the surface acidity of the catalysts [26,27]. Catalytic materials comprising B, Al, Ga, and In elements have been tested in various catalytic reactions of petrochemicals (e.g., aromatization of light alkanes [35–37] and dehydrogenation of long-chain alkanes [38,39]) and environmental interest (e.g., selective catalytic reduction of nitrogen oxides [40–43]). Pure alumina is widely used as a catalyst for several reactions where it activates hydrogen–hydrogen, carbon–hydrogen, and carbon–carbon bonds [44]. Also a variety of soluble systems of Al have very good catalytic activity for olefin polymerization in homogeneous catalysis [14,15]. Boria-based catalysts can be effectively used in the selective oxidation of hydrocarbons such as ethane [45], while gallium- or indium-supported oxides are promising catalysts for combustion reactions and NO_x abatement processes, such as the selective catalytic reduction of NO_x by hydrocarbons [24,46–48].

The group IIIA elements are also used to replace some silicon atoms in the solid matrix of micro- or mesoporous silicates or aluminosilicates in order to modify the surface acidity of such solids [49]. Acid catalysis is particularly important in a number of industrially catalytic reactions, such as catalytic reforming, cracking, isomerization, alkylation, and dealkylation [50]. Also, insertion of gallium with a lower charge density than aluminum in the lattice framework of different types of zeolites affects acid sites strength and distribution generating more coke-selective materials [51]. The utility of coke-selective materials is important because the coke deposition and deactivation are the major factors limiting the industrial utility of this kind of material and are thought to be closely related to the strong acidity [52]. The systems such as active aluminas, silica-aluminas, Y zeolites, and Ga-modified zeolites were found to be useful in several hydrocarbon conversion processes including fluid catalytic cracking (FCC), dealkylation, C_8–C_{10} separation, and partial oxidation reaction [53]. Fluid catalytic cracking (FCC) is one of the most important processes in industrial refineries for the production of gasoline, diesel, liquefied petroleum gas (LPG), and light olefins [54,55]. It is generally accepted that the cracking of heavy hydrocarbons to lighter ones is catalyzed by acidic sites through the formation of carbonium and/or carbenium intermediates [56,57]. Consequently, alumina plays an important role in the performances of FCC catalysts. For example, due to the limited access of the heavy hydrocarbon molecules in residue-containing feeds into the pores of Y zeolites, an acidic component with large pores and pore volume is needed for residue cracking. Acidic binders can play such a role. Incorporation of alumina produces acidic binders with large pores that allow the access of the large molecules contained in residue feeds.

In all above mentioned applications, the surface properties of group IIIA elements based solids are of primary importance in governing the thermodynamics of the adsorption, reaction, and desorption steps, which represent the core of a catalytic process. The method often used to clarify the mechanism of catalytic action is to search for correlations between the catalyst activity and selectivity and some other properties of its surface as, for instance, surface composition and surface acidity and basicity [58–60]. Also, since contact catalysis involves the adsorption of at least one of the reactants as a step of the reaction mechanism, the correlation of quantities related to the reactant chemisorption with the catalytic activity is necessary. The magnitude of the bonds between reactants and catalysts is obviously a relevant parameter. It has been quantitatively confirmed that only a fraction of the surface sites is active during catalysis, the more reactive sites being inhibited by strongly adsorbed species and the less reactive sites not allowing the formation of active species [61].

Previously mentioned properties can be conveniently investigated by studying the adsorption of a suitably chosen probe molecule on the solid. Adsorption influences all phenomena depending on the properties of the surface (e.g., it constitutes the primary step in corrosion as well as the prerequisite for every catalytic reaction involving solid catalysts).

Measurement of heat of adsorption by means of microcalorimetry has been used extensively in heterogeneous catalysis to gain more insight into the strength of gas–surface interactions and the catalytic properties of solid surfaces [61–65]. Microcalorimetry coupled with volumetry is undoubtedly the most reliable method, for two main reasons: (i) the expected physical quantities (the heat evolved and the amount of adsorbed substance) are directly measured; (ii) no hypotheses on the actual equilibrium of the system are needed. Moreover, besides the provided heat effects, adsorption microcalorimetry can contribute in the study of all phenomena, which can be involved in one catalyzed process (activation/deactivation of the catalyst, coke production, pore blocking, sintering, and adsorption of poisons in the feed gases) [66].

On the basis of research activity in the intervening years, this chapter will focus on the applications of microcalorimetry as a powerful tool for studying the catalysts involving oxides of the group IIIA elements. The main protagonists will be boron, aluminum, gallium, and indium elements in different materials: oxides, nitrides, and zeolites. The purpose of this chapter is to give the reader a short summary of the latest accomplishments in the area of group IIIA elements compounds from the point of view of the relationship between their acid–basic and adsorptive properties and their catalytic performances. Besides the role of group IIIA elements in the industrially very important processes such as FCC will be given.

This chapter begins with a general discussion about acid–base character of solid surfaces. It continues with the information gained from microcalorimetry technique and experimental considerations in interpreting calorimetric data. This will be followed by a review of literature on the use of this technique to study catalysts comprising group IIIA elements (oxides, nitrides, zeolites, and related materials).

13.2 ACID–BASE CHARACTER OF SOLID SURFACES

13.2.1 Definitions

Acidity and basicity are paired concepts that are very often invoked to explain the catalytic properties of divided metal oxides and zeolites. The concept of acids and bases has been important since ancient times. It has been used to correlate large amounts of data and to predict trends. During the early development of acid–base theory, experimental observations included the sour taste of acids and the bitter taste of bases, color changes in indicators caused by acids and bases, and the reaction of acids with bases to form salts.

Although many other acid–base definitions have been proposed and have been useful in particular types of reactions, only a few have been widely adopted for general use. Among them are those attributed to Arrhenius (based on hydrogen and hydroxide ion formation), Brønsted–Lowry (hydrogen ion donors and acceptors), and Lewis (electron pair donors and acceptors) [6,67–70].

S. A. Arrhenius defined an acid as any hydrogen-containing species able to release protons and a base as any species able to form hydroxide ions [71]. The aqueous acid–base reaction is the reaction between hydrogen ions and hydroxide ions with water formation. The ions accompanying the hydrogen and hydroxide ions form a salt, so the overall Arrhenius acid–base reaction can be written:

$$acid + base \rightarrow salt + water (HA + B = A - B + water).$$

This explanation works well in aqueous solution, but it is inadequate for nonaqueous solutions and for gas and solid phase reactions in which H^+ and OH^- may not exist. Definitions by Brønsted and Lewis are more appropriate for general use.

According to the concepts independently proposed by J. M. Brønsted and T. M. Lowry in 1923, an acid is any hydrogen-containing species able to release a proton and a base is any species capable of combining with a proton. This definition does not exclusively imply water as the reaction medium. In this view, acid–base interactions consist in the equilibrium exchange of a proton from an acid HA to a base B giving rise to the conjugated base of HA, A^-, plus the conjugated acid of B, HB^+ [72,73a,b]:

$$HA + B = A^- + HB^+.$$

In the same year (1923) G. N. Lewis first proposed a different approach. In this view, an acid is any species that, because of the presence of an incomplete electronic grouping, can accept an electron pair to give rise to a dative or coordination bond. Conversely, a base is any species that possesses a nonbonding electron pair that can be donated to form a dative or coordination bond. The Lewis-type acid–base interaction can be depicted as follows [74]:

$$B: + A = {}^{\delta+}B \rightarrow A^{\delta-}.$$

This definition is completely independent from water as the reaction medium and is more general than the previous ones. In terms of Lewis acidity, the Brønsted-type acid HA is the result of the interaction of the Lewis-type acid species H^+ with the base A^-. According to the definitions given, Lewis-type acids (typically, but not only, coordinatively unsaturated cations) do not correspond to Brønsted-type acids (typically species with acidic hydroxyl groups). On the contrary, Lewis basic species are also Brønsted bases.

The Lewis acid–base interaction can also be described taking into account the molecular orbital theory: the molecular orbitals that will be of greatest interest for reactions between molecules are *the highest occupied molecular orbital* (HOMO) and *the lowest unoccupied molecular orbital* (LUMO), collectively named as frontier orbitals because they lie at the occupied-unoccupied frontier. Thus, the Lewis definition of acids and bases can be also formulated in terms of frontier orbitals: a base has an electron pair in an occupied orbital (HOMO) of suitable symmetry to interact with the vacant orbital (LUMO) of the acid (although lone pair orbitals with the wrong geometry may need to be ignored). The better the energy match between the base's HOMO and the acid's LUMO, the stronger the interaction. In most acid–base reactions, a HOMO–LUMO combination forms new HOMO and LUMO orbitals of the product.

Later on, Pearson [75] introduced the concept of hard and soft acid and bases (HSABs): hard acids (defined as small-sized, highly positively charged, and not easily polarizable electron acceptor) prefer to associate with hard bases (i.e., substances that hold their electrons tightly as a consequence of large electronegativities, low polarizabilities, and difficulty of oxidation of their donor atoms) and soft acids prefer to associate with soft bases, giving thermodynamically more stable complexes. According to this theory, the proton is a hard acid, whereas metal cations may have different hardnesses.

13.2.2 Acid–Base Character of Oxides

The oxides represent one of the most important and widely employed classes of solid catalysts, either as active phases or as supports. They are used for both their acid–base and ReDox properties and constitute the largest family of catalysts in heterogeneous catalysis [76].

The oxides are the compounds of any element with oxygen: the oxides of nonmetals are defined as acidic oxides, while the oxides of metals are denoted as basic oxides. Some oxides have both acidic and basic characters and are consequently defined as amphoteric [4,77].

13.2.2.1 Acidic Oxides: The Example of Boria

Tanabe [78] defined a solid acid as a solid on whose surface a basic indicator changes its color or a base is chemically adsorbed. Several acidic properties originate from local surface defects. More strictly, following both the Brønsted and Lewis definitions, a solid acid shows a tendency to donate a proton or to accept an electron pair.

Oxygen has a highly electronegative character and its bond with nonmetals is covalent, due to the high electronegativity of the nonmetal elements [76,79]. By decreasing the element electronegativity, covalent network structures are formed for semimetal oxides (e.g., B_2O_3 in their room pressure forms). The bulk acidity of these nonmetal solid oxides is associated with the acidity of the product of their reaction with water, which is an oxo-acid. The acidity of oxo-acids is associated with the strong covalent character of the bond between the nonmetal and oxygen that subtracts electronic charge from the oxygen atom so allowing the O–H bond to be highly polarized and partly stabilizing the negative charge of the anion produced by dissociation. On the other hand, the acid strength of oxo-acids is primarily enhanced by the presence of double-bonded oxygen atoms around the nonmetal atom, which allows the delocalization of the negative charge produced by acid dissociation. For this reason, the higher the oxidation state of the nonmetal, the more double-bonded oxygen atoms there are and the stronger the acidity of the oxo-acid [67,80]. Between the group IIIA elements, B_2O_3 belongs to this category.

13.2.2.2 Basic Oxides

Typical basic species have electron pairs in nonbonding orbitals. These doublets can be used to produce a dative bond with species having empty orbitals, such as protons or coordinatively unsaturated cations [69].

The oxides of low-valency metals (i.e., with cations in oxidation number $\leq +4$) are typically ionic compounds [76]. They are most frequently easily obtained in crystalline forms. In ionic metal oxides the coordination of the cations (four to eight) is generally higher than their valency (one to four) and this also occurs for the coordination of O^{2-} oxide ions (three to six). The bulk basic nature of the ionic metal oxides is associated with the strong polarization of the metal-oxygen bond, to its tendency to be dissociated by water and to the basic nature of the products of their reaction with water (i.e., the metal hydroxides) [67].

The basic oxides act during a catalytic reaction as base either by abstraction of a proton from the reactants (Brønsted base) or by donation of an electron pair to the reactants (Lewis base) to form anionic intermediates that undergo a catalytic cycle [81].

13.2.2.3 Amphoteric Oxides: The Example of Alumina

The surface of fully dehydroxylated ionic oxides can be considered as a spatially organized distribution of coordinatively unsaturated positively (cations) and negatively (oxygen) charged centers [70]. The cations behave as Lewis acids while lattice oxygen anions are Lewis bases. The acidic character of the cations depends on their positive charge and size while the basic character of the lattice oxygen anions depends on the ionic character of the metal–oxygen bonds [76,82].

While a fully dehydrated surface presents only Lewis acid and Lewis base centers of variable strength, the real oxide surfaces are variably covered by hydroxyl groups [44,83,84].

In fact, oxide surfaces usually terminate in oxide ions due to the larger size of the O^{2-} anion as compared with the M^{n+} cation and low-polarizing power [85]. In the crystal of a MO_x oxide, the symmetry and coordination of the M^{n+} cations are lost at the surface. This surface unsaturation tends to be compensated by reaction with gases, and particularly with moisture. Upon exposure to moist environments at low temperature (<230°C [86]), the surface of a metal oxide undergoes a series of chemical reactions that are largely dictated by the chemistry of the cations. The first step in surface hydration involves the formation of surface hydroxyl groups (dissociative adsorption) followed by molecular adsorption. Protons attach to bridging oxygen sites and behave as Brønsted acids, whereas the OH⁻ fragments adsorb to the cation sites and behave as Brønsted bases.

Aluminum oxide or alumina (Al_2O_3) exhibiting a remarkable series of crystallographic modifications, is used in industrial processes primarily as catalyst support, whereas only a few processes apply aluminas as the catalyst. In addition to the thermodynamically stable oxide α-Al_2O_3 (corundum), there exist a large variety of metastable forms of alumina. The transformation between alumina phases strongly depends on the precursors and the thermal treatment used in their stabilization. For instance, calcination of boehmite at increasing temperatures gives rise to the sequence γ-$Al_2O_3 \rightarrow \delta$-$Al_2O_3 \rightarrow \theta$-$Al_2O_3 \rightarrow \alpha$-$Al_2O_3$, while from calcination of bayerite the sequence obtained is η-$Al_2O_3 \rightarrow \theta$-$Al_2O_3 \rightarrow \alpha$-$Al_2O_3$ [80,87–89].

The importance of aluminas is due to their availability in large quantities and in high purity presenting high thermal stability and surface areas (in the 100–250 m²/g range and even more). Their pore volumes can be controlled during fabrication and bimodal pore size distributions can be achieved. However, besides these textural aspects, the surface chemical properties of aluminas play a major role, since these are involved in the formation and stabilization of catalytically active components supported on their surfaces. Despite the widespread interest in catalytic aluminas there is still only a limited understanding about the real nature of the alumina surface [44,80,90].

Aluminas are amphoteric, hence, they possess acidic and basic properties that are controlled by the surface groups or ions which terminate the microcrystallites [91]. The acidic and basic properties of these materials can be modified by the heat treatment conditions and by incorporating additives, such as halogen or alkali.

The catalytic activity of aluminas are mostly related to the Lewis acidity of a small number of low coordination surface aluminum ions, as well as to the high ionicity of the surface Al–O bond [67,92]. The number of such very strong Lewis sites present on aluminum oxide surfaces depends on the dehydroxylation degree and on the particular phase and preparation. Depending on the activation temperature, the density of the strongest Lewis acid sites tends to decrease as the calcination temperature of the alumina increases (i.e., upon the sequence $\gamma \rightarrow \delta \rightarrow \theta$, which is also a sequence of decreasing surface area and increasing catalyst stability).

Among different aluminas, γ-Al_2O_3 (mostly obtained by decomposition of the boehmite oxyhydroxide γ-AlOOH) is the most used material in any field of

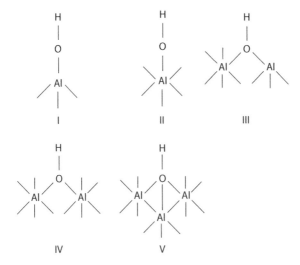

FIGURE 13.1   (I) a terminal OH group is coordinated to a single tetrahedral Al^{3+} cation; (II) the OH group is coordinated to a single cation in an octahedral interstice if possible vacant cation positions are existing; (III) a bridging OH group links a tetrahedral and an octahedral cation; (IV) the OH group links two cations in octahedral positions; (V) the OH group is coordinated to three cations in octahedral interstices. (From Knözinger, H. and Ratnasamy, P., *Catal. Rev. Sci. Eng.*, 17, 31–70, 1978.)

technologies. However, the details of its structure are still a matter of controversy. It is well known that γ-alumina is always hydroxylated; dehydroxylation occurring only at a temperature where conversion to other alumina forms is obtained [93,94]. There is general consensus that the adsorptive and reactive properties of alumina surfaces are governed by the surface hydroxyl groups (OH species). Several models of the structure and multiplicity of the surface hydroxyl groups have been proposed in the literature [44,95,96].

Among the different surface models and site configurations reported in the literature, Knözinger and Ratnasamy [44] proposed five possible OH configurations as presented in Figure 13.1.

The nature of hydroxyl species on alumina surface has been comprehensively reviewed also by Morterra and Magnacca [97] and by Lambert and Che [93]. Additional advances in understanding the intricate surface chemistry of γ-Al_2O_3 have been made by Busca and coworkers [98,99] and Tsyganenko and Mardilovich [100].

The most important result of the above studies is that a minimum of five different OH configurations should be expected to exist on the surface of γ-alumina. The OH groups in these various configurations bear slightly different net charges. As a consequence, they should possess different properties: the protonic acidity of the OH groups will decrease as the net charge on them becomes more negative while their basicity will increase at the same time. This should explain the amphoteric character of alumina.

Thus, the pretreatment of the solid is probably the most important experimental parameter. During the pretreatment, the regular dehydroxylation process leads to the formation of coordinatively unsaturated (*cus*) oxygens (Lewis base sites) and of anion vacancies (Lewis acid sites) that expose *cus* aluminum ions. In fact, the surface dehydroxylation could be described by condensation of adjacent OH groups, whereby the more negatively charged (more basic) groups would combine with the proton provided by the more positively charged (acidic) groups [44]. In the temperature range below 400°C, dehydroxylation will form surface oxide ions with a small negative charge ($-O^-$), anion vacancies that expose coordinatively unsaturated (*cus*) Al species and part of the hydroxyl groups (–OH) will remain on the surface. The degree of unsaturation of the Al^{3+} (*cus*) ion is determined by the configuration of the –OH groups that are removed as water. The Al^{3+} (*cus*) sites have Lewis acid (electron pair acceptor) character, whereas $-O^-$ sites should function as Lewis base (electron pair donor) sites, while the –OH species may principally develop basic or proton acidic properties. During dehydroxylation, an anion vacancy (Lewis acid site) and a *cus* oxygen atom (Lewis base site) are formed for every two hydroxyls that leave the surface as water.

Although it is clear that surface Lewis acid sites on alumina are due to coordinatively unsaturated Al^{3+} ions, it is not fully clear what is the coordination of such surface ions. Most authors agree that at least three different types of Lewis acid sites (weak, medium, strong) exist on alumina, arising in some way from the two or three coordinations of the ions in the bulk spinel-type structure, namely octahedral and tetrahedral (normal spinel positions) and trigonal [101]. The acid strength of the Lewis sites depends on the degree of unsaturation of the Al^{3+} ion (e.g., tetrahedral Al^{3+} exposed in a vacancy is a stronger site than octahedral Al^{3+}).

It has to be pointed out that Lewis acid and base sites produced during the regular dehydroxylation process can hardly be all involved in catalytic reactions as active sites. It has already been indicated [44] that only defect sites can be considered as active sites because of their low site density. The configuration of such defect sites can hardly be predicted from idealizing model considerations.

The strongest Lewis acidic oxides in normal circumstances are alumina and gallia, which are oxides of elements at the limit of the metallic character.

13.2.3 Acid–Base Character of Zeolites

Zeolites are crystalline aluminosilicates with a regular pore structure. These materials have been used in major catalytic processes for a number of years. The application using the largest quantities of zeolites is FCC [102]. The zeolites with significant cracking activity are dealuminated Y zeolites that exhibit greatly increased hydrothermal stability, and are accordingly called ultrastable Y zeolites (USY), ZSM-5 (alternatively known as MFI), mordenite, offretite, and erionite [103].

Humphries et al. [104] have given an excellent description of the surface acidity of zeolites and its influence on FCC. This topic has been comprehensively reviewed also by Shen and Auroux [105] and Auroux [103].

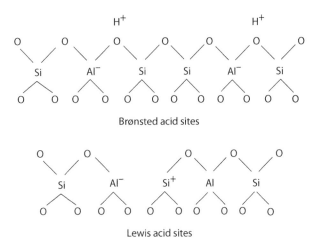

Brønsted acid sites

Lewis acid sites

FIGURE 13.2   Example of Brønsted and Lewis acid sites in zeolites. (From Humphries, A., Harris, D. H., and O'Connor, P., *Stud. Surf. Sci. Catal.*, 76, 41–82, 1993.)

The catalytic activity of zeolites has its origin in the fact that some of the silicon atoms in the crystalline framework of the solids are replaced by an aluminum atom. Since aluminum is trivalent, the replacement of the tetravalent silicon results in the introduction of a negative charge into the zeolite lattice. This negative charge has to be compensated by cations and particularly by protons, the latter resulting in the so-called Brønsted acidity (Figure 13.2) that plays an important role in the catalytic activity of zeolites.

Brønsted acid sites can be further dehydroxylated to form Lewis acid sites as shown also in Figure 13.2. The elimination of water by dehydroxylation should lead to the creation of one Lewis acid site from every two Brønsted sites.

Factors influencing the acid properties of zeolites include the method of preparation, temperature of dehydration, the silica to alumina ratio, and the distribution of the framework atoms [106–108].

For example, Brønsted sites of different strengths can be observed. These Brønsted acid sites may be Si–OH–Al species having different numbers of next-nearest neighbors Al-centered tetrahedra or tetrahedral Al. It was found [109] that isolated Al framework atoms (having no next-nearest Al neighbors) have the highest strength and that, as the number of next-nearest Al atoms increases, the acid strength is likely to decrease.

The basicity of zeolites is due to anions such as O^{2-}, AlO_4^- or OH^-. In Si-Al zeolites, the framework oxygen bears the negative charge, and when this charge is compensated by cations with low electronegativity (such as alkali cations), the charge may become high enough to create basic properties. The cation then acts as a Lewis acid while the associated framework oxygens act as Lewis bases. Basic strength and the density of the basic sites decrease with an increase in the framework Si/Al ratio, while the basic strength increases with an increase in electropositivity of the countercation in zeolites [110].

13.2.4 Measurement of Acid–Base Interactions

The characterization of the acidic and basic properties of solids is relatively complicated compared to that of liquids. Interaction between acids and bases can be measured in many ways [6,80,111,112]:

1. Titration method (consisting in the study of the interaction of indicator dyes with solids from solutions) is a technique for both qualitative and quantitative characterization of solid surfaces. If a basic indicator B is used, the proton acidity of the surface is expressed by the Hammett acidity function. Similarly, the basicity can be defined when an acid is converted by its conjugated base. This allows defining acidity and basicity in the same scale.

2. Direct calorimetric method or temperature dependence of equilibrium constants can be used to measure enthalpies and entropies of acid–base reactions. Calorimetric techniques allow obtaining an interesting quantification and evaluation of the gas–solid interactions and more details on use of data from these measurements will be given in the following section.

3. Gas phase measurements of the formation of protonated species can provide similar thermodynamic data.

4. Infrared (IR) spectroscopy can be also used to characterize the acidic and basic properties of solids by studying adsorbed probe molecules. This technique today finds wide application due to its well-established principle and moderate cost of the FT-IR instruments. By using appropriate bases as probes, IR spectroscopy allows the separated characterization of Lewis and Brønsted acid sites and (with more difficulty) of basic sites. For example, three different types of Lewis sites on the alumina surfaces were observed by FT-IR characterization of the surface cationic sites using pyridine as probe molecule [113]. These Lewis sites interact via a coordination bonding with pyridine, forming three adsorbed species characterized by the $\upsilon 8a$ bands near 1600–1590 cm^{-1} (site III), 1615 cm^{-1} (site II), and 1625 cm^{-1} (site I). Sites I and II have been assigned by Morterra et al. [114,115] to coordinatively unsaturated (*cus*) tetrahedral Al^{3+} and to a pair of *cus* Al^{3+} ions in octahedral and tetrahedral coordination, respectively, while site III is assigned to *cus* octahedral Al^{3+} cations. Moreover, the analysis of the IR spectrum of the pure catalyst allows the detection of the vibrational modes of the surface hydroxyl groups (OH stretchings in the region of 3800–2500 cm^{-1} [69]), which are potential Brønsted sites.

5. Nuclear magnetic resonance (NMR) is another powerful technique to study solid acid catalysts. Advanced NMR methods such as magic-angle spinning (MAS) of solids have increased the capability of this technique to study acid sites in solid acid catalysts [80]. For example, ^{1}H MAS NMR technique performed on the solid catalysts after activation and upon adsorption allows the detection of the signals due to the magnetic resonance of the protons

of the surface hydroxyl groups, the position of which is indicative of their environment [69].

6. Ultraviolet or visible spectra can show changes in energy levels in the molecules as they combine.

7. Temperature-programmed desorption (TPD) of adsorbed acid/basic probe molecules and thermogravimetric analysis (TGA) are also appropriate techniques. Coupled TPD–TGA technique provides information about the nature of acid/basic sites according to the different products evolved during TPD while TGA gives quantitative amounts for the corresponding types of acid/basic sites when the gas-chromatography or mass spectrometry methods are used for analysis of the gases evolved from a surface upon a temperature-programmed heating ramp after adsorption of probes.

8. Catalytic test reactions represent an important tool for acid–basic characterization. Conversion of secondary alcohols such as isopropanol, 2-butanol, and cyclohexanol either to olefins or to ketones, is considered to be evidence of acidic and basic behavior, respectively [104].

9. X ray photoelectron spectroscopy (XPS) is powerful in identifying species present at the surface/interface and atoms or functional groups involved in acid–base interactions [116]. Since XPS measures the kinetic energy of photoelectrons emitted from the core levels of surface atoms upon X ray irradiation of the uppermost atomic layers, it can be used to characterize surface acid sites, in combination with base probe molecules adsorption.

Different methods of measuring acid–base strength yield different results, which is not surprising when the physical properties being measured are considered. An interesting and important feature of solid acid–base catalysts is that in many cases, both acidic and basic sites exist simultaneously on the solid surface. Considerable interest has been directed to the possible correlation between catalytic activity and the acidic and/or basic properties of the catalyst. The search for correlations has been implemented through appropriate measurements of the number, nature, strength, location, and environment of the acidic (or basic) active sites. From numerous papers that have been published in the literature [62,64,117–125], it is apparent that adsorption microcalorimetry technique is the most reliable method in providing not only the heats evolved during adsorption but also the strength and distribution of acidic and basic sites of catalysts.

13.3   ADSORPTION MICROCALORIMETRY

13.3.1   Fundamentals in Adsorption and Calorimetry

The surface of a solid exerts an attractive force on chemical species coming into contact with it owing to incomplete saturation of the coordination sphere of atoms, ions, or molecules at the surface. Adsorption is thus an accumulation of *the adsorptive* (probe molecules) on the surface of *the adsorbent* (the solid), giving rise to *the adsorbate* (or adsorbed phase).

Adsorption influences all phenomena depending on surface properties, since it constitutes the primary step for every catalytic reaction involving solid catalysts. According to the relationship:

$$\Delta G_{ads} = \Delta H_{ads} - T\Delta S_{ads}, \tag{13.1}$$

adsorption is generally exothermic ($\Delta H_{ads} < 0$), as it occurs spontaneously ($\Delta G_{ads} < 0$) and leads to a more ordered state ($\Delta S_{ads} < 0$). The heat evolved is called heat of adsorption [126] and is related to the ability of the sites to interact with the probe molecule (i.e., to their basic or acidic character). This heat that represents also a measure of the strength of the interaction can be determined experimentally by the calorimetric technique. The strength of the binding forces between the adsorbed molecules or atoms and the adsorbing surface has been frequently used as a criterion to distinguish *chemisorption* from *physisorption*. Thus, if the heat of adsorption exceeds ~50 kJ/mol, a true chemical bond is formed between the adsorbate and the adsorbent; this case is referred to as *chemisorption*. When the heat of adsorption is lower than about 50 kJ/mol, adsorption involves secondary (electrostatic or Van der Waals) forces and this case is referred to as *physisorption* [127,128].

A calorimeter suitably adapted to heat of adsorption measurements is required to present high sensitivity and thermal stability and large interval of utilization temperature. Bruzzone [129] and Hansen and Russell [130] reviewed and compared various types of calorimeters and calorimetric methods.

Microcalorimetry has gained importance as one of the most reliable method for the study of gas–solid interactions due to the development of commercial instrumentation able to measure small heat quantities and also the adsorbed amounts. There are basically three types of calorimeters sensitive enough (i.e., microcalorimeters) to measure differential heats of adsorption of simple gas molecules on powdered solids: isoperibol calorimeters [131,132], constant temperature calorimeters [133], and heat-flow calorimeters [134,135]. During the early days of adsorption calorimetry, the most widely used calorimeters were of the isoperibol type [136–138] and their use in heterogeneous catalysis has been discussed in [134]. Many of these calorimeters consist of an inner vessel that is imperfectly insulated from its surroundings, the latter usually maintained at a constant temperature. These calorimeters usually do not have high resolution or accuracy.

An apparatus with high sensitivity is the heat-flow microcalorimeter originally developed by Calvet and Prat [139] based on the design of Tian [140]. Several Tian-Calvet type microcalorimeters have been designed [141–144]. In the Calvet microcalorimeter, heat flow is measured between the system and the heat block itself. The principles and theory of heat-flow microcalorimetry, the analysis of calorimetric data, as well as the merits and limitations of the various applications of adsorption calorimetry to the study of heterogeneous catalysis have been discussed in several reviews [61,118,134,135,141,145]. The Tian-Calvet type calorimeters are preferred because they have been shown to be reliable, can be used with a wide variety of solids, can follow both slow and fast processes, and can be operated over a reasonably broad temperature range [118,135]. The apparatus is composed by an experimental vessel, where the system is located, which is contained into a calorimetric block (Figure 13.3 [146]).

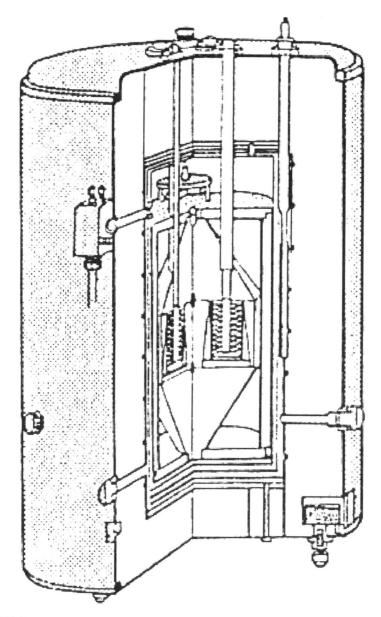

FIGURE 13.3   Calvet microcalorimeter. (From Solinas, V. and Ferino, I., *Catal. Today,* 41, 179–89, 1998.)

The temperature of the block, which works as a heat sink, is controlled very precisely. The heat generated in the system flows to the heat sink and is accurately measured by means of a detector. This is made up of a large number of identical conductive thermocouples (a thermopile) that surround the vessel and connect it to the block in such a way that the vessel and the block temperatures are always close to each other.

In the adsorption microcalorimetry technique, the sample is kept at a constant temperature, while a probe molecule adsorbs onto its surface, and a heat-flow detector emits a signal proportional to the amount of heat transferred per unit time.

Undesired signals due to external temperature fluctuations in the calorimetric block are minimized by connecting in opposition two heat-flow detectors from two identical vessels, one of which is used to perform the experiment, the other being used as a reference. Heat related to the introduction of the probe molecule and other parasitic phenomena are thus compensated.

In general, the heat generated by a sample inside the vessel is transferred to the surroundings by heat conduction through the thermopile, radiation, conduction along the wall of the sample cell, conduction and convection through the fluid phase in the sample cell. If the heat loss by means other than conduction through the thermopile is minimized by the design of the apparatus and is a constant fraction of the total heat flow, it can be shown that the total heat produced in an event, Q, is equal to

$$Q = K \int E \mathrm{dt}, \tag{13.2}$$

where K is the instrument constant, E is the voltage output of the thermopile, and the integral is over the time of the thermal event. Typically, the end of the time interval is chosen to be a point when the thermopile output has returned to the baseline.

This highly sensitive calorimeter needs to be connected to a sensitive volumetric system in order to determine accurately the amounts of gas or vapor adsorbed. A schematic representation of the whole assembly is shown in Figure 13.4 [147]. The volumetric determination of the adsorbed amount of gas is performed in a constant-volume vessel linked to a vacuum pump. The apparatus consists of two parts: the measuring section equipped with a capacitance manometer, and the vessels section that includes the cells placed in the calorimeter (a sample cell in which the adsorbent solid is set, and an empty reference cell).

The experiments are carried out isothermally by admitting stepwise increasing doses of the probe gas or vapor to the solid. As adsorption of water has a strong effect on the distribution of Lewis and Brønsted acid sites, prior to adsorption, the samples are pretreated by heating at the desired temperature under high vacuum (~ 0.1 mPa). After cooling down to the adsorption temperature and establishing the thermal equilibrium of the calorimeter, a dose of gaseous probe molecule is brought into contact with the catalyst sample, and both the pressure and heat signal are monitored until equilibrium is reached. Then, successive new doses are added and the new equilibrium pressures are recorded together with the corresponding evolved heats. The heat evolved by each dose is measured and the corresponding amount adsorbed is obtained by the pressure drop in the known volume of the apparatus. This volume is determined by the expansion of a known quantity of gas, contained in the measuring part of the assembly, into the previously evacuated vessels section. This calibration must be made with the same gas and at the same temperature as the proposed study.

For each dose thermal equilibrium must be attained before the pressure p_i, the adsorbed amount $\Delta n_{a,i}$ and the integral heat evolved $\Delta Q_{int.,i}$ are measured.

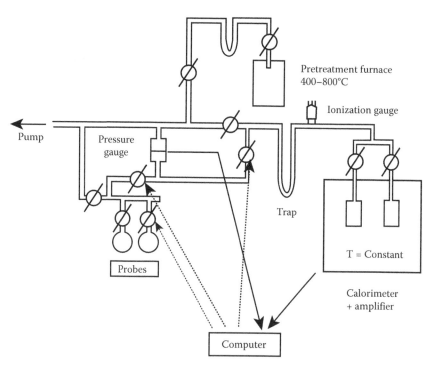

FIGURE 13.4   The volumetric-calorimetric line. (From Bennici, S. and Auroux, A., *Metal Oxides Catalysis,* Wiley-VCH Verlag GmbH & Co., Weinheim, Germany, 391–441, 2009.)

The adsorption experiment is conducted until a relatively high pressure is reached without significant evolution of heat and the adsorbed amount becomes negligible. Owing to the high sensitivity of the method, only small quantities of sample are required. The error may be around 1%, as for the Tian-Calvet calorimeter [129].

13.3.2   HEAT OF ADSORPTION AND THE DATA OBTAINED FROM ADSORPTION CALORIMETRY

As already mentioned, the first step in any heterogeneous catalytic reaction is the adsorption of a gas molecule onto a solid surface. Adsorption heat measurements can provide information about the adsorption process not available using other surface analytical tools. For example, differential heat measurements can provide valuable insights into sites distribution on the catalyst surface as well as quantitative information on the changes in catalyst particle surface chemistry that result from changes in particle size or catalyst support material [148–150].

Heats of adsorption are usually determined in two ways: either by direct calorimetric determination at a chosen temperature, or by calculating the isosteric heats from adsorption isotherms measured at different temperatures and using the Clausius–Clapeyron equation. Thus, isosteric heats of adsorption are calculated from the temperature dependence of the adsorption isotherms (the isosteres). Indeed, q_{st} can be computed from the experimental isosteres for each average temperature according to the equation

$$q_{st} = -RT^2 \left(\frac{\partial (\ln P)}{\partial T} \right)_{n_r},$$
(13.3)

where T is the absolute temperature, R the gas constant, and n_r the number of reversibly adsorbed molecules. This method is limited to true reversible adsorption processes and has been little utilized. Direct calorimetric measurements provide more reliable data.

In the direct calorimetric determination ($-\Delta H = f(n_a)_T$), the amount adsorbed (n_a) is calculated either from the variations of the gas pressure in a known volume (volumetric determination) or from variations of the weight of the catalyst sample in a static or continuous-flow apparatus (gravimetric determination), or from variations of the intensity of a mass spectrometer signal [151].

The average errors in evaluation of the differential heats of adsorption, as estimated by Stach et al. [152], are 1–2% only for the direct measurement and around 5% for the isosteric measurements.

Presently, calorimetry linked to the volumetric technique is still the most commonly used method to study the gas–solid interactions [123]. A complete description of the technique and valuable information provided is given in the different reviews by Cardona-Martinez and Dumesic [118], Auroux [145], Andersen and Kung [153] and Farneth and Gorte [154].

If the surface can be considered a priori as heterogeneous, the adsorption heat, the amount adsorbed, and the kinetics of adsorption must be measured by very small successive doses of the adsorbate so as to obtain information on the variation of these quantities as a function of the coverage. The volumetric apparatus gives the adsorbed quantity and the equilibrium gas pressure (and thus the adsorption isotherm can be plotted), while the variations in the thermal signal indicate the amount of heat evolved. The adsorption of n_a moles of gas is accompanied by the liberation of the total (integral) heat of adsorption, Q_{int}. If heats are measured isothermally at particular coverage (θ) values, in such a way that no external work is transferred to the calorimeter as heat during the adsorption, the true differential heat of adsorption Q_{diff} is obtained as defined by $Q_{diff} = \partial Q_{int}/\partial n_a$.

The data obtained directly from adsorption calorimetry measurements can be expressed in different ways (Figure 13.5 [155]) as follows:

1. The raw data obtained for each dose of probe molecule; that is, the evolution of the pressure above the sample (P) and the exothermic heat evolved signal (Q) as a function of time (Figure 13.5a). The study of the time constant of the heat evolution for each dose provides a description of the kinetics of the adsorption process.
2. The amount of gas adsorbed at constant temperature plotted as a function of the equilibrium pressure (adsorption isotherm I).

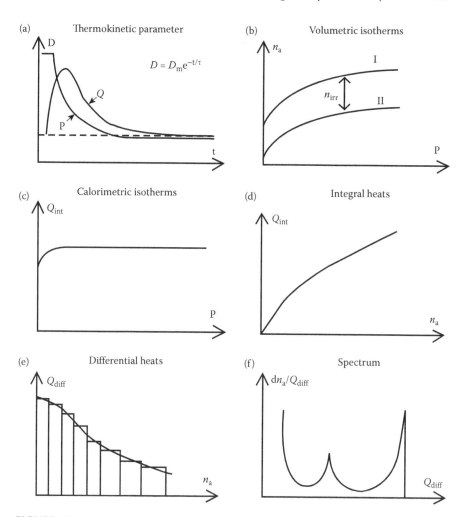

FIGURE 13.5 Calorimetric and volumetric data obtained from adsorption calorimetry measurements: Raw pressure and heat flow data obtained for each dose of probe molecule and Thermokinetic parameter (a), Volumetric isotherms (b), Calorimetric isotherms (c), Integral heats (d), Differential heats (e), Site Energy Distribution Spectrum (f). (From Damjanović, Lj. and Auroux, A., *Handbook of Thermal Analysis and Calorimetry, Further Advances, Techniques and Applications,* Elsevier, Amsterdam, 387–438, 2007. With permission.)

In order to accurately determine the chemisorbed amount from the overall adsorption isotherm, the sample can be further outgassed at the same temperature to remove the physically adsorbed amount, after which a new adsorption procedure is carried out to obtain isotherm II. The difference between the first and second isotherm gives the extent of irreversible adsorption (n_{irr}) at a given temperature (Figure 13.5b), and can be considered as a measurement of the amount of strong sites in the catalyst. However, in the first approximation, the magnitude of the heat of adsorption can be considered as a simple criterion to distinguish between physical and chemical adsorption.

3. The corresponding calorimetric isotherms (Q_{int} versus P; Figure 13.5c).
4. The integral heats (Q_{int}) as a function of the adsorbed quantities (n_a; Figure 13.5d).

This representation leads to the detection of coverage ranges with constant heat of adsorption, those for which the evolved heat is a linear function of the coverage.

5. The differential heat $Q_{diff} = \partial Q_{int}/\partial n_a$ (molar adsorption heat for each dose of adsorbate) as a function of n_a (Figure 13.5e).

The ratio of the amount of heat evolved for each increment to the number of moles adsorbed (in the same period) is equal to the average value of the differential enthalpy of adsorption in the interval of the adsorbed quantity considered. The curve showing the differential heat variations in relation to the adsorbed amount is traditionally represented by histograms. However, for simplification, the histogram steps are often replaced by a continuous curve connecting the centers of the steps.

According to the ideal Langmuir model [156] the heats of adsorption should be independent of coverage, but this requirement is seldom fulfilled in real systems because the effects of surface heterogeneity and sorbate–sorbate interactions are generally significant.

Differential heats of chemisorption usually fall with increasing volume adsorbed. The way in which the heat of chemisorption falls with increasing coverage varies both with the adsorbate and with the adsorbent. Information concerning the magnitude of the heat of adsorption and its variation with coverage can provide useful data concerning the nature of the surface and the adsorbed phase. The shape of the variation of Q_{diff} with coverage (θ) often takes the shape of one of those shown in Figure 13.6 [146,153] where:

I. *Curve I* is for samples that possess distinctly different types of sites; each type adsorbs the molecule with a characteristic value of Q_{diff}. The most energetically favored sites (i.e., those with the highest heat of adsorption) are occupied first, giving rise to the plateau labeled 1. Occupation of the second energetically most favored sites occurs only after sites of the first type are saturated and gives rise to the plateau 2, and so on. Thus the $Q_{diff} - \theta$ curve is characterized by the presence of well-defined steps, with each step corresponding to a specific type of sites. The sharpness of the break between plateaus is limited theoretically by thermal equilibration between sites as dictated by the Boltzmann distribution and, in practice, by whether sites are occupied sequentially according to the magnitude of Q_{diff} [146,153].

II. *Curve II* is for samples that contain both sites of characteristic heats of adsorption as well as those with a continuous variation of Q_{diff}. Samples represented by this curve would contain a small number of sites that adsorb the molecule with high values of heats of adsorption and that are occupied first. As θ increases, sites of lower values of Q_{diff} are occupied. Thus, region 1 in this curve shows a declining slope. After these sites of

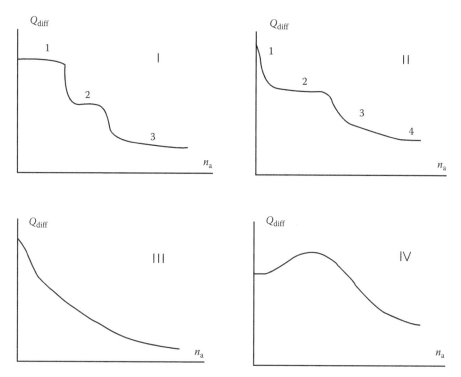

FIGURE 13.6 Types of generalized thermograms obtained in isothermal adsorption microcalorimetry. (From Solinas, V. and Ferino, I., *Catal. Today*, 41, 179–89, 1998 and Andersen, P. J. and Kung, H. H., *Catalysis*, 11, 441–66, 1994. With permission.)

high adsorption heats are occupied, further adsorption takes place on sites with a characteristic Q_{diff}. Thus, the plateau 2 is obtained. After these sites are occupied, adsorption begins to take place on sites of a range of decreasing Q_{diff}, and region 3 is obtained. Finally, adsorption takes place on the weak sites of region 4 [146,153].

III. *Curve III* shows the situation where the sample only contains sites of a range of heats of adsorption and no sites of characteristic heats. Thus, a continuous curve of decreasing Q_{diff} is obtained as θ increases [146,153].

IV. In order to obtain one of these curves, it is implicitly assumed that the adsorbate molecule occupies sites in the order of decreasing Q_{diff}. However, for porous samples, the adsorbate molecules are always exposed to sites near the pore mouth first. Exposure to the sites inside the pores would require desorption and readsorption or surface diffusion into the pore. Thus, if the sites near the pore mouth adsorb the molecules strongly (i.e., have a sufficiently high Q_{diff}), then desorption from the sites or surface diffusion would be too slow for equilibration. As a result, the sites first occupied might not be the strongest sites that are inside the pores. In such a case, the sample might show a $Q_{diff} - \theta$ curve like *curve IV* [153]. A similar shape of $Q_{diff} - \theta$ curve like *curve IV* was also observed for the competitive adsorption of two different gases [157,158]. Rakic et al. [158] studied the adsorption properties of copper-exchanged ZSM5 (Si/Al = 20) by N_2O and CO adsorption microcalorimetry at 30°C. When carbon monoxide was adsorbed on the sample previously contacted with N_2O, a competitive adsorption was found between adsorbed N_2O and incoming CO from the gas phase having as result a profile of differential heats versus CO uptake as presented in Figure 13.6, type IV.

6. The distribution of the energies of the adsorption sites (Figure 13.5f).

As shown in Figure 13.6, in some cases the variation of the adsorption heats with progressive coverage corresponds to step-shaped curves. Such a behavior may be associated with a discrete surface heterogeneity due to the existence of several energetic levels [159]. In such cases, to describe the change in the adsorption heats with coverage, another approach is to plot energy spectra (Figure 13.5f). Assuming that the variation in the adsorption heats coincides with energy distributions, one may wish to measure the number of sites with the same energy (i.e., sites that give rise to the same differential heat). This is achieved upon plotting $-dn_a/dQ_{diff}$ as a function of Q_{diff}. The area below the curve included between Q_{diff} and $Q_{diff} + dQ_{diff}$ represents the population of sites of identical strength estimated via Q_{diff}.

7. Estimate of the entropy of adsorption from the adsorption equilibrium constants obtained from adsorption isotherms and heat of adsorption data obtained microcalorimetrically [160].

The differential molar entropies can be plotted as a function of the coverage. Adsorption is always exothermic and takes place with a decrease in both free energy ($\Delta G < 0$) and entropy ($\Delta S < 0$). With respect to the adsorbate, the gas–solid interaction results in a decrease in entropy of the system. The cooperative orientation of surface-adsorbate bonds provides a further entropy decrease. The integral molar entropy of adsorption S^a and the differential molar entropy S^a_{diff} are related by the formula $S^a_{diff} = \partial(n_a S^a)/\partial n_a$ for the particular adsorbed amount n_a. The quantity S^a can be calculated from

$$\Delta_a S = S^a - S^{g,P} = \frac{Q_{int}}{Tn_a} + \frac{R}{n_a} \int_0^P n_a d(\ln P), \tag{13.4}$$

where S^a is the molar entropy of the adsorbed phase, $S^{g,P}$ the molar entropy of the gaseous phase (available from tables), and n_a the adsorbed amount. After integration of the plot of n_a versus $\ln(P)$ between 0 and P from the adsorption isotherm at the temperature T, and of the plot of Q_{diff} versus n_a between the same boundaries to obtain Q_{int}/n_a, the value of S^a can be obtained, and then that of S^a_{diff}.

8. Plot of the variation of the thermokinetic parameter as a function of the adsorbed amount of probe.

Heat conduction microcalorimetric output consists of power versus time and hence can undergo analysis to produce not only thermodynamic but also kinetic data. The kinetics of heat release during adsorption can be monitored by the change in the thermokinetic parameter τ [58,161]. The calorimetric signal decreases exponentially with the adsorption time after the maximum of each adsorption peak. This can be approximated by $D = D_m \exp(-t/\tau)$, where D and D_m are the deviation at time t and the maximum deviation of the calorimetric signal, respectively (see Figure 13.5a). In this expression, the thermokinetic parameter τ, known also as time constant, can thus be calculated as the reciprocal of the slope of the straight line obtained upon plotting $\log(D)$ as a function of time [161]. This thermokinetic parameter is not constant and varies with coverage. Thus, the variations of the thermokinetic parameter can be plotted versus the amount of adsorbed probe (Figure 13.7).

Measurement of the thermokinetic parameter can be used to provide a more detailed characterization of the acid properties of solid acid catalysts, for example, differentiate reversible and irreversible adsorption processes. For example, Auroux et al. [162] used volumetric, calorimetric, and thermokinetic data of ammonia adsorption to obtain a better definition of the acidity of decationated and boron-modified ZSM5 zeolites (Figure 13.7).

Because chemisorption may be a slow, irreversible process involving activation of the adsorbate, a longer time and, therefore, a broader thermogram would distinguish such a process from a faster, reversible physisorption process. This feature was exploited to monitor the change in adsorption with coverage. The adsorption process was initially slow and became slower, reaching a minimum, before a significant

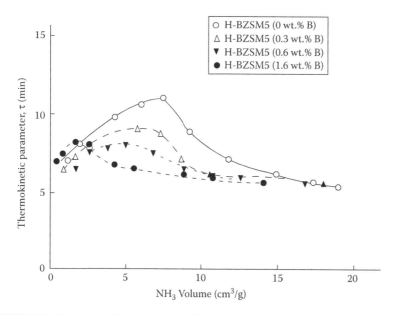

FIGURE 13.7 Variation of the thermokinetic parameter (in minutes) versus the ammonia uptake for a H-ZSM5 sample and boron-modified H-ZSM5. (From Auroux, A., Sayed, M. B., and Védrine, J. C., *Thermochim. Acta*, 93, 557–60, 1985. With permission.)

acceleration of the process was observed on approaching the physisorbed state at high coverage. The minimum rate appears as a maximum in a plot of the thermokinetic parameter as a function of the surface coverage, indicative of a change from irreversible to reversible adsorption (Figure 13.7).

The number of strong sites can be estimated directly from the number of sorbed basic molecules defined by the peak maximum, provided that one basic molecule interacts with one acidic site. Auroux and colleagues [162] observed that ammonia adsorption shifts from strong chemisorption for H-ZSM5 to a process controlled by physisorption (shorter τ) for boron-modified zeolites.

When the thermokinetic parameter was plotted versus the amount of NH_3 adsorbed for samples of H-ZSM5 (Si/Al = 10.3) pretreated at 400 and 800°C it was found that the maximum time constant is higher for the sample pretreated at 800°C than for that pretreated at 400°C [103]. In fact, the increase of the pretreatment temperature caused dealumination; extra-framework aluminum species were created that restricted the access to the channels and created diffusional limitations.

13.3.3 Possible Limiting Factors

Although it is a powerful and informative technique, adsorption calorimetry presents several inherent limitations that require its use in combination with other characterization methods [103,154].

1. The nature of the binding sites cannot be known through calorimetry alone. Adsorption may occur at Brønsted sites, Lewis sites, or as a result of any combination of surface/vapor attractive forces. In many cases this technique fails to distinguish between cations and protonic sites and so to discriminate Lewis and Brønsted sites, due to the insufficient selectivity of the adsorption if no complementary techniques are used. As no exact information can be obtained regarding the nature of the adsorption centers from the calorimetric measurements, suitable IR, MAS NMR, and/or XPS investigations are necessary to identify these sites. However due to the complex nature of the acid strength distribution, it is currently still not possible to establish a detailed correlation between sites of different nature and their strength.

   Even when it can be demonstrated that binding results from proton transfer (adsorption at Brønsted sites on the surface of the solid), the heat of adsorption is not a measurement of the proton affinity of the site. It is, in fact, a convolution of the proton affinity of the acid site on the solid, the proton affinity of the reference base, and the heat of interaction of the resulting ion pair.

2. Although it is common to talk about using calorimetry to measure distributions of site strengths, the differential heats are a more complicated thermochemical quantity that, in the general case, will have contributions specific to the structure of the probe molecule, the local geometry of the binding site, and the coverage of the surface. The way in which the differential heat of adsorption depends on the adsorbed amount is affected

both by the adsorbent and by the adsorbed species. The influence of the former depends on crystallographic and energetic heterogeneity and chemical composition. The contributions from the probes in general reflect their bonding character to the solid, their physical interactions (attractive and repulsive), and their chemical interactions. To establish the contributions of the individual factors to the heat from the shape of the heat curve versus coverage is, of course, very difficult even for the simplest gases [163].

3. A temperature must be chosen that allows sufficient mobility so that all available binding sites are sampled within the time scale of the experiment. Otherwise, the differential heats represent some type of kinetically averaged distribution rather than a binding strength distribution that can be interpreted by equilibrium thermodynamics. Generally, the adsorption temperature should not be too low, in order to allow the detection of differences among the sites; otherwise, under certain circumstances the measured evolved heat can be just an average value. Another important issue is that the temperature used during calorimetric measurement must ensure that *chemisorption* predominates over *physisorption*. On the other hand, at temperatures that are too high, chemical reactions can occur and the measured heats become a complex mixture of enthalpies of adsorption and reaction. Therefore, the site strength distribution should be determined at a sufficiently high temperature that the adsorption process is specific, but at a low enough temperature ensuring that the adsorption equilibrium constant is favorable. For example, the temperature dependence of the heat of adsorption for alumina is characteristic of a strong acidic surface, with the initial differential heat increasing and the adsorption capacity decreasing with increasing adsorption temperature [118].

4. On the heterogeneous surface there will be a tendency for the most energetic sites to be covered first, both because adsorption is likely to proceed more rapidly on them, and also because, even if there has been random coverage initially, a spreading to the most active sites will subsequently take place. Thus, as the coverage increases, sites of weaker energy will be covered so that the heat of adsorption continuously decreases. However there is some evidence that under certain experimental conditions, simultaneous adsorption on all types of sites can occur. So when a plateau of essentially constant heats of adsorption is observed, one may wonder whether it is due to nonspecific adsorption of the probe molecule on sites of varying strength, giving rise to an apparently constant heat of adsorption that is an average value. The presence of distinct regions of strong sites can be proof of the contrary, but the issue of site equilibration should be addressed separately for each adsorption system [62,154].

5. The adsorption of the gas should not be limited by diffusion, neither within the adsorbent layer (external diffusion) nor in the pores (internal diffusion). Should diffusion limitations occur, then adsorption on active but less accessible sites may only occur after better exposed but less active sites have interacted. Diffusion may thus cause the "smoothing out" of significant details in the energy spectrum, and the differential heat curves determined

under the influence of diffusion phenomena may indicate less surface heterogeneity than actually exists on the adsorbent surface [135].

6. Wall adsorption, although significant mainly in metallic volumetric systems, appears not to result in major problems if a suitably careful calibration is performed. Rapid transfer of heat from the sample to the cell wall is recommended [164].

7. Careful heat-flow calibrations have to be performed. Chemical calibrations present many disadvantages: they rely on prior results, with no general agreement and no control of rate, and are generally available only at a single temperature. On the contrary, electrical calibrations (Joule effect) provide many advantages and they are easy to perform at any temperature [103].

8. The absence of a plateau of constant heat in the differential heat curve can be the result of molecular interactions between molecules adsorbed at neighboring sites rather than a true indication of differences between sites [165]. This matter can be checked by varying the probe size or the site density.

9. The accuracy with which the differential heats of adsorption Q_{diff} could be measured is ca. 2%. Rapid collection of evolved heat is an important criterion and sometimes, the calorimeter response has to be corrected from the instrumental distortion due to thermal lags. The peak width at half maximum of the signal from the thermal fluxmeters allows comparing the various calorimeters responses [62].

13.3.4 PROBE MOLECULES

On solids, the amount and strength of acid or basic sites are quite independent parameters, so both of them must be analyzed independently for a complete characterization. Additionally, several different families of acid sites may occur in the same solid surface, so their distribution must be characterized. The key to the effective utilization of microcalorimetry in heterogeneous catalysis is the judicious choice of gas-phase molecules for study.

Adsorption calorimetry allows the total number of adsorption sites and potentially catalytically active centers to be estimated; the values obtained depend on the nature and size of the probe molecule. Appropriate probe molecules to be selected for adsorption microcalorimetry should be stable over time and with temperature. The probe adsorbed on the catalyst surface should also have sufficient mobility to equilibrate with active sites at the given temperature [103].

The acid sites strength can be determined by measuring the heats of adsorption of basic probe molecules. The basic probes most commonly used are NH_3 (pKa = 9.24, proton affinity in gas-phase = 857.7 kJ/mol) and pyridine (pKa = 5.19, proton affinity in gas-phase = 922.2 kJ/mol). The center of basicity of these probes is the electron lone pair on the nitrogen. When chemisorbed on a surface possessing acid properties, these probes can interact with acidic protons, electron acceptor sites, and hydrogen from neutral or weakly acidic hydroxyls.

Different factors should be taken into consideration when comparing the adsorption heats obtained using different probe molecules and adsorption conditions. For example, since pyridine has a higher gas phase proton affinity than ammonia, it is usually observed that the heat of pyridine adsorption is higher that the heat of ammonia adsorption over the same acidic surface. Moreover, the site strength distributions derived from the adsorption of pyridine and ammonia are usually different. This is because of the different sizes of pyridine and ammonia that allow them to reach different locations in the pores and because of their different Van der Waals interactions with the walls of pores. The Van der Waals interaction strongly depends on the diameter of pores in which the probe molecules are adsorbed. It is usually difficult to subtract the Van des Waals interaction from the heat of adsorption in order to account for the pure interaction energy of a probe molecule with a surface acid site. Ammonia may have the advantage as a probe molecule because it is among the smallest strongly basic molecules, and its diffusion is hardly affected by the porous structure. This makes it the most commonly used probe in calorimetry. Piperidine ($pKa = 11.1$), n-butylamine ($pKa = 10.6$), triethylamine ($pKa = 10.8$) are also used occasionally [146,153]. When these bases are compared in terms of their respective proton affinities in gas phase, the order of basic strength is ammonia $< n$-butylamine $<$ pyridine $<$ trimethylamine $<$ piperidine [118].

For basicity measurements, the number of acidic probes able to cover a wide range of strength is rather small [166]. The most common acidic probe molecules used are CO_2 ($pKa = 6.37$) and SO_2 ($pKa = 1.89$). Carboxylic acids such as acetic acid can also be used but dimmers can be formed, particularly at high coverage. Pyrrole may also be used, particularly at low adsorption temperature, but has sometimes shown some amphoteric character [103]. Hexafluoroisopropanol has also been used to characterize the surface basicity of some solids [145].

13.3.5 METHODOLOGY OF ADSORPTION MICROCALORIMETRY MEASUREMENTS

Most of the adsorption experiments performed by the authors were carried out in a heat-flow microcalorimeter linked to a conventional volumetric apparatus (see Figure 13.4). The heat flow generated during adsorption was measured by a microcalorimeter of the Tian-Calvet type, which can control the temperature during adsorption within 0.01°C. The volumetric system includes a gas handling system with probe molecule reservoir, the calibrated dosing section, and the measuring cells. The adsorbate is first dosed into the dosing section, which consists of the volume between three valves: the dosing valve, the gas-supply valve, and the pump valve. The quartz measuring cells comprise a sample cell and a reference cell, both placed inside the thermopile block of the microcalorimeter. The probe molecules can be inserted into the sample cell through its upper end. Gas pressures are measured with a capacitance manometer and its signal is recorded continuously and transferred to a computer. The measuring cells and the dosing section can be evacuated by a turbomolecular pump linked to a roughing pump. The leakage rate of the whole volumetric system should be as small as possible.

The differential heats of adsorption are measured at a certain temperature (ranging from room temperature to 300°C and depending on the calorimeter used) as a function of coverage by repeatedly sending small doses of probe molecule over the solid until an equilibrium pressure of about 67 Pa was reached. The sample was then outgassed for 30 minutes at the same temperature and a second adsorption was performed until an equilibrium pressure of about 27 Pa was attained, in order to calculate the amount of irreversibly chemisorbed probe at this pressure. The difference between the amounts of gas adsorbed at 27 Pa during the two adsorption runs corresponds to the number of strong adsorption sites. The adsorbed amounts are calculated from the difference between the admission pressure and the equilibrium pressure after the adsorption process.

The duration of each dosing experiment is about 15–50 minutes (depending on the sample and of the time constant of the calorimeter), which was long enough to yield well-resolved heat-flow peaks and a stable horizontal baseline of the microcalorimeter. For all catalysts presented here, adsorption always reached thermodynamic equilibrium. Prior to adsorption measurements, the samples were pretreated in the calorimetric cell by heating overnight under vacuum.

13.4 MEASUREMENT OF ACID–BASE INTERACTIONS IN GROUP IIIA CONTAINING SAMPLES

The measurement of heats of adsorption by means of microcalorimetry has been used extensively in heterogeneous catalysis in the past few decades to gain more insight into the nature of gas–surface interactions and the catalytic properties of solid surfaces. Specific attention will be focused on group IIIA containing samples in this section.

13.4.1 BULK OXIDES OF B, AL, GA AND IN ELEMENTS

In metal oxides, the capacity of combined oxygen to act as electron donor is expected to be related to the partial charge on the oxygen: the higher the negative charge, the better the donor properties. Oxides may also serve as electron acceptors, the metal atom seeking for electrons. Therefore, oxides of group IIIA metals may be amphoteric. It is known that when the partial charge on oxygen is greater than –0.50, no acidic behavior can be evidenced and the oxides are strongly basic. When the partial charge on oxygen is less than –0.10, the oxides appear to be exclusively acidic, with no basicity. Between –0.10 and –0.50 lays a wide area within which a few oxides appear to be exclusively acidic or exclusively basic, but most are amphoteric. Amphoteric oxides may favor acidity or basicity, and it is not possible to determine their predominant behavior a priori in terms of the partial charge on oxygen only. The group IIIA oxides lie in this area, as can be seen in Table 13.1 which lists the electronegativity, relative partial charge on oxygen, and cation radius of B, Al, Ga, and In oxides [167].

Bulk alumina and india are isostructural, with a linear structure OMOMO, while B_2O_3 molecule is V-shaped. The Ga_2O_3 can present the both types of isomers, the V-shaped structure being a little more stable than the linear one [1]. These very different structural features (shape, electronegativities, etc.) of group IIIA oxides may help explain their specific properties that fail to strictly follow any simple rule. Their amphoteric character (except for boria) that is not easy to evaluate, has been confirmed and quantified by the experimental microcalorimetric results.

TABLE 13.1
Physical Characteristics of Group IIIA Oxides

Cation	Electronegativity	Relative Partial Charge on Oxygen	Ion Radius (A)
B^{3+}	2.84	−0.24	0.20
Al^{3+}	2.25	−0.31	0.50
Ga^{3+}	3.23	−0.19	0.62
In^{3+}	2.86	−0.23	0.81

Source: From Sanderson, R. T., *Inorganic Chemistry,* Reinhold Publishing Corporation, New York, 1967. With permission.

Bulk boron oxide was found to be much more acidic than basic [168]. When SO_2 adsorption microcalorimetry was used, no basic sites were observed, but some physisorption occurred. Ammonia and pyridine adsorption microcalorimetry were used to characterize the acidity of B_2O_3. Boron oxide displays an initial heat for NH_3 adsorption of 80 kJ/mol and can adsorb irreversibly a large amount of ammonia. The number of active sites determined by pyridine adsorption and the corresponding integral heats were found to be much lower than those determined by using ammonia.

The surface of alumina presents strong acid and basic sites, as demonstrated by the differential heats of adsorption of basic probe molecules such as ammonia [169–171] and pyridine [169,172] or of acidic probe molecules such as SO_2 [169,171] and CO_2 [173,174]. Table 13.2 presents a survey of microcalorimetric studies performed for Al_2O_3.

As can be seen in Table 13.2, the heats of NH_3, pyridine, CO_2 or SO_2 adsorption clearly show that these molecules are chemisorbed on all aluminas (heats of adsorption higher than 100 kJ/mol) in spite of the different origins of Al_2O_3 and different pretreatment and adsorption temperatures used.

The pretreatment temperature is an important factor that influences the acidic/basic properties of solids. For Brønsted sites, the differential heat is the difference between the enthalpy of dissociation of the acidic hydroxyl and the enthalpy of protonation of the probe molecule. For Lewis sites, the differential heat of adsorption represents the energy associated with the transfer of electron density toward an electron-deficient, coordinatively unsaturated site, and probably an energy term related to the relaxation of the strained surface [147,182]. Increasing the pretreatment temperature modifies the surface acidity of the solids. The influence of the pretreatment temperature, between 300 and 800°C, on the surface acidity of a transition alumina has been studied by ammonia adsorption microcalorimetry [62]. The number and strength of the strong sites, which should be mainly Lewis sites, have been found to increase when the temperature increases. This behavior can be explained by the fact that the Lewis sites are not completely free and that their electron pair attracting capacity can be partially modified by different OH group environments. The different pretreatment temperatures used affected the whole spectrum of adsorption heats

TABLE 13.2

Literature Survey of Calorimetric Measurements on Alumina

Phase	Probe Molecules	Adsorption Temperature (°C)	Q_{init} (kJ/mol)	Reference
γ-Al$_2$O$_3$	NH$_3$	80	195	169
ns-Al$_2$O$_3$	NH$_3$	150	222	170
γ-Al$_2$O$_3$	NH$_3$	80	217	171,175
γ-Al$_2$O$_3$	NH$_3$	150	125	176
γ-Al$_2$O$_3$	NH$_3$	150	165	177
γ-Al$_2$O$_3$	NH$_3$	150	150	178
γ-Al$_2$O$_3$	NH$_3$	80	213	160
γ-Al$_2$O$_3$	NH$_3$	150	206	179
porous-Al$_2$O$_3$	NH$_3$	80	230	168
nonporous-γ-Al$_2$O$_3$	NH$_3$	80	180	168
γ-Al$_2$O$_3$	Pyridine	150	175	169
γ-Al$_2$O$_3$	Pyridine	150	152	180
ns-Al$_2$O$_3$	Pyridine	150	226	170
γ-Al$_2$O$_3$	Pyridine	80	180	160
nonporous-γ-Al$_2$O$_3$	Pyridine	80	243	168
γ-Al$_2$O$_3$	SO$_2$	80	194	171,175
γ-Al$_2$O$_3$	SO$_2$	80	174	160
porous γ-Al$_2$O$_3$	SO$_2$	80	191	168
nonporous-γ-Al$_2$O$_3$	SO$_2$	80	196	168
γ-Al$_2$O$_3$	CO$_2$	80	176	160
γ-Al$_2$O$_3$	CO$_2$	23	180	173
γ-Al$_2$O$_3$	CO$_2$	150	132	176
γ-Al$_2$O$_3$	CO$_2$	30	150	180
γ-Al$_2$O$_3$	CO$_2$	150	115	178
γ-Al$_2$O$_3$	H$_2$O	80	265	179
γ-Al$_2$O$_3$	H$_2$O	30	155	181

at various coverages, proving the complex nature of the alumina dehydration process. The initial heat increases and the adsorption capacity decreases with increasing pretreatment temperature.

Two different temperatures (80 and 150°C) for ammonia adsorption were used by Carniti et al. [183] to obtain the acid site strength distribution of γ-Al$_2$O$_3$. When comparing the experimental results of NH$_3$ adsorption measured at 80 and 150°C, each point of the isotherm at 80°C lay above the corresponding one at 150°C at the same pressure. This is in agreement with thermodynamics, as ammonia chemisorption is less favored at higher temperature, being an exothermic reaction. At both adsorption temperatures, the differential heats decreased fast and almost monotonically with increasing ammonia uptake. The strength distribution of the acid sites obtained from adsorption at different temperatures was found to be different. When the temperature and the time employed in measurements ensure that the adsorption equilibrium has been obtained on all sites, the differences among the observed distributions have to be ascribed to the influence of temperature on the thermodynamic parameters of adsorption, particularly on the adsorption constants. To obtain the acid site distribution in terms of energy of adsorption, Carniti and colleagues [183] used a mathematical model taking into account at the same time the volumetric and calorimetric data collected at the two temperatures. Alumina showed a significant amount of acid sites with ammonia heats of adsorption at around 40 kJ/mol corresponding to hydrogen-bonded NH_3 and three types of more energetic sites, with adsorption heats higher than 100 kJ/mol.

The influence of the probe molecule on various thermodynamic parameters of the adsorption on $\gamma\text{-}Al_2O_3$ was studied by Gervasini and Auroux [160]. Thus, the isotherm, the differential heat, the integral heat of adsorption, and the entropy of adsorbate were determined as a function of the coverage for a series of 15 probe molecules with pKa varying over the full range of the pKa scale: basic (i.e., piperidine, diethylamine, n-butylamine, ammonia, pyridine, aniline), amphoteric (i.e., pyrrole, water, methylalcohol *tert*-butyl alcohol, acetonitrile), and acidic (i.e., m-cresol, carbon dioxide, acetic acid, sulphur dioxide) molecules. Figure 13.8 presents the results obtained by Gervasini and Auroux [160] for NH_3, pyridine, CO_2, and SO_2 adsorption. For these probes, the variation of differential adsorption heats versus coverage were found to be roughly composed of a sharp decrease in Q_{diff} at the very beginning, which was assigned to the adsorption on few very strong sites possibly arising from surface defects. Then, in the intermediate region, a plateau or a continuous and monotonous decrease in Q_{diff} values was observed, corresponding to the heats released during adsorption on the predominant surface sites and to the saturation of the chemisorption sites before a final plateau corresponding to the formation of small amounts of physisorbed or liquid-like species.

The introduction of guest ions in the surface of alumina leads to modifications of the acid–basic character of its surface and can also vary to a more or less extent its catalytic properties [184]. Gervasini et al. [171] observed that the chemical properties of $\gamma\text{-}Al_2O_3$ have been changed substantially by doping with metal ions such as lithium, nickel, and sulphate ions. The authors have shown by using adsorption microcalorimetry of ammonia and sulphur dioxide that the change in alumina acid/base properties depends on the nature and amount of the introduced ion. For example, an increase of medium and weak acidity upon the addition of nickel was observed, while alkali ion (Li^+) above a certain loading decreased the total acidity in all strength domains. Basicity of modified aluminas was much more affected than acidity. On increasing the concentration of lithium, the strength of very strong basic centers increased enormously. A high concentration of nickel provoked a huge decrease of the number and strength of basic sites. The presence of sulphate ions decreased both the number and strength of the basic sites.

The physicochemical characterization of the acidity of the doped alumina performed by microcalorimetry, has been complementarized with the catalytic test of 2-propanol decomposition [175]. It was found that the modification of $\gamma\text{-}Al_2O_3$ surface properties with small amounts of Ca^{2+}, Li^+, Nd^{3+}, Ni^{2+}, SO_4^{2-}, Zr^{4+} ions changed moderately its amphoteric properties. The catalytic test of 2-propanol

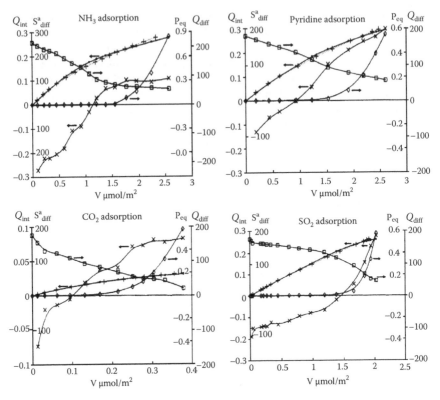

FIGURE 13.8 Differential heat (kJ/mol; □), integral heat (J/m²; ■), calculated integral heat (J/m²;+), differential entropy (J/mol K; x), equilibrium pressure (Torr; ◊) vs. the adsorbed volume (μmol/m²) on γ-Al₂O₃ for ammonia, pyridine, SO₂, and CO₂ at 80°C. (From Gervasini, A. and Auroux, A., *J. Phys. Chem.*, 97, 2628–39, 1993. With permission.)

conversion led to results in agreement with the acidity determination performed by NH_3 adsorption microcalorimetry. The intermediate acidity strength played a central role in 2-propanol dehydration. Gervasini and colleagues [175] found that the acidity of the catalysts correlated with the charge/radius ratio and with the generalized electronegativity of the doping ions. Doping a strongly amphoteric surface as alumina with small amounts of different ions created a very weak variation in its basic character. For example, the initial heat of SO_2 adsorption slightly increased when adding Ca^{2+} ($Q_{init} = 228$ kJ/mol) or more with Nd^{3+} ($Q_{init} = 281$ kJ/mol) compared to alumina ($Q_{init} = 194$ kJ/mol). The medium basic site strength appeared to be lowered by addition of sulphate, and in that case, the strength distribution was more homogeneous with a plateau around 150 kJ/mol.

As already mentioned, in spite of the widespread use of alumina in industry as adsorbent, catalyst, or catalyst support, there is only a limited understanding about the relationship between its surface properties and dehydroxylation–rehydroxylation behavior. The rehydration–dehydration behavior of transition aluminas containing controlled amounts of pentahedral Al has been investigated by Coster et al. [185].

Microcalorimetric measurements at the early stage of surface rehydration suggested that the differential heat of water chemisorption rises with increasing crystallinity. The heat released during the dissociative chemisorption varied between 250 and 85 kJ/mol and was larger than that corresponding to bulk rehydroxylation (65 kJ/mol). On the average, 5.4 OH per nm^2 were generated during this process. Bulk defects were cured when aluminas were contacted with saturated vapor pressure and thereafter calcined at high temperatures (600 or 750°C). While the surface area decreased, the crystallinity increased and fivefold coordinated Al was transformed into a more stable coordination.

In contrast to the large number of studies dealing with alumina in the literature, very few studies are devoted to boron, indium, and gallium bulk oxides. The differential heats of adsorption of NH_3 and CO_2 over 18 oxides have been determined by Auroux and Gervasini [173]. Higher heats of ammonia adsorption were observed for Ga_2O_3 when compared with Al_2O_3 in all coverage domains. The chemisorbed amounts varied in the same direction. Ga_2O_3 showed a plateau of homogeneous strong sites before a drastic decrease and a discontinuous heterogeneity in the field of medium-strength sites (≈ 100 kJ/mol). During CO_2 adsorption, alumina surface exhibited a remarkable chemical heterogeneity with initial heat of 180 kJ/mol proving its amphoteric character. Contrary to the results of NH_3 adsorption, smaller heats of CO_2 adsorption were observed for Ga_2O_3 in comparison with those found for alumina. The basic sites are related to O^{2-} ions, and this is consistent with electron-donor properties. Gallium oxide exhibits weaker basic sites (in strength and number) than those observed for Al_2O_3.

Another study examined the acidity and basicity of bulk Ga_2O_3 by NH_3 and SO_2 adsorptions microcalorimetry performed at 150°C. As alumina, Ga_2O_3 is amphoteric, with heats higher than 100 kJ/mol for both NH_3 and SO_2 adsorption, respectively [186]. The amphoteric character of bulk gallium oxides and strong heterogeneity of the surface acidic and basic sites were proved also by Petre et al. [179] using microcalorimetry of pyridine adsorption at 150°C and CO_2 adsorption at 30°C.

The amphoteric indium oxide can be considered as more basic than acidic when comparing the adsorption heats and irreversible adsorbed amounts, which are clearly higher for SO_2 adsorption than for ammonia adsorption [40,47]. The heats of NH_3 adsorption decreased continuously with coverage, while the SO_2 adsorption heat remained constant over a wide range of coverage.

13.4.2 SUPPORTED AND MIXED OXIDES

Studies on the nature of the interaction between the dispersed metal oxide species and the support have shown that their catalytic behavior and their acid–base properties are strongly affected by the inductive effect of the metal ions in the solids [187,188]. It has also been established in the literature that the support influences strongly the nature and the extent of metal oxide-support interaction and that the physicochemical properties of dispersed metal oxides are usually very different from those of the corresponding bulk phases [118,189]. Besides the effect of the support, the loading amount of metal oxide has also a very strong impact on the nature of the interaction between metal oxide and support [190] and on the acid–basic properties of these systems [191,192].

The surface acid–base properties of supported oxides can be conveniently investigated by studying the adsorption of suitably chosen basic–acidic probe molecules on the solid. As shown, acidic and basic sites are often present simultaneously on solid surfaces. The knowledge of the detailed amphoteric character of supported metal oxides is of extreme interest due to the possibility of using them as catalysts in different reactions in which acidity governs the reaction mechanism.

13.4.2.1   The Influence of the Amount of Loaded Oxide

Petre et al. [193] have studied the modifications of acid–base properties of γ-Al_2O_3 by depositing variable amounts (3–25 wt.%) of B_2O_3, Ga_2O_3, and In_2O_3 using NH_3 and SO_2 adsorption microcalorimetry experiments at 80°C. From the adsorption of NH_3 and SO_2, it was found that the addition of B_2O_3 on alumina led to an increase of the number of acid sites, while no real basicity could be evidenced. Small amounts of boron oxide deposited on Al_2O_3 covered the acid and basic sites of the amphoteric alumina, generating acid sites of medium strength. Increasing the amount of supported B_2O_3 increased strongly the number of acidic sites, majoritarily in the domain of weak acidic strength ($100 > Q > 50$ kJ/mol), by creating mainly Brønsted acid sites.

Similar results were obtained for supported boria catalysts prepared using two different methods, a classical impregnation method and chemical vapor deposition on porous and nonporous γ-alumina [45,168,194]. The acidity studied by NH_3 adsorption microcalorimetry has been shown to increase in sites number but not in strength with boron oxide content. A large number of weak acid sites were created on the catalyst surface when the boria amount was greater than the theoretical monolayer. However the number of acid sites determined when using pyridine as probe molecule was lower than using ammonia. Ammonia was shown to cover all types of sites from strong to weak acid sites, while pyridine only titrated the stronger sites of the samples, perhaps because of steric hindrance. At low loadings, new acid sites with medium strength were generated by coating of the strong acid sites of alumina. At high boron oxide loadings, weak acid sites were generated by formation of oxide agglomerates. The basicity of the system, measured by sulfur dioxide adsorption, decreased progressively with the increase in boron oxide content. It was also shown that the basic sites of the amphoteric alumina support were neutralized by 10 wt.% of boron oxide on a nonporous alumina support and 20 wt.% of B_2O_3 on a porous alumina. Moreover, the catalytic activity for partial oxidation of ethane increased with acidity and reached a maximum constant value for the monolayer [45,168].

Sato et al. [195] have studied the surface borate structures and the acidic properties of alumina-boria (3–20 wt.%) catalysts prepared by impregnation method using ^{11}B(MAS)-NMR measurements and TPD of pyridine, as well as their catalytic properties for 1-butene isomerization. The number of Brønsted acid sites was found to increase with increasing boria content, and the catalytic activity was explained by the strong Brønsted acid sites generated by BO_4 species on the surface of alumina.

Concerning gallium and indium oxides, Petre et al. [193] observed that coating alumina with Ga_2O_3 decreased slightly the acidity and did not affect markedly the basicity, while depositing In_2O_3 on Al_2O_3 decreased both the acidity and basicity of alumina. Alumina-supported Ga_2O_3 and In_2O_3 samples displayed a well-preserved amphoteric character, which decreased in the order $Al_2O_3 > Ga_2O_3/Al_2O_3 > In_2O_3/Al_2O_3$. The strength of the amphoteric character seemed to decrease concomitantly with increasing ionic radius (see Table 13.1).

The influence of gallium oxide loading on the adsorptive properties of alumina was also characterized by Gergely et al. [186,196] using NH_3, SO_2, and NO adsorption microcalorimetry. The authors have shown that SO_2 and NH_3 are strongly adsorbed on catalysts surface. The deposition of Ga_2O_3 was reported to cause a slight decrease in the acidity as determined by NH_3 adsorption. The number of sites titrated by SO_2 on the supported gallium oxide samples increased with increasing amounts of Ga_2O_3. This behavior suggested that gallium oxide is more bonded to the acidic sites than to the basic sites of alumina, creating a loss of acidic sites in alumina, or it may also reflect the presence of new basic sites provided by Ga_2O_3.

The structure and Lewis acidic properties of gallia-alumina catalysts prepared by impregnation have been investigated using IR spectroscopy (with CO as probe) and ESR spectroscopy (with anthraquinone) by Pushkar et al. [197]. The modification by gallia was a source of significant changes in the alumina hydroxyl cover. $Ga_2O_3–Al_2O_3$ systems with low gallia content (1–3 mol.%) were found to form solid solutions exhibiting a behavior close to that of pure γ-Al_2O_3. It was shown that, when the gallia content reached 20 mol.%, the conditions on the alumina surface were favorable (more than in the case of pure gallia) for the formation of two types of Lewis acid sites (due to Ga^{3+} *cus* formation).

A detailed study of a supported In_2O_3 on γ-Al_2O_3 system at different In loadings (2 < wt.% < 22) has been presented by Perdigon-Melon et al. [40]. As already shown, indium oxide can be considered as basic rather than acidic: the overall acidity was slightly decreased and the basicity significantly increased with increasing indium oxide loading on the alumina support. The irreversible amounts of SO_2 adsorbed were greater than the irreversible amounts of NH_3 adsorbed suggesting the presence of stronger basic sites on the indium loaded samples. Alumina was found to be completely unable to reduce any NO_x. The addition of a very low amount of In was sufficient to impart new catalytic activity to the Al_2O_3 support. The NO_x conversion increased with In_2O_3 loading up to 13%; from this point on, further indium addition caused a decrease in conversion. Thus, a balanced presence of indium centers and acidic sites of the alumina support has been found to be very important for achieving optimal catalytic performances in NO_x reduction by hydrocarbons. The high activity for de-NO_x process on In_2O_3/A_2O_3 catalyst was explained by the bifunctional mechanism involved in the reaction: the well-dispersed indium sites activate the hydrocarbons into partially oxygenated compounds, and the acidic alumina sites readily use the oxygenated hydrocarbons to reduce NO_x.

13.4.2.2   The Influence of the Support

Recently, certain group IIIA oxides (Al_2O_3, Ga_2O_3, In_2O_3) have attracted much interest as catalysts for selective catalytic reduction (SCR) of NO_x by hydrocarbons in the presence of excess oxygen [28,198] and for the dehydrogenation or aromatization of light alkanes [199]. It was proved that the catalytic performances in these processes were highly affected by the support effect, hence the importance of the choice of support in the development of highly efficient catalysts. Therefore, the acid–base properties of different supports may be affected by the deposition of Al, Ga, or In amphoteric oxides. It is expected that, by studying the changes in the acid–base properties, one can gain useful information on the guest oxides, including their dispersion on different support surfaces, thus allowing one to enhance the catalytic properties.

The influence of the oxide support (i.e., Al_2O_3, Nb_2O_5, SiO_2, ZrO_2, CeO_2, TiO_2) or nonoxide support such as BN on the acid–base properties of supported boron, aluminum, indium, and gallium oxide catalysts has been investigated by adsorption microcalorimetry [47,179,180,200].

The acid–base properties of supported gallium oxide catalysts have been investigated by Petre et al. [179,180] using pyridine and CO_2 adsorption. The studied supports covered a wide range of acidity as measured by pyridine adsorption, from the weakly acidic silica to the strongly acidic titania and γ-alumina, and sites of different strengths were involved. The order of the number of acid sites, as determined from the chemisorption uptakes, was the following: $SiO_2 < ZrO_2 < Al_2O_3 < TiO_2$. Pyridine adsorption microcalorimetry has shown that loading supports such as γ-Al_2O_3, TiO_2, and ZrO_2 with gallium oxide in a surface concentration close to the theoretical monolayer, resulted in a decrease of the surface acidity of the catalysts compared to that of the supports, while in the case of SiO_2, new Lewis acid sites were created. The number of acid sites for the supported and bulk Ga_2O_3 catalysts was observed to be in the order: $Ga_2O_3/SiO_2 < Ga_2O_3/ZrO_2 < Ga_2O_3/TiO_2 \approx Ga_2O_3 < Ga_2O_3/Al_2O_3$. This order was found to be related to the degree of dispersion of gallium oxide on the surface of each support and to the interaction between guest and host oxides.

Two types of Lewis acid sites were identified by pyridine adsorption, corresponding to the support and the supported gallium oxide, respectively. Lewis acidity depends on the existence of exposed metal cations at the surface and is influenced by factors such as ionic charge, degree of coordinative unsaturation, and bandgap. The sites with strong or medium acid strength are more affected by the deposition of gallium oxide than the weak ones. The addition of gallium oxide decreased also the hydrophilic properties of alumina, titania, and zirconia, but increased the amount of water adsorbed on silica [179]. In fact, the ordering of the studied samples according to their hydrophilic character was close to that for acidity, both in the case of supports ($SiO_2 \ll ZrO_2 \sim TiO_2 < Al_2O_3$) and supported gallium oxide catalysts ($Ga_2O_3/TiO_2 < Ga_2O_3/ZrO_2 < Ga_2O_3/SiO_2 < Ga_2O_3/Al_2O_3$).

The NH_3 and H_2O are both donors with lone-pair electrons able to interact with cation surface sites explaining why water adsorption defined the same scale of acidity as ammonia. The comparison of the CO_2 uptakes on studied supports and supported gallium oxide catalysts showed that amphoteric Ga_2O_3 is covering some of the basic sites of alumina, titania, and zirconia. After depositing Ga_2O_3 on ZrO_2, an amphoteric support that displays the largest and strongest population of basic sites, the CO_2 uptake decreased much more than in the case of alumina-supported catalyst.

On silica, the deposited Ga_2O_3 did not create a specific basicity. $Ga_2O_3/\gamma\text{-}Al_2O_3$ presented the highest acidity and was the only sample displaying a relatively strong basicity. $Ga_2O_3/\gamma\text{-}Al_2O_3$ presented a specific local structure (superficial spinel) different from bulk gallium oxide and was revealed as highly effective in SCR of NO_x. The catalytic performances in the selective catalytic reduction of NO by C_2H_4 in excess oxygen were decreasing in the order: $Ga_2O_3/Al_2O_3 > Ga_2O_3/TiO_2 > Ga_2O_3/ZrO_2 > Ga_2O_3/SiO_2$ [179].

The influence of the oxide support (i.e., Al_2O_3, Nb_2O_5, SiO_2, and TiO_2) on the surface acid–base properties and catalytic performances of supported indium oxide catalysts were also studied [47]. Indium oxide was found to be responsible for the acidic–basic and adsorptive properties of In_2O_3/SiO_2 catalyst as silica chemisorbed neither ammonia nor SO_2. The poor dispersion, due to the inertia of the silica surface made the adsorbed quantities very small. The heats of ammonia adsorption presented an interesting behavior when indium oxide was supported on $\gamma\text{-}Al_2O_3$, with lower values for $In_2O_3/\gamma\text{-}Al_2O_3$ than on the bare support. The deposition of indium oxide decreased both the acidity and basicity of alumina.

The acidity of TiO_2 and Nb_2O_5 supports was lowered by In_2O_3 deposition while their basic properties increased after indium oxide deposition. Supports able to disperse the In_2O_3 aggregates with high In stabilization gave rise to active catalytic systems. Among the studied oxide supports, Al_2O_3 and to a lower extent TiO_2 were found to be the best supports for obtaining indium based active and selective de-NO_x catalysts.

Similar results were found when the acid/base properties of a series of group IIIA oxides (Al_2O_3, Ga_2O_3 and In_2O_3) supported on niobia were studied by adsorption microcalorimetry using ammonia and sulfur dioxide as probe molecules at 80°C [200]. The changes found by Petre and colleagues in acidic and basic properties of niobia by the deposition of alumina, gallia, and india are presented in Figures 13.9 and 13.10, respectively. As it can be seen in Figure 13.9, the differential heats of NH_3 adsorption of Al_2O_3/Nb_2O_5 and Nb_2O_5 are almost identical except for the first part of the curve where the Al_2O_3/Nb_2O_5 sample presents a range of higher adsorption heats (coverage $\approx$ 42 µmol/g). This could be attributed to adsorption on newly created alumina acid sites, knowing that alumina presents Lewis acid sites stronger than the acid sites found in the niobia used as support [173]. Also, the amount of ammonia adsorbed by the Al_2O_3/Nb_2O_5 sample is larger than for niobia, the difference being more important at higher pressures when the adsorption is due mainly to physisorption. By contrast, the chemisorbed amounts are identical for both samples, indicating that the acidic properties of the two samples are very similar. It was found that gallium and indium oxides were preferentially deposited on the acid sites of niobia decreasing the acidity of the support.

The acid site densities were found to vary in the following order: $Ga_2O_3/Nb_2O_5 < In_2O_3/Nb_2O_5 < Al_2O_3/Nb_2O_5$. The more important contribution of the bare Nb_2O_5 to total acidity of In_2O_3/Nb_2O_5 in spite of the more pronounced basic character of In_2O_3 was explained by the low dispersion of the indium oxide when supported on niobia. Since no basicity could be observed for Nb_2O_5, the basicity observed for the three supported samples should be attributed to the supported oxides only (Figure 13.10). The order of basic strength as determined by SO_2 adsorption was:

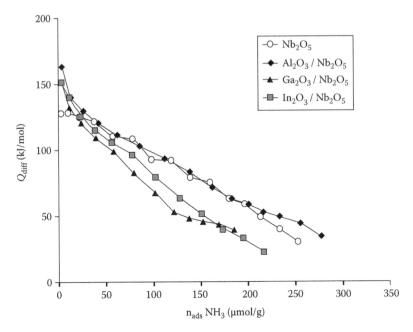

FIGURE 13.9 Differential heats of ammonia adsorption at 80°C versus coverage for Nb_2O_5, Al_2O_3/Nb_2O_5, Ga_2O_3/Nb_2O_5, and In_2O_3/Nb_2O_5. (From Petre, A. L., Perdigon-Melon, J. A., Gervasini, A., and Auroux, A., *Catal. Today,* 78, 377–86, 2003. With permission.)

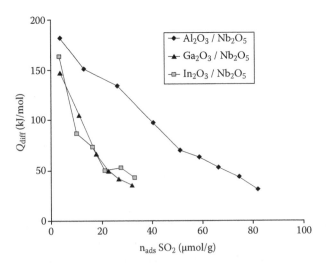

FIGURE 13.10 Differential heats of SO_2 adsorption at 80°C versus coverage for Nb_2O_5, Al_2O_3/Nb_2O_5, Ga_2O_3/Nb_2O_5, and In_2O_3/Nb_2O_5. (From Petre, A. L., Perdigon-Melon, J. A., Gervasini, A., and Auroux, A., *Catal. Today,* 78, 377–86, 2003. With permission.)

$Al_2O_3/Nb_2O_5 \gg Ga_2O_3/Nb_2O_5 \approx In_2O_3/Nb_2O_5$. Since it is commonly accepted that indium oxide is the most basic of the three amphoteric group IIIA oxides, the low basicity observed for In_2O_3/Nb_2O_5 has once more been attributed to a very poor dispersion [200].

As it can be observed in Figures 13.9 and 13.10, the total amounts of SO_2 adsorbed are markedly smaller than the corresponding NH_3 amounts, indicating that the surfaces of all supported catalysts can be considered as mostly acidic. Ga_2O_3/Nb_2O_5 presented the best conversion and selectivity toward dehydrogenation of propane between studied samples. The measured propane conversions varied in the order $Ga_2O_3/Nb_2O_5 > In_2O_3/Nb_2O_5 > Al_2O_3/Nb_2O_5 \approx Nb_2O_5$, exactly the inverse of the order observed in the acidity measurements. In fact, the ratio of acidic to basic sites was the main criteria that determined the reactivity of these samples.

Calorimetric measurements of NH_3 and SO_2 at 80°C on a nonoxidic support as BN and Ga_2O_3/BN and In_2O_3/BN samples have provided evidence that the acid sites are predominant on their surfaces, while no real basicity could be evidenced [201]. BN presented a heterogeneous surface for NH_3 adsorption with heats of adsorption ranging from 150 to 30 kJ/mol. The influence of the oxide additives was evident: In_2O_3 and Ga_2O_3 increased sharply the number of acidic sites and the ammonia adsorption gave rise to adsorption heats ranging between 180 and 30 kJ/mol. The number of acidic sites was found to vary in the following order: $In_2O_3/BN > Ga_2O_3/BN > BN$. For the BN support only physical adsorption of SO_2 was observed. The addition of oxides as active phases resulted in the appearance of basic sites due to the amphoteric character of Ga and In oxides. The most basic catalyst was Ga_2O_3/BN, but even for this sample the basic character was not significant when compared to the acidic character.

Another study examined the NH_3 and SO_2 adsorption on (B_2O_3, Al_2O_3, Ga_2O_3, In_2O_3)–CeO_2 mixed oxides, differing in their preparation procedure and in the loading amount of supported oxides [202,203]. When the mixed oxides were prepared by a *coprecipitation method*, only boria created significant acidity, whereas the basicity has been found to be dependent on the nature and amount of group IIIA element [202]. The differential heats of SO_2 adsorption showed a wide range of variability, displaying either a plateau of constant heat for Al_2O_3-CeO_2, Ga_2O_3-CeO_2, and In_2O_3-CeO_2 samples, or a continuous decrease indicative of adsorption heterogeneity for B_2O_3–CeO_2 samples. The smallest adsorption of SO_2 was found for boria-containing samples; the increase of boria content produced the decrease of the amount of adsorbed gas.

The low amount of alumina did not modify the basicity of ceria; however, with increasing Al_2O_3 content a slight decrease in basicity has been observed for alumina-ceria sample. The additions of gallium and indium oxides created some additional basic sites on ceria surface, enhancing the basic character of CeO_2. By coupling the microcalorimetric results with an additional method such as XPS, where NH_3 and SO_2 adsorptions were investigated through the recording of N1s and S2p lines, respectively, it was observed that the basicity comes from Brønsted (mainly) and Lewis sites, whereas the acidity is only of the Lewis type for all investigated mixed oxides. The (B_2O_3, Al_2O_3, Ga_2O_3, In_2O_3)–CeO_2 mixed oxides prepared by *sol-gel method* expressed surface amphoteric character; however the surface basicity was more pronounced than surface acidity [203]. All sol-gel mixed oxides presented only weak acidity lower than pure ceria, while basic character was diminished for boria-ceria and alumina-ceria formulations and remained similar to that of ceria for gallia-ceria and india-ceria.

This result was unexpected after those found for the mixed oxides prepared by coprecipitation: for example, the significant acidity noticed for boria (6–17 wt.%)-ceria samples. Even if the B_2O_3–CeO_2 prepared by sol-gel method had a higher amount of boria (36 wt.%), this sample showed only insignificant acidity. Bonnetot and colleagues [203] found that the high amount of boria in this sample and the preparation method were the factors that produced an unfavorable distribution of B_2O_3 particles in the mixed oxides. B_2O_3 particles could be coated by ceria. In that way, the sites active in NH_3 adsorption were blocked. Indeed, the acid–base properties were also influenced by the applied preparation method: the acid–base features of the samples prepared by *sol-gel method* were found to be quite different from those of the samples prepared by *coprecipitation*. Sol-gel Ga_2O_3–CeO_2 sample for example, with the same composition as for its homologue prepared by coprecipitation presented an increased basicity. The applied procedure is in fact decisive for the dispersion/mixing of both group IIIA oxides and ceria. In the case of Ga_2O_3–CeO_2, by applying the sol-gel procedure, the contribution of acidic sites is decreased, which means that the influence of ceria in the mixture is minimized.

13.4.3 NITRIDED PHOSPHATES COMPOUNDS

The substitution of oxygen by nitrogen in PO_4 tetrahedron has allowed the synthesis of a new family of solids with original properties: the nitrided phosphates. These systems (e.g., AlPON, AlGaPON) with tunable acid–base properties are used in a growing number of intermediate and fine chemistry production processes [204] as well as supports in heterogeneous catalysis (e.g., dehydrogenation reactions) [205].

On nitrided aluminophosphates, AlPON, Massinon et al. [206] observed on a series of six samples with increasing nitrogen contents a good correlation between the catalytic activity in the Knoevenagel condensation reaction and the amount of surface NH_x species ($1 \leq x \leq 4$) quantified by the Kjeldahl method. The authors suggest that those species are not the only active species and evoke an additional role of the nitride ions in the reaction [206]; on the other hand, Benitez et al. [207] suggest hydroxyls linked to aluminum cations in the vicinity of terminal $P-NH_2$ groups as basic centers.

Mixed nitrided galloaluminophosphates are obtained by heating an amorphous $Al_{0.5}Ga_{0.5}PO_4$ phosphate under ammonia. This treatment allows substituting three oxygen atoms by two nitrogen atoms in the anionic network and generates NH_4^+, coordinated NH_3, $-NH_2$ and $-NH-$ species on the surface of the AlGaPON solids [208]. Substituting part of the aluminum atoms by gallium allows to lower the minimum nitridation temperature and facilitates the nitrogen enrichment [209]. Those changes confer specific properties and applications to AlGaPON, as compared to those of the phosphate precursor AlGaPO: their acidity decreases while their basicity increases with the nitrogen enrichment. IR, microcalorimetry, and chemisorption studies have highlighted the evolution of the number, nature, and strength of the acidic and basic sites with the nitrogen content, leading to the full mastering of the acid–base properties of phosphates through nitridation [204,210,211]. The replacement of acidic –OH by NH_x species decreases the surface Brønsted acidity, while the replacement of coordinatively unsaturated metals by $M–NH_x$ species reduces the number of Lewis acidic sites [212].

Sulfur dioxide was chosen as a probe molecule to perform calorimetric and volumetric gas–solid titration of the basic sites of the AlGaPON samples with different nitrogen content (0–24 wt.% N) [211]. It was observed that the total amount of SO_2 adsorption sites increased progressively with the nitrogen enrichment. On the sample without nitrogen, the adsorption of SO_2 is weak due to the interaction of the probe with the surface –OH groups, while on the oxynitrides, SO_2 is adsorbed on $–NH_2$ groups. The number of sites that irreversibly adsorb SO_2 constituted between 44 and 70% of the total number of sites, the highest values being obtained for the sample with the highest nitrogen content (23.3 wt.%; see Figure 13.11).

The strength of the basic sites also increases with the total nitrogen content: the initial adsorption heats rise from 83 kJ/mol on the least nitrided sample to 142 kJ/mol for the most nitrided sample.

For all samples, the differential molar heats of adsorption decreased rapidly with SO_2 coverage. However the steepness of the curve decreased with the nitrogen enrichment, suggesting greater sites homogeneity.

The AlGaPON samples were used as catalysts of the Knoevenagel condensation reaction and the authors [211] found that the $–NH_2$ groups present at the surface of the samples were the basic sites responsible for the condensation properties of the catalysts. The catalytic performances of the studied samples increased with their basic character observed by SO_2 adsorption microcalorimetry.

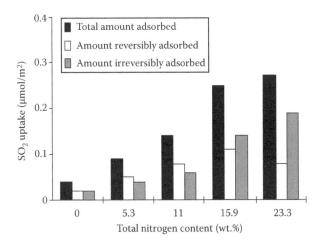

FIGURE 13.11   Evolution of the number of sites that reversibly and irreversibly adsorb SO_2 at 80°C with the total nitrogen content of AlGaPON samples. (From Delsarte, S., Auroux, A., and Grange, P., *Phys. Chem. Chem. Phys.*, 2, 2821–27, 2002. With permission.)

13.4.4 ZEOLITES AND RELATED MATERIALS

Zeolites consist of linked tetrahedra of SiO_4 and AlO_4. The substitution of the aluminum by other trivalent atoms such as B, Ga, and In in order to modify the surface acidity of such solids has aroused considerable interest [34,213–215].

Zeolites, or crystalline aluminosilicates, differ from more conventional crystalline materials in that the anhydrous crystal has a large, regular pore structure, making the internal surface available for adsorption or catalysis [104]. Since their successful introduction some 40 years ago, zeolite catalysts have been the subject of considerable academic and industrial research efforts. By far the major industrial process that utilizes zeolites is the catalytic cracking of petroleum. It is interesting to note that silica by itself has no activity for cracking and little acidity [216,217]. Gayer [218] observed that by introducing small amounts of alumina, both the activity and the acidity of the mixture began to rise and Whitmore [219] proposed acid sites as the active centers. The correlation between the acidity of a zeolite and its catalytic properties is a difficult task. Three factors are important here: the total number of acidic sites, the ratio of Brønsted to Lewis sites, and the acid strength distribution (and density) of each type of site. For example, Y zeolites present a maximum in strong acid sites and cracking activity when silica to alumina ratios vary from about 7 to 15 [220]. In contrast, for ZSM5, hexane cracking ability increases linearly with increasing aluminum content [104] leading to the conclusion that the maximum in acidity is a function not only of the zeolite structure but also the surroundings of the individual aluminum atoms in the framework.

13.4.4.1 Acidity of the Main Components in Fluid Cracking Catalysts (FCC)

The main components in fluid cracking catalysts (FCCs) are USY (dealuminated Y zeolite), a binding matrix, and an acidic component consisting of a small amount of H-ZSM5 zeolite in order to enhance the octane number of gasoline. The catalytic activity of such materials is due to the presence of acidic sites and is determined by the zeolite content and by the types of zeolite and matrix in the FCC catalyst. The catalytic selectivity is among other factors determined by the zeolite type, the nature (Brønsted or Lewis), strength, concentration, and distribution of the acid sites, the pore size distribution, the matrix surface area, and the presence of additives or contaminants. Stability is affected by both the composition and the structural characteristics of the catalyst components. The acidity of FCCs is designed to meet specific requirements, and a full characterization of the acidity is necessary; this gives a great deal of importance to the information gathered by direct methods such as the monitoring by microcalorimetry of the adsorption of gaseous bases, particularly ammonia or pyridine.

The determination of acidity in FCCs from adsorption microcalorimetry of probe molecules was the object of a review article by Shen and Auroux [105]. Adsorption microcalorimetry results obtained using ammonia as a probe molecule revealed that, as long as Lewis acid sites with strength greater than 100 kJ/mol are present and as long as these sites are available to gas oil, FCCs can retain their useful cracking activity and selectivity properties [221].

The effect of aging and regeneration on the acidity of FCC catalysts has also been studied by Occelli et al. [222,223] using ammonia and pyridine as probe molecules. The results obtained have shown that after aging, either under microactivity test conditions or in a fluid cracking catalyst unit, a fresh FCC undergoes severe losses in acid sites density while retaining most of the strength of its strongest Lewis acid sites. The presence of these sites and the retention of an open micro- and mesoporosity are believed to be responsible for the cracking activity of aged FCCs. Moreover, an increase in acidity and microporosity is consistent with the observed enhanced cracking activity under microactivity test conditions of a regenerated FCC [223].

13.4.4.1.1 *Acidity of H-Y and Dealuminated H-Y/H-USY*

Adsorption microcalorimetry has been used to measure the acid strength on a variety of original and dealuminated HY zeolites [103,224–227]. The removal of aluminum from HY zeolite crystals leads to products with high framework Si/Al ratios and calorimetric investigation of NH_3 adsorption showed changes in the initial values and coverage dependence of differential heats of adsorption. These changes were interpreted in terms of the presence within the zeolite porous matrix of an alumina type phase that is behaving as an additional acidic component. The differential heats of ammonia adsorption on the parent HY sample (Si/Al = 2.4) exhibited a plateau within 137–140 kJ/mol, which expanded over a large domain of coverage indicating a uniform strength distribution within the structural acid sites of this material. The differential heats then declined progressively as the coverage increased. This irregular distribution among the weaker sites was also reported in the literature and was attributed to the effect of coverage [228]. Microcalorimetric investigation of the acidity of dealuminated HY showed that they are in fact composite materials associating the dealuminated zeolite lattice with a guest aluminum oxide phase. The lattice retained most of its acidic properties unchanged, though the number of acid sites was closely related to the residual aluminum tetrahedral.

A detailed study of the acidity of different commercial HY zeolites (both nondealuminated and dealuminated by hydrothermal treatment) was reported by Boréave et al. [226]. Using NH_3 adsorption microcalorimetry coupled with FTIR, XPS, and TPD techniques, the authors found three families of sites in terms of acid strength for the studied zeolites: weak, intermediate, and strong (beyond 160 kJ/mol). The weak sites (associated with NH_3 adsorption heats below 130 kJ/mol) were mostly of Lewis type resulting from the degradation of the crystalline structure of the zeolite. This degradation was more important as the zeolite was more dealuminated, since the aluminum-containing debris become especially concentrated at the surface of the crystallites. Part of these weak sites was of the Brønsted type and corresponds to the protons surrounded by three aluminum atoms as second-nearest neighbors. These sites were found to be numerous in the nondealuminated zeolite but vanish after thorough dealumination. The sites of intermediate strength (NH_3 adsorption heats between 130 and 150 kJ/mol) were found to be mostly of the Brønsted type, and were associated with the bridged OH groups of the structure that are surrounded by less than three second-nearest neighbor aluminum atoms. The strong (NH_3 adsorption heat between 150 and 180 kJ/mol) and very strong (NH_3 adsorption heat between 180 and 230 kJ/mol) sites were found to correspond to both structural and extra-framework OH groups, as well as aluminum atoms that are either extra-framework or associated with framework defects. The high strength of these sites is due to interactions between Brønsted and Lewis sites and/or the neighborhood of cationic Al species.

NH_3 adsorption microcalorimetry was used by Shannon et al. [225] to follow the changes in acid sites of a HY zeolite during dehydroxylation, framework dealumination, and the formation of nonframework aluminum species.

Acid site distributions obtained by adsorption microcalorimetry showed that dehydroxylation of HY zeolite at 650°C resulted in the destruction of most of the weak and medium strength acid sites (75–140 kJ/mol) and replacement by fewer but strong (150–180 kJ/mol) sites, as shown in Figure 13.12 where the differential heat plots of ammonia adsorption for HY, dehydroxylated HY, pseudo-boehmite and γ-Al_2O_3 are presented. This acid site distribution was similar to that found in γ-Al_2O_3 heated to 600°C suggesting that boehmite-like species have developed surface defects like γ-Al_2O_3 or that remaining Brønsted acid strength has been modified by the presence of the boehmite-like phase.

Similar results were also obtained by Stach et al. [229] who studied the acidity of HY zeolites with different Si/Al ratios (2.4, 5.6, and 12.0) by calorimetric measurements

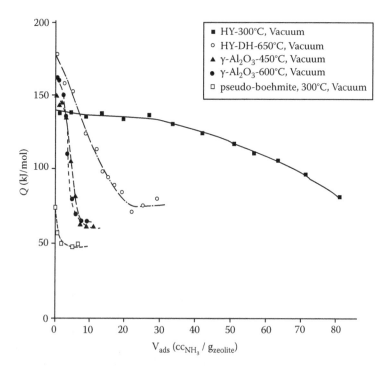

FIGURE 13.12 Differential heats of ammonia adsorption on HY, dehydroxylated HY (HY-DH), pseudo-boehmite, and γ-Al_2O_3 at 150°C. (From Shannon, R. D., Gardner, K. H., Staley, R. H., Bergeret, G., Gallezot, P., and Auroux, A., *J. Phys. Chem.*, 89, 4778–88, 1985. With permission.)

of ammonia adsorption at 150°C. Dealumination of Y zeolites was found to decrease the total number of acid sites and to generate some very strong acid sites.

Macedo et al. [227] studied HY zeolites dealuminated by steaming, and found that the strength of intermediate sites decreased with increasing dealumination for Si/Al ratios varying from 8 to greater than 100. For comparison, isomorphously substituted HY, which is free of extra-framework cationic species, possesses more acid sites than conventionally dealuminated solids with a similar framework Si/Al ratio [227]. This is because some of the extra-framework aluminum species act as charge-compensating cations and therefore decrease the number of potential acid sites.

Shi et al. [230] measured the strength of HY zeolites dealuminated by treatment with $SiCl_4$ for Si/Al ratios ranging from 4.2 to 37.1, and they also found a decrease in the number of sites possessing intermediate strength with increasing dealumination. As the Si/Al ratio increased, the initial heat of adsorption first increased and then passed through a maximum at a Si/Al ratio equal to 36. The heats of adsorption at the plateau of Brønsted acid sites passed through a maximum at a Si/Al ratio about 7.

The acid properties of nondealuminated and dealuminated commercial HY were also determined by Colon et al. [231] and Ferino et al. [58] using pyridine adsorption microcalorimetry at 150°C or by Biaglow et al. [165] and Chen et al. [232] using NH_3 adsorption microcalorimetry at 150°C.

It is known that the activation temperature can influence the acid strength distribution. For example, measurements of the differential heats of ammonia adsorbed at 150°C for a HY zeolite have led to the conclusion that stronger acid sites, in the 150–180 kJ/mol range, are formed upon increasing the activation temperature from 300 to 650°C. Dehydroxylation at high temperature resulted in the formation of strong Lewis acid sites and the disappearance of intermediate and weak Brønsted sites [62].

NH_3 adsorption microcalorimetry has been used to characterize the acid sites of an H-USY zeolite and another USY sample in which the strong Lewis acid sites were poisoned with ammonia. It was found that poisoning of the Lewis acid sites did not affect the rate of deactivation, the cracking activity and the distribution of cracked products during 2-methylpentane cracking. Thus, strong Lewis acid sites did not seem to play any important role in the cracking reactions [233].

13.4.4.1.2 Acidity of ZSM5

Extensive studies of the acidity and basicity of zeolites by adsorption calorimetry have been carried out over the past decades, and many reviews have been published [62,64,103,118,120,121,145,146,153,154]. For a given zeolite, different factors can modify its acidity and acid strength: the size and strength of the probe molecule, the adsorption temperature, the morphology and crystallinity, the synthesis mode, the effect of pretreatment, the effect of the proton exchange level, the Si/Al ratio and dealumination, the isomorphous substitution, chemical modifications, aging, and coke deposits.

Ammonia adsorption experiments have shown that there exists a strong acidity in H-ZSM5 zeolite, which exceeds that present in HY zeolite [234] while no electron donor (basic) sites have been evidenced [235].

The acidity of ZSM5 zeolites with various SiO_2/Al_2O_3 ratios (28, 50, 75, 100, 150), synthesized with and without the aid of a template, has been investigated by adsorption microcalorimetry [236]. The integral heats of adsorption were found to increase with the Al content per unit cell, but the nontemplated zeolites gave rise to higher adsorption heats compared to the analogous templated zeolites for aluminum contents greater than 3 Al atoms per unit cell ($SiO_2/Al_2O_3 < 75$). This can be explained by the possible presence of more Lewis acid sites in the nontemplated zeolites. Narayanan and colleagues [236] used aniline alkylation and cumene cracking reactions to evaluate the catalytic properties involving the acid sites present on ZSM5 zeolites. The nontemplated zeolites showed good catalytic activity for aniline alkylation. The presence of more Lewis sites in nontemplated zeolites than in templated ones was found to be responsible for high aniline alkylation and low-cumene-cracking activity.

The effect of the Si/Al ratio of H-ZSM5 zeolite-based catalysts on surface acidity and on selectivity in the transformation of methanol into hydrocarbons has been studied using adsorption microcalorimetry of ammonia and *tert*-butylamine. The observed increase in light olefins selectivity and decrease in methanol conversion with increasing Si/Al ratio was explained by a decrease in total acidity [237].

The effect of temperature on ammonia adsorption by ZSM5 samples has been investigated by microcalorimetry, varying the adsorption temperature from 150 to 400°C [235]. The initial heats of adsorption were independent of temperature up to 300°C. When the adsorption temperature increased, there was a competition between the formation of ammonium ions on Brønsted sites and their decomposition. The total number of titrated sites decreased with increasing adsorption temperature. It appeared that an adsorption temperature between 150 and 300°C is appropriate for these calorimetric experiments.

The microcalorimetry of NH_3 adsorption coupled with infrared spectroscopy was used to study the effect of the synthesis medium (OH^- or F^-) on the nature and amount of acid sites present in Al,Si-MFI zeolites [103]. Both techniques revealed that H-MFI (F^-) with Si/Al < 30 contained extra-framework aluminum species. Such species were responsible for the presence of Lewis acid sites and poisoning of the Brønsted acidity. In contrast, MFI (F^-) characterized by Si/Al > 30 presented the same behavior as H-MFI (OH^-).

13.4.4.2 Acidity of Other Zeolites and Related Materials

H-ZSM11 samples were found to be also very strongly acidic, but slightly less than the corresponding H-ZSM5 zeolites [235,238]. The acid strength distributions of H-offretites (fresh catalyst and aged samples) have been determined at 150°C using ammonia as a basic probe and the results obtained were compared to those of other zeolites (ZSM5, ZSM11, mordenite) [239]. The acidity of offretite dealuminated by hydrothermal treatment has also been characterized by microcalorimetry of NH_3 adsorption. It was observed that dealumination up to ca. 50% primarily decreased the number of sites of medium acid strength of offretite and left part of the strong acid sites unaffected [240]. The parent and dealuminated samples displayed very similar initial heats of NH_3 adsorption of about 180 kJ/mol when the Si/Al ratio increased from 4.3 to 9.7, followed by a plateau of heats between 170 and 150 kJ/mol and then a continuous decrease to ca. 70–50 kJ/mol. Two kinetic regimes of heat evolution were observed for the starting offretite, which perhaps arose from two differently accessible types of sites presumably located in the cages and channels. Only one kinetic regime was observed for dealuminated samples.

The adsorption microcalorimetry has been also used to measure the heats of adsorption of ammonia and pyridine at 150°C on zeolites with variable offretite-erionite character [241]. The offretite sample (Si/Al = 3.9) exhibited only one population of sites with adsorption heats of NH_3 near 155 kJ/mol. The presence of erionite domains in the crystals provoked the appearance of different acid site strengths and densities, as well as the presence of very strong acid sites attributed to the presence of extra-framework Al. In contrast, when the same adsorption experiments were repeated using pyridine, only crystals free from stacking faults, such as H-offretite, adsorbed this probe molecule. The presence of erionite domains in offretite drastically reduced pyridine adsorption. In crystals with erionite character, pyridine uptake could not be measured. Thus, it appears that chemisorption experiments with pyridine could serve as a diagnostic tool to quickly prove the existence of stacking faults in offretite-type crystals [241].

The acidity of a series of dealuminated mazzites with Si/Al ratios varying from 5 to 23, prepared by combined steam-acid leaching and ion-exchange procedures, all obtained from the same parent zeolite (Si/Al = 4), has been studied using adsorption microcalorimetry [242] with particular emphasis on the evaluation of the influence of framework and nonframework aluminum species on the nature, strength, and accessibility of the acid sites. It was shown that the strong acid sites present in dealuminated mazzite were associated with the framework aluminum atoms, whereas nonframework species contributed essentially to the weaker acidity. Initial and intermediate heats of adsorption of ammonia were with 10–20 kJ/mol higher than those usually reported for dealuminated zeolites. The samples presenting nonframework aluminum displayed a heterogeneous distribution of acid sites, with a continuous decrease of the differential heats versus coverage, whereas the samples presenting no extra-framework species displayed a very homogeneous heat of adsorption.

Since their relatively recent discovery, ordered mesoporous materials have attracted much interest because of their high surface area and uniform distribution of mesopore diameters. Because of its hexagonal array of uniform one-dimensional mesopores, varying in diameter from 1.5 to 10 nm, MCM41 is a potentially interesting catalyst for converting large molecules of nondistillable feeds to fuels and other products [103].

The methods for measuring the acidity of nanoporous aluminosilicates such as MCM41 have been reviewed by Zheng et al. [243], including microcalorimetry measurements of probe molecules adsorption.

The influence of the nature of the aluminum source on the acidic properties of mesostructured materials (MCM41) has also been studied in the literature [244]. Microcalorimetry experiments using ammonia as a probe molecule have shown that Al insertion into the mesoporous silicate framework affected acid site strength and distribution in a manner controlled by the synthesis conditions (materials prepared with $Al(OH)_3$, $Al(iPrO)_3$ or $NaAlO_2$ in the presence of a surfactant). The initial heats of ammonia adsorption varied from 165 to 120 kJ/mol depending on the source of aluminum used for synthesis and on the Al(VI)/Al(IV) ratio in the material.

A combined microcalorimetry and adsorption study was used by Meziani et al. [49] to characterize the surface acidity of a series of MCM41 aluminosilicates (referred to as $SiAl_xC_n$, where x is the molar Si:Al ratio and n the chain length of the surfactant template). With the exception of $H-SiAl_{32}C_{14}$ and $SiAl_8C_{14}$, all samples were found to present low surface acidity.

The acidity of thermally stable mesoporous aluminophosphates (AlPO) and silicoaluminophosphates (SAPO) has also been studied by microcalorimetry [245]. By contrast with microporous crystalline aluminophosphate molecular sieves, mesoporous compounds are amorphous and characterized by Al/P ratios greater than 1. These particularities are responsible for a strong Lewis acidity, making these mesoporous materials more acidic than the microporous analogues, with an amount of strong acid sites that increases with the silicon content.

13.4.4.3 The Influence of B, Ga, and In on the Acidity of Zeolites

The acidic/basic properties of zeolites can be changed by introduction of B, In, Ga elements into the crystal framework. For example, a coincorporation of aluminum and boron in the zeolite lattice has revealed weak acidity for boron-associated sites [246] in boron-substituted ZSM5 and ZSM11 zeolites. Ammonia adsorption microcalorimetry gave initial heats of adsorption of about 65 kJ/mol for H-B-ZSM11 and showed that B-substituted pentasils have only very weak acidity [247]. Calcination at 800°C increased the heats of NH_3 adsorption to about 170 kJ/mol by creation of strong Lewis acid sites as it can be seen in Figure 13.13. The lack of strong Brønsted acid sites in H-B-ZSM11 was confirmed by poor catalytic activity in methanol conversion and in toluene alkylation with methanol.

A microcalorimetric investigation of NH_3 adsorption at 150°C was also applied to characterize the modified acidity of ZSM5 zeolite impregnated with increasing amounts of H_3BO_3 and pretreated at two different temperatures (400 and 800°C) [248]. The former pretreatment was responsible for removing part of the Brønsted sites, while the latter also induced pore plugging, which therefore drastically reduced the indicated overall zeolite acidity. Zeolite impregnation with increasing amounts of H_3BO_3 contributed to an increased loss of acidity.

The isomorphous replacement of aluminum by gallium in the framework structure of zeolites (beta, MFI, offretite, faujasite) offers new opportunities for modified acidity and subsequently modified catalytic activity such as enhanced selectivity toward aromatic hydrocarbons [249,250]. The Ga^{3+} ions in zeolites can occupy tetrahedral framework sites (T) and nonframework cationic positions.

Auroux et al. [251] used adsorption microcalorimetry of different alkanes to investigate Ga and Al substituted MFI zeolites used as catalysts in dehydrogenation and cracking reactions.

The variation of cracking selectivity in the conversion of alkanes over substituted H-Ga-MFI and H-Al-MFI zeolites has been correlated with the basicity of the C–C bond of the alkane, while the selectivity toward dehydrogenation was found to be related to the attenuation of the acid strength of the zeolite [251].

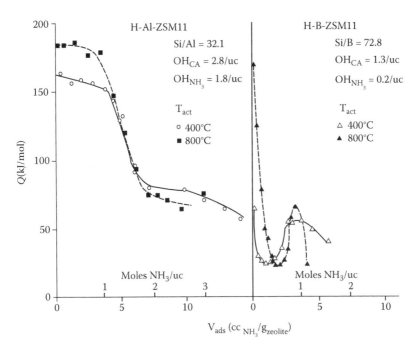

FIGURE 13.13   Differential heats of ammonia adsorption on H-Al-ZSM11 and H-B-ZSM11 at 150°C. (From Coudurier, G., Auroux, A., Védrine, J. C., Farlee, R. D., Abrams, L., and Shannon, R. D., *J. Catal.*, 108, 1–14, 1987. With permission.)

Microcalorimetric experiments of NH_3 adsorption have shown that the isomorphous substitution of Al with Ga in various zeolite frameworks (offretite, faujasite, beta) leads to reduced acid site strength, density, and distribution [250,252,253]. To a lesser extent, a similar behavior has also been observed in the case of a MFI framework [51,254]. A drastic reduction in the acid site density of H,Ga-offretites has been reported, while the initial acid site strength remained high [248,250].

Microcalorimetry experiments with NH_3 and pyridine as probe molecules indicated that insertion of Ga into the offretite aluminosilicate structure increased the overall acid sites strength of the crystals while decreasing its acid sites density [255]. The observed heterogeneity of acid site strength distribution of H,Ga,Al-offretites was attributed to some extra-framework Al(VI) and Ga(VI) species generated during the ion exchange and calcination procedures used to prepare H-offretite crystals.

With Ga-Beta it was found that, when the Si/Ga ratio increased from 10 to 40, the number of strong sites decreased drastically for Si/Ga between 10 and 25 and then reached a plateau above Si/Ga = 25 [53]. The strength and density of acid sites in H(Ga, La)-Y were also found to be lower than those in HY crystals of the type used in FCC preparation (LZY-82) [250]. Similar catalytic selectivities were obtained for both Ga-ZSM5 and Al-ZSM5 in Prins condensation of isobutylene with formaldehyde. Catalytic tests coupled with microcalorimetric measurements have shown that medium to weak acid strength sites favor the selectivity to isoprene [254].

Several publications have been dedicated to the study of In-based zeolites as catalysts in different catalytic processes (e.g., SCR of NO_x, alkylaromatic transformation, etc.) and the role played by the acidic character in their good performances was followed mostly by IR technique [256–259] or NH_3-TPD [260]. It was observed [260] that incorporation of indium in the framework of H-ZSM5 resulted in a strong interaction between the indium species and the protonic acid sites of the zeolite framework producing a great decrease of protonic acidity and the formation of highly dispersed indium species, which were suggested to be the active centers for the SCR of NO with CH_4 in the presence of O_2. In spite of the evident correlation between the acidity and catalytic performances of indium-based zeolites, the microcalorimetric studies are covered in a very limited extent in the literature.

The catalytically active sites of isomorphous substituted MFI structures (Al-Sil and In-Sil) have been characterized by infrared spectroscopy and microcalorimetric measurements using ammonia and acetonitrile as probes [261]. The first derivative of the heat of adsorption curves, dQ/da, as function of the loading, a, gave maxima at about 140 and 100 kJ/mol for Al-Sil and In-Sil, respectively. Jänchen and colleagues [261] explained these decreasing values by decreasing acidic strength of the bridging OH. The incorporation of trivalent cations with increasing ion radius into the MFI lattice resulted in decreasing acidic Brønsted centers.

13.5   CONCLUSIONS

In this chapter, a brief summary of studies that made use of calorimetry to characterize compounds comprising group IIIA elements (zeolites, nitrides, and oxides catalysts) was presented. It was demonstrated that adsorption microcalorimetry can be used as an efficient technique to characterize the acid–base strength of different types of materials and to provide information consistent with the catalytic data.

It was proven that microcalorimetry technique is quite well developed and very useful in providing information on the strength and distribution of acidic and basic sites of catalysts. When interpreting calorimetric data, caution needs to be exercised. In general, one must be careful to determine if the experiments are conducted under such conditions that equilibration between the probe molecules and the adsorption sites can be attained. By itself, calorimetry only provides heats of interaction. It does not provide any information about the molecular nature of the species involved. Therefore, other complementary techniques should be used to help interpreting the calorimetric data. For example, IR spectroscopy needs to be used to determine whether a basic probe molecule adsorbs on a Brønsted or Lewis acid site.

ACKNOWLEDGMENT

Georgeta Postole gratefully thank Professor Vesna Rakic for a fruitful discussion.

REFERENCES

1. Archibong, E. F., and Sullivan, R. *J. Phys. Chem.* 100 (1996): 18078–82.
2. Downs, A. J. *Coord. Chem. Rev.* 189 (1999): 59–100.

3. Roesky, H. W., and Shravan Kumar, S. *Chem. Commun.* (2005): 4027–38.
4. Silberberg, M. S. *CHEMISTRY, The Molecular Nature of Matter and Change.* New York: McGraw-Hill, 2006.
5. Buchin, B., Steinke, T., Gemel, C., Cadenbach, T., and Ficher, R. A. *Z. Anorg. Allg. Chem.* 631 (2005): 2756–62.
6. Miessler, G. L., and Tarr, D. A. *Inorganic Chemistry.* Upper Saddle River, NJ: Pearson Prentice Hall, 2003.
7. Vaucher, S., Kuebler, J., Beffort, O., Biasetto, L., Zordan, F., and Colombo, P. *Compos. Sci. Tech.* 68 (2008): 3202–7.
8. Shaula, A. L., Yaremchenko, A. A., Kharton, V. V., Logvinovich, D. I., Naumovich, E. N., Kovalevsky, A. V., Frade, J. R., and Marques, F. M. B. *J. Membr. Sci.* 221 (2003): 69–77.
9. Gogotsi, Y. G., Yaroshenko, V. P., and Porz, F. *J. Mat. Sci. Lett.* 11 (1992): 308–10.
10. Yang, W., Xie, Z., Ma, J., Miao, H., Luo, J., Zhang, L., and An, L. *J. Am. Ceram. Soc.* 88 (2005): 485–87.
11. Zhang, Y., He, X., Han, J., and Du, S. *J. Mater. Proc. Technol.* 116 (2001): 161–64.
12. Yoon, S. J., and Jha, A. *J. Mater. Sci.* 30 (1995): 607–14.
13. Kumar, A., and Pawar, S. S. *J. Molec. Catal. A: Chem.* 208 (2004): 33–37.
14. Harlan, C. J., Bott, S. G., and Barron, A. R. *J. Am. Chem. Soc.* 117 (1995): 6465–74.
15. Watanabi, M., McMahon, N., Harlan, C. J., and Barron, A. R. *Organometallics* 20 (2001): 460–67.
16. Liacha, M., Yous, S., Poupaert, J. H., Depreux, P., and Aichaoui, H. *Monatsh. Chem. Chem. Mon.* 130 (1999): 1393–97.
17. Komon, Z. J. A., Bazan, G. C., Fang, C., and Bu, X. *Inorg. Chim. Acta* 345 (2003): 95–102.
18. Li, J., Hao, J., Fu, L., Zhu, T., Liu, Z., and Cui, X. *React. Kinet. Catal. Lett.* 80 (2003): 75–80.
19. Satsuma, A., Segawa, Y., Yoshida, H., and Hattori, T. *Appl. Catal. A: Gen.* 264 (2004): 229–36.
20. Gerritsen, G., Duchateau, R., Van Santen, R. A., and Yap, G. P. A. *Organometallics* 22 (2003): 100–10.
21. Basilio de Caland, L., Soares Santos, L. S., Rodarte de Moura, C. V., and Miranda de Moura, E. *Catal. Lett.* 128 (2009): 392–400.
22. Gebauer-Henke, E., Grams, J., Szubiakiewicz, E., Farbotko, J., Touroude, R., and Rynkowski, J. *J. Catal.* 250 (2007): 195–208.
23. Zakharchenko, N. I. *Russ. J. Appl. Chem.* 76 (2003): 1455–60.
24. Park, P. W., Ragle, C. S., Boyer, C. L., Balmer, M. L., Engelhard, M., and McCready, D. *J. Catal.* 210 (2002): 97–105.
25. Mao, D., Lu, G., Chen, Q., Xieb, Z., and Zhang, Y. *Catal. Lett.* 77 (2001): 119–24.
26. Li, C., Chen, Y.-W., Yang, S.-J., and Wu, J.-C. *Ind. Eng. Chem. Res.* 32 (1993): 1573–78.
27. Ramirez, J., Castillo, P., Cedeno, L., Cuevas, R., Castillo, M., Palacios, J. M., and Lopez-Agudo, A. *Appl. Catal. A: Gen.* 132 (1995): 317–34.
28. Shimizu, K., Satsuma, A., and Hattori, T. *Appl. Catal. B: Env.* 16 (1998): 319–26.
29. El-Hakam, S. A., and El-Sharkawy, E. A. *Mater. Lett.* 36 (1998): 167–73.
30. Passos, F. B., Schmal, M., and Vannice, M. A. *J. Catal.* 160 (1996): 118–24.
31. Passos, F. B., Aranda, D. A. G., and Schmal, M. *J. Catal.* 178 (1998): 478–88.
32. Wang, X., Zhang, T., Sun, X., Guan, W., Liang, D., and Lin, L. *Appl. Catal. B: Env.* 24 (2000): 169–73.
33. Berndt, H., Schütze, F.-W., Richter, M., Sowade, T., and Grünert, W. *Appl. Catal. B: Env.* 40 (2003): 51–67.
34. Li, Y., and Armor, J. N. *J. Catal.* 145 (1994): 1–9.

35. Lukyanov, D. B., and Vazhnova, T. *Appl. Catal. A: Gen.* 316 (2007): 61–67.
36. Montes, A., and Giannetto, G. *Appl. Catal. A: Gen.* 197 (2000): 31–39.
37. Grunert, W., Saffert, W., Feldhaus, R., and Anders, K. *J. Catal.* 99 (1986): 149–58.
38. Wu, Z., and Stair, P. C. *J. Catal.* 237 (2006): 220–29.
39. Joshi, Y. V., and Thomson, K. T. *Catal. Today* 105 (2005): 106–21.
40. Perdigon-Melon, J. A., Gervasini, A., and Auroux, A. *J. Catal.* 234 (2005): 421–30.
41. Burch, R., and Watling, T. C. *Appl. Catal. B: Env.* 11 (1997): 207–16.
42. Liu, Z., Hao, J., Fu, L., Zhu, T., Li, J., and Cui, X. *Appl. Catal. B: Env.* 48 (2004): 37–48.
43. Mečárová, M., Miller, N. A., Clark, N. C., Ott, K. C., and Pietraß, T. *Appl. Catal. A: Gen.* 282 (2005): 267–72.
44. Knözinger, H., and Ratnasamy, P. *Catal. Rev. Sci. Eng.* 17 (1978): 31–70.
45. Colorio, G., Védrine, J. C., Auroux, A., and Bonnetot, B. *Appl. Catal. A: Gen.* 137 (1996): 55–68.
46. Ozaki, T., Masui, T., Machida, K.-I., Adachi, G.-Y., Sakata, T., and Mori, H. *Chem. Mater.* 12 (2000): 643–49.
47. Gervasini, A., Perdigon-Melon, J. A., Guimon, C., and Auroux, A. *J. Phys. Chem. B* 110 (2006): 240–49.
48. Zahir, Md. H., Katayama, S., and Awano, M. *Mater. Chem. Phys.* 86 (2004): 99–104.
49. Meziani, M. J., Zajac, J., Jones, D. J., Partyka, S., Roziere, J., and Auroux, A. *Langmuir* 16 (2000): 2262–68.
50. Peil, K. P., Galya, L. G. and Marcelin, G. *J. Catal.* 115 (1989): 441–51.
51. Ducourty, B., Occelli, M. L., and Auroux, A. *Thermochim. Acta* 312 (1998): 27–32.
52. Wolker, A., Hudalla, C., Eckert, H., Auroux, A., and Occelli, M. L. *Solid State NMR* 9 (1997): 143–53.
53. Occelli, M. L., Eckert, H., Wolker, A., and Auroux, A. *Micropor. Mesopor. Mater.* 30 (1999): 219–32.
54. O'Connor, P., Hakuli, A., and Imhof, P. *Stud. Surf. Sci. Catal.* 149 (2004): 305–22.
55. Harding, R. H., Peters, A. W., and Nee, J. R. D. *Appl. Catal. A: Gen.* 221 (2001): 389–96.
56. Sanchez-Castillo, M. A., Agarwal, N., Miller, C., Cortright, R. D., Madon, R. J., and Dumesic, J. A. *J. Catal.* 205 (2002): 67–85.
57. Gates, B. C. *Catalytic Chemistry.* New York: Wiley, 1992.
58. Ferino, I., Monaci, R., Rombi, E., and Solinas, V. *J. Chem. Soc., Faraday Trans.* 94 (1998): 2647–52.
59. Cutrufello, M. G., Ferino, I., Solinas, V., Primavera, A., Trovarelli, A., Auroux, A., and Picciau, C. *Phys. Chem. Chem. Phys.* 1 (1999): 3369–75.
60. Auroux, A., Artizzu, P., Ferino, I., Solinas, V., Leofanti, G., Padovan, M., Messina, G., and Mansani, R. *J. Chem. Soc., Faraday Trans.* 91 (1995): 3263–67.
61. Gravelle, P. C. *Thermochim. Acta* 96 (1985): 365–76.
62. Auroux, A. *Top. Catal.* 4 (1997): 71–89.
63. Xia, X., Naumann d'Alnoncourt, R., Strunk, J., Litvinov, S., and Muhler, M. *J. Phys. Chem. B* 110 (2006): 8409–15.
64. Spiewak, B. E., and Dumesic, J. A. *Thermochim. Acta* 290 (1996): 43–53.
65. Ostrovskii, V. E. *Thermochim. Acta* 489 (2009): 5–21.
66. Llewellyn, P. L., and Maurin, G. *C. R. Chimie* 8 (2005): 283–302.
67. Busca, G. *Phys. Chem. Chem. Phys.* 1 (1999): 723–36.
68. Luder, W. F. *Chem. Rev.* 27 (1940): 547–83.
69. Busca, G. *Chem. Rev.* 107 (2007): 5366–5410.
70. Zecchina, A., Lamberti, C., and Bordiga, S. *Catal. Today* 41 (1998): 169–77.
71. Arrhenius, S. A. *Z. Phys. Chem.* 1 (1887): 631–48.
72. Brønsted, J. N. *Recl. Trav. Chim. Pays-Bas* 42 (1923): 718–28.

73a. Lowry, T. M. *Chim. Ind.* (London) 42 (1923): 43–47.
73b. Lowry, T. M. *Trans. Faraday Soc.* 20 (1924): 13–15.
74. Lewis, G. N. *Valency and Structure of Atoms and Molecules.* New York: Wiley, 1923.
75. Pearson, R. G. *J. Am. Chem. Soc.* 85 (1963): 3533–39.
76. Védrine, J. C. *Top. Catal.* 21 (2002): 97–106 and references therein.
77. Smith, D. W. *J. Chem. Educ.* 64 (1987): 480–81.
78. Tanabe, K., Misono, M., Ono, Y., and Hattori, H. *Stud. Surf. Sci. Catal.* 51 (1989): 1–213.
79. Reddy B. M. *Metal Oxides: Chemistry and Applications.* Edited by J. L. G. Fierro, 215–46. Boca Raton, FL: CRC Press, Taylor & Francis, 2006.
80. Busca, G. *Metal Oxides: Chemistry and Applications.* Edited by J. L. G. Fierro, 247–318. Boca Raton, FL: CRC Press, Taylor & Francis, 2006.
81. Hattori, H. *Appl. Catal. A: Gen.* 222 (2001): 247–59.
82. Kung, H. H. *Stud. Surf. Sci. Catal.* 45 (1989): 1–277.
83. Jehng, J.-M., Deo, G., Weckhuysen, B. M., and Wachs, I. E. *J. Molec. Catal. A: Chem.* 110 (1996): 41–54.
84. Caldararu, M., Postole, G., Hornoiu, C., Bratan, V., Dragan, M., and Ionescu, N. I. *Appl. Surf. Sci.* 181 (2001): 255–64.
85. Pawelec, B. *Metal Oxides: Chemistry and Applications.* Edited by J. L. G. Fierro, 111–31. Boca Raton, FL: CRC Press, Taylor & Francis, 2006.
86. Wachs, I. E. *Metal Oxides: Chemistry and Applications.* Edited by J. L. G. Fierro, 1–30. Boca Raton, FL: CRC Press, Taylor & Francis, 2006.
87. Levin, I., and Brandon, D. *J. Am. Ceram. Soc.* 81 (1998): 1995–2012.
88. Wolverton, C., and Hass, K. C. *Phys. Rev. B* 63 (2001): 24102–16.
89. Wilson, S. J., and Mc Connell, J. D. C. *J. Solid State Chem.* 34 (1980): 315–22.
90. Peri, J. B. *J. Phys. Chem.* 69 (1965): 220–30.
91. Knozinger, H. *Stud. Surf. Sci. Catal.* 20 (1985): 111–25.
92. Busca, G. *Metal Oxide Catalysis.* Edited by S. D. Jackson and J. S. J. Hargreaves, 95–175. Weinheim, Germany: Wiley-VCH Verlag GmbH & Co., 2009.
93. Lambert, J. F., and Che, M. *J. Molec. Catal. A: Chem.* 162 (2000): 5–18.
94. Boehm, H. P. *Avd. Catal.* 16 (1966): 179–274.
95. Peri, J. B. *J. Phys. Chem.* 69 (1965): 211–20.
96. Tsyganenko, A., and Filimonov, V. N. *Spectrosc. Lett.* 5 (1972): 477–87.
97. Morterra, C., and Magnacca, G. *Catal. Today* 27 (1996): 497–532.
98. Busca, G., Lorenzelli, V., Escribano, V. S., and Guidetti, R. *J. Catal.* 131 (1991): 167–77.
99. Busca, G., Lorenzelli, V., Ramis, G., and Willey, R. J. *Langmuir* 9 (1993): 1492–99.
100. Tsyganenko, A. A., and Mardilovich, P. P. *J. Chem. Soc., Faraday Trans.* 92 (1996): 4843–52.
101. Liu, X., and Truitt, R. E. *J. Am. Chem. Soc.* 119 (1997): 9856–60.
102. Stockenhuber, M. *Metal Oxide Catalysis.* Edited by S. D. Jackson and J. S. J. Hargreaves, 299–322. Weinheim, Germany: Wiley-VCH Verlag GmbH & Co., 2009.
103. Auroux, A. *Molecular Sieves—Science and Technology: Acidity and Basicity, Determination by Adsorption Microcalorimetry,* 45–152. New York: Springer Verlag, 2008.
104. Humphries, A., Harris, D. H., and O'Connor, P. *Stud. Surf. Sci. Catal.* 76 (1993): 41–82.
105. Shen, J., and Auroux, A. *Stud. Surf. Sci. Catal.* 149 (2004): 35–70.
106. Barthomeuf, D. *J. Phys. Chem.* 83 (1979): 249–56.
107. Jacobs, P. A. *Catal. Rev. Sci. Eng.* 24 (1982): 415–40.
108. Pine, L. A., Maher, P. J., and Wachter, W. A. *J. Catal.* 85 (1984): 466–76.
109. Chen, D. T., Sharma, S. B., Filimonov, I., and Dumesic, J. A. *Catal. Lett.* 12 (1992): 201–11.

110. Huang, M., Kaliaguine, S., Muscas, M., and Auroux, A. *J. Catal.* 157 (1995): 266–69.
111. Sun, C., and Berg, J. C. *Adv. Colloid Interface Sci.* 105 (2003): 151–75.
112. Corma, A. *Chem. Rev.* 95 (1995): 559–614.
113. Abbattista, F., Delmastro, S., Gozzelino, G., Mazza, D., Vallino, M., Busca, G., Lorenzelli, V., and Ramis, G. *J. Catal.* 117 (1989): 42–51.
114. Morterra, C., Chiorino, A., Ghiotti, G., and Garrone, E. *J. Chem. Soc., Faraday Trans.* I 75 (1979): 271–88.
115. Morterra, C., Coluccia, S., Chiorino, A., and Boccuzzi, F. *J. Catal.* 54 (1978): 348–64.
116. Guimon, C., and Martinez, H. *Recent Res. Devel. Catal.* 2 (2003): 99–120.
117. Gorte, R. J. *Catal. Lett.* 62 (1999): 1–13.
118. Cardona-Martinez, N., and Dumesic, J. A. *Adv. Catal.* 38 (1992): 149–244.
119. Handy, B. H., Sharma, S. B., Spiewak, B. E., and Dumesic, J. A. *Meas. Sci. Technol.* 4 (1993): 1350–56.
120. Spiewak, B. E., and Dumesic, J. A. *Thermochim. Acta* 312 (1998): 95–104.
121. Auroux, A. *Top. Catal.* 19 (2002): 205–13.
122. Babitz, S. M., Williams, B. A., Kuehne, M. A., Kung, H. H., and Miller, J. T. *Thermochim. Acta* 312 (1998): 17–25.
123. Parrillo, D. J., and Gorte, R. J. *Thermochim. Acta* 312 (1998): 125–32.
124. Gorte, R. J., and White, D. *Top. Catal.* 4 (1997): 57–69.
125. Cardona-Martinez, N., and Dumesic, J. A. *J. Catal.* 127 (1991): 706–18.
126. Everett, D. H. *Pure Appl. Chem.* 31 (1972): 579–638.
127. Cerofolini, G. F., and Rudzifiski, W. *Stud. Surf. Sci. Catal.* 104 (1997): 1–104.
128. Della Gatta, G. *Thermochim. Acta.* 96 (1985): 349–63.
129. Bruzzone, G. *Thermochim. Acta* 96 (1985): 239–58.
130. Hansen, L. D., and Russell, D. J. *Thermochim. Acta* 450 (2006): 71–72.
131. Gale, R. J., Haber, J., and Stone, F. S. *J. Catal.* 1 (1962): 32–38.
132. Vass, M. I., and Budrugeac, P. *J. Catal.* 64 (1980): 68–73.
133. Zarifyanz, Y. A., Kiselev, V. F., Lezhner, N. N., and Nikitina, O. V. *Carbon* 5 (1967): 127–35.
134. Gravelle, P. C. *Catal. Rev. Sci. Eng.* 16 (1977): 37–110.
135. Gravelle, P. C. *Adv. Catal.* 22 (1972): 191–262.
136. Hsieh, P. Y. *J. Catal.* 2 (1963): 211–22.
137. Wahba, M., and Kemball, C. *Trans. Fataday Soc.* 49 (1953): 1351–60.
138. Stone, F. S., and Whalley, L. *J. Catal.* 8 (1967): 173–82.
139. Calvet, E., and Prat H. *Microcalorimétrie, Applications Physicochimiques et Biologiques.* Paris: Masson, 1956.
140. Tian, A. *C. R. Hebd. Seances Acad. Sci.* 178 (1924): 705.
141. Sircar, S., Mohr, R. J., Ristic, C., and Rao, M. B. *J. Phys. Chem.* 103 (1999): 6539–46.
142. Dunne, J., Mariwala, R., Rao, M. B., Sircar, S., Gorte, R. J., and Myers, A. L. *Langmuir* 12 (1996): 5888–5904.
143. Dunne, J., Rao, M. B., Sircar, S., Gorte, R. J., and Myers, A. L. *Langmuir* 12 (1996): 5896–5904.
144. Dunne, J., Rao, M. B., Sircar, S., Gorte, R. J., and Myers, A. L. *Langmuir* 13 (1997): 4333–41.
145. Auroux, A. *Catalyst Characterization: Physical Techniques for Solid Materials.* Edited by B. Imelik and J. C. Védrine, 611–50. New York: Plenum Press, 1994.
146. Solinas, V., and Ferino, I. *Catal. Today* 41 (1998): 179–89.
147. Bennici, S., and Auroux, A. *Metal Oxides Catalysis.* Edited by S. D. Jackson and J. S. J. Hargreaves, 391–441. Weinheim, Germany: Wiley-VCH Verlag GmbH & Co., 2009.
148. O'Neil, M., Lovrien, R., and Phillips, J. *Rev. Sci. Instrum.* 56 (1985): 2312–18.
149. Mason, M. G., and Baetzold, R. C. *J. Chem. Phys.* 64 (1976): 271–76.
150. Baetzold, R. C. *J. Phys. Chem.* 80 (1976): 1504–9.

151. Siril, P. F., Davison, A. D., Randhawa, J. K., and Brown D. R. *J. Molec. Catal. A: Chem.* 267 (2007): 72–78.
152. Stach, H., Lohse, U., Thamm, H., and Schirmer, W. *Zeolites* 6 (1986): 74–90.
153. Andersen, P. J., and Kung, H. H. *Catalysis* 11 (1994): 441–66.
154. Farneth, W. E., and Gorte, R. J. *Chem. Rev.* 95 (1995): 615–35.
155. Damjanović, Lj., and Auroux, A. *Handbook of Thermal Analysis and Calorimetry, Further Advances, Techniques and Applications.* Edited by M. Brown and P. Gallagher, 387–438. Amsterdam: Elsevier, 2007.
156. Ruthven, D. M. *Principles of Adsorption and Adsorption Processes.* New York: Wiley, 1984.
157. Postole, G., Bennici, S., and Auroux, A. *Appl. Catal. B: Env.* 92 (2009): 307–17.
158. Rakic, V., Rac, V., Dondur, V., and Auroux, A. *Catal. Today* 110 (2005): 272–80.
159. Klyachko, A. L., Brueva, T. R., Mishin, I. V., Kapustin, G. I., and Rubinshtein, A. M. *Acta Phys. Chem.* 24 (1978): 183–88.
160. Gervasini, A., and Auroux, A. *J. Phys. Chem.* 97 (1993): 2628–39.
161. Auroux, A., Huang, M., and Kaliaguine, S. *Langmuir* 12 (1996): 4803–7.
162. Auroux, A., Sayed, M. B., and Védrine, J. C. *Thermochim. Acta* 93 (1985): 557–60.
163. King, D. A., and Woodruff, D. P. *Adsorption on Solid Surfaces.* Amsterdam: Elsevier, 1983.
164. Coker, E. N., and Karge, H. G. *Rev. Sci. Instrum.* 68 (1997): 4521–24.
165. Biaglow, A. I., Parrillo, D. J., Kokotailo, G. T., and Gorte, R. J. *J. Catal.* 148 (1994): 213–23.
166. Barthomeuf, D. *Mater. Chem. Phys.* 18 (1988): 553–75.
167. Sanderson, R. T. *Inorganic Chemistry.* New York: Reinhold Publishing Corporation, 1967.
168. Cucinieri Colorio, G., Auroux, A., and Bonnetot, B. *J. Therm. Anal. Cal.* 40 (1993): 1267–76.
169. Le Bars, J., Védrine, J. C., Auroux, A., Trautmann, S., and Baerns, M. *Appl. Catal. A: Gen.* 119 (1994): 341–54.
170. Occelli, M. L., Biz, S., Auroux, A., and Iyer, P. S. *Appl. Catal. A: Gen.* 179 (1999): 117–29.
171. Gervasini, A., Fenyvesi, J., and Auroux, A. *Langmuir* 12 (1996): 5356–64.
172. Auroux, A., Monaci, R., Rombi, E., Solinas, V., Sorrentino, A., and Santacesaria, E. *Thermochim. Acta* 379 (2001): 227–31.
173. Auroux, A., and Gervasini, A. *J. Phys. Chem.* 94 (1990): 6371–79.
174. Gervasini, A., and Auroux, A. *J. Therm. Anal. Cal.* 37 (1991): 1737–44.
175. Gervasini, A., Bellussi, G., Fenyvesi, J., and Auroux, A. *J. Phys. Chem.* 99 (1995): 5117–25.
176. Zou, H., Ge, X., and Shen, J. *Thermochim. Acta* 397 (2003): 81–86.
177. Spiewak, B. E., Handy, B. E., Sharma, S. B., and Dumesic, J. A. *Catal. Lett.* 23 (1994): 207–13.
178. Zou, H., and Shen, J. *Thermochim. Acta* 351 (2000): 165–70.
179. Petre, A. L., Auroux, A., Gervasini, A., Caldararu, M., and Ionescu, N. I. *J. Therm. Anal. Cal.* 64 (2001): 253–60.
180. Petre, A. L., Auroux, A., Gélin, P., Caldararu, M., and Ionescu, N. I. *Thermochim. Acta* 379 (2001): 177–85.
181. Bailbs, M., and Stone, F. S. *Catal. Today* 10 (1991): 303–13.
182. Auroux, A., Muscas, M., Coster, D. J., and Fripiat, J. J. *Catal. Lett.* 28 (1994): 179–86.
183. Carniti, P., Gervasini, A., and Auroux, A. *J. Catal.* 150 (1994): 274–83.
184. Gervasini, A., Bellussi, G., Fenyvesi, J., and Auroux, A. New Frontiers in Catalysis. In *Proceedings of the 10th International Congress on Catalysis.* Edited by L. Guczi, et al., 2047–50. Amsterdam: Elsevier, 1993.

185. Coster, D. J., Fripiat, J. J., Muscas, M., and Auroux, A. *Langmuir* 11 (1995): 2615–20.
186. Gergely, B., and Auroux, A. *Res. Chem. Intermed.* 25 (1999): 13–24.
187. Tanabe, K. *Catalysis, Science and Technology.*, Edited by J. R. Anderson and M. Boudart, 232–71. Berlin: Springer, 1987.
188. Chen, Y., and Zhang, L. *Catal. Lett.* 12 (1992): 51–62.
189. Berteau, P., and Delmon, B. *Catal. Today* 5 (1989): 121–37.
190. Turek, A. M., Wachs, I. E., and DeCanio, E. *J. Phys. Chem.* 96 (1992): 5000–5007.
191. Youssef, N. A., and Youssef, A. M. *Bull. Soc. Chim. Fr.* 128 (1991): 864.
192. Gervasini, A., and Auroux, A. *J. Catal.* 131 (1991): 190–98.
193. Petre, A. L., Perdigon-Melon, J. A., Gervasini, A., and Auroux, A. *Top. Catal.* 19 (2002): 271–81.
194. Cucinieri Colorio, G., Auroux, A., and Bonnetot, B. *J. Therm. Anal. Cal.* 38 (1992): 2565–73.
195. Sato, S., Kuroki, M., Sodesawa, T., Nozaki, F., and Maciel, G. E. *J. Molec. Catal. A: Chem.* 104 (1995): 171–77.
196. Gergely, B., Redey, A., Guimon, C., Gervasini, A., and Auroux, A. *J. Therm. Anal. Cal.* 56 (1999): 1233–41.
197. Pushkar, Y. N., Sinitsky, A., Parenago, O. O., Kharlanov, A. N., and Lunina, E. V. *Appl. Surf. Sci.* 167 (2000): 69–78.
198. Maunula, T., Kintaichi, Y., Inaba, M., Haneda, M., Sato, K., and Hamada, H. *Appl. Catal. B: Env.* 15 (1998): 291–304.
199. Nakagawa, K., Kajita, C., Ide, Y., Okamura, M., Kato, S., Kasuya, H., Ikenaga, N., Kobayashi, T., and Suzuki, T. *Catal. Lett.* 64 (2000): 215–21.
200. Petre, A. L., Perdigon-Melon, J. A., Gervasini, A., and Auroux, A. *Catal. Today* 78 (2003): 377–86.
201. Postole, G., Gervasini, A., Caldararu, M., Bonnetot, B., and Auroux, A. *Appl. Catal. A: Gen.* 325 (2007) 227–36.
202. Yuzhakova, T., Rakic, V., Guimon, C., and Auroux, A. *Chem. Mater.* 19 (2007): 2970–81.
203. Bonnetot, B., Rakic, V., Yuzhakova, T., Guimon, C., and Auroux, A. *Chem. Mater.* 20 (2008): 1585–96.
204. Delsarte, S., Maugé, F., Lavalley, J.-C., and Grange, P. *Catal. Lett.* 68 (2000): 79–83.
205. Delsarte, S., Maugé, F., and Grange, P. *J. Catal.* 202 (2001): 1–13.
206. Massinon, A., Guéguen, E., Conanec, R., Marchand, R., Laurent, Y., and Grange, P. *Stud. Surf. Sci. Catal.* 101 (1996): 77–85.
207. Benitez, J. J., Diaz, A., Laurent, Y., and Odriozola, J. A. *Catal. Lett.* 54 (1998): 159–64.
208. Centeno, M.-A., Delsarte, S., and Grange, P. *J. Phys. Chem. B* 103 (1999): 7214–21.
209. Peltier, V., Conanec, R., Marchand, R., Laurent, Y., Delsarte, S., Guéguen, E., and Grange, P. *Mater. Sci. Eng. B* 47 (1997): 177–83.
210. Delsarte, S., Florea, M., Maugeé, F., and Grange, P. *Catal. Today* 116 (2006): 216–25.
211. Delsarte, S., Auroux, A., and Grange, P. *Phys. Chem. Chem. Phys.* 2 (2002): 2821–27.
212. Delsarte, S., Peltier, V., Laurent, Y., and Grange, P. *Stud. Surf. Sci. Catal.* 118 (1998): 869–78.
213. Howden, M. G. *Zeolites* 5 (1985): 334–38.
214. Woolery, G. L., Kuehl, G. H., Timken, H. C., Chester, A. W., and Vartuli, J. C. *Zeolites* 19 (1997): 288–96.
215. Mavrodinova, V. P., Popova, M. D., Neinska, Y. G., and Minchev, C. I. *Appl. Catal. A: Gen.* 210 (2001): 397–408.
216. Janin, A., Maache, M., Lavalley, J. C., Joly, J. F., Raatz, F., and Szydlowski, N. *Zeolites* 11 (1991): 391–96.
217. Cardona-Martinez, N., and Dumesic, J. A. *J. Catal.* 125 (1990): 427–44.

218. Gayer, F. H. *J. Ind. Eng. Chem.* 25 (1933): 1122–27.
219. Whitmore, F. C. *J. Ind. Eng. Chem.* 26 (1934): 94–95.
220. Beagley, B., Dwyer, J., Fitch, F. R., Mann, R., and Walters, J. *J. Phys. Chem.* 88 (1984): 1744–51.
221. Occelli, M. L., Olivier, J. P., Petre, A., and Auroux, A. *J. Phys. Chem. B* 107 (2003): 4128–36.
222. Occelli, M. L., Auroux, A., Baldiraghi, F., and Leoncini, S. *Fluid Cracking Catalysts.* Edited by M. L. Occelli and P. O'Connor, 203–16. New York: Marcel Dekker, 1998.
223. Occelli, M. L., Kalwei, M., Wölker, A., Eckert, H., Auroux, A., and Gould, S. A. C. *J. Catal.* 196 (2000): 134–48.
224. Auroux, A., and Ben Taarit, Y. *Thermochim. Acta* 122 (1987): 63–70.
225. Shannon, R. D., Gardner, K. H., Staley, R. H., Bergeret, G., Gallezot, P., and Auroux, A. *J. Phys. Chem.* 89 (1985): 4778–88.
226. Boréave, A., Auroux, A., and Guimon, C. *Micropor. Mater.* 11 (1997): 275–91.
227. Macedo, A., Auroux, A., Raatz, F., Jacquinot, E., and Boulet, R. *Perspectives in Molecular Sieve Science. ACS Symposium Series 368.* Edited by W. H. Flank and T. E. Whyte, 98–116. Washington, DC: ACS, 1988.
228. Mitani, Y., Tsutsumi, K., and Takahashi. H. *Bull. Chem. Soc. Japan* 56 (1983): 1921–23.
229. Stach, H., Wendt, R., Lohse, U., Jänchen, J., and Spindler, H. *Catal. Today* 3 (1988): 431–36.
230. Shi, Z. C., Auroux, A., and Ben Taarit, Y. *Can. J. Chem.* 66 (1988): 1013–17.
231. Colon, G., Ferino, I., Rombi, E., Selli, E., Forni, L., Magnoux, P., and Guisnet, M. *Appl. Catal. A: Gen.* 168 (1998): 81–92.
232. Chen, D., Sharma, S., Cardona-Martinez, N., Dumesic, J. A., Bell, V. A., Hodge, G. D., and Madon, R. J. *J. Catal.* 136 (1999): 392–402.
233. Babitz, S. M., Kuchne, M. A., Kung, H. H., and Miller, J. T. *Ind. Eng. Chem. Res.* 36 (1997): 3027–31.
234. Auroux, A., Bolis, V., Wierzchowski, P., Gravelle, P. C., and Védrine, J. C. *J. Chem. Soc., Faraday Trans II* 75 (1979): 2544–55.
235. Védrine, J. C., Auroux, A., and Coudurier, G. *Catalytic Materials, Relationship Between Structure and Reactivity. ACS Symposium Series 248.* Edited by T. E. Whyte, Jr., R. A. Dalla Betta, E. G. Derouane, and R. T. K. Baker, 254–73. Washington, DC: ACS, 1984.
236. Narayanan, S., Sultana, A., Le, Q. T., and Auroux, A. *Appl. Catal. A: Gen.* 168 (1998): 373–84.
237. Gayubo, A. G., Benito, P. L., Aguayo, A. T., Olazar, M., and Bilbao, J. *J. Chem. Tech. Biotechnol.* 65 (1996): 186–92.
238. Védrine, J. C., Auroux, A., Coudurier, G., Engelhard, P., Gallez, J. P., and Szabo G. *Proceedings of the 6th International Zeolite Conference.* Edited by D. Olson and A. Bisio, 497–507. Surrey: Butterworths, Guildford, 1983.
239. Auroux, A., Védrine, J. C., and Gravelle, P. C. *Stud. Surf. Sci. Catal.* 10 (1982): 305–22.
240. Fernandez, C., Auroux, A., Védrine, J. C., Grosmangin, J., and Szabo G. *Stud. Surf. Sci. Catal.* 28 (1986): 345–50.
241. Auroux, A., and Occelli, M. L. *Stud. Surf. Sci. Catal.* 84 (1994): 693–99.
242. McQueen, D., Chiche, B. H., Fajula, F., Auroux, A., Guimon, C., Fitoussi, F., and Schulz, P. *J. Catal.* 161 (1996): 587–96.
243. Zheng, J., Song, C., Xu, X., Turaga, U. T., and Zhao, X. S. *Ser. Chem. Eng.* 4 (2004): 464–86.
244. Occelli, M. L., Biz, S., Auroux, A., and Ray, G. J. *Micropor. Mesopor. Mater.* 26 (1998): 193–213.

245. d'Arbonneau, S., Tuel, A., and Auroux, A. *J. Therm. Anal. Cal.* 56 (1999): 287–96.
246. Sayed, M., Auroux, A., and Védrine, J. C. *J. Catal.* 116 (1989): 1–10.
247. Coudurier, G., Auroux, A., Védrine, J. C., Farlee, R. D., Abrams, L., and Shannon, R. D. *J. Catal.* 108 (1987): 1–14.
248. Sayed, M. B., Auroux, A., and Védrine, J. C. *Appl. Catal.* 23 (1986): 49–61.
249. Occelli, M. L., Eckert, H., Hudalla, C., Auroux, A., Ritz, P., and Iyer, P. S. *Stud. Surf. Sci. Catal.* 105 (1997): 1981–87.
250. Occelli, M. L., Schwering, G., Fild, C., Eckert, H., Auroux, A., and Iyer, P. S. *Micropor. Mesopor. Mater.* 34 (2000): 15–22.
251. Auroux, A., Tuel, A., Bandiera, J., Ben Taarit, Y., and Guil, J. M. *Appl. Catal. A: Gen.* 93 (1993): 181–90.
252. Wolker, A., Hudalla, C., Eckert, H., Auroux, A., and Occelli, M. L. *Solid State NMR* 9 (1997): 143–53.
253. Occelli, M. L., Schweizer, A. E., Fild, C., Schwering, G., Eckert, H., and Auroux, A. *J. Catal.* 192 (2000): 119–27.
254. Dumitriu, E., Hulea, V., Fechete, I., Catrinescu, C., Auroux, A., Lacaze, J. F., and Guimon, C. *Appl. Catal. A: Gen.* 181 (1999): 15–28.
255. Occelli, M. L., Eckert, H., Hudalla, C., Auroux, A., Ritz, P., and Iyer, P. S. *Micropor. Mater.* 10 (1997): 123–35.
256. Requejo, F. G., Ramallo-López, J. M., Ledea, E. J., Miró, E. E., Pierella, L. B., and Anunziata, O. A. *Catal. Today* 54 (1999): 553–58.
257. Halfász, J., Kónya, Z., Fudala, Á., Béres, A., and Kiricsi, I. *Catal. Today* 31 (1996): 293–304.
258. Mihalyi, R. M., Beyer, H. K., Mavrodinova, V., Minchev, Ch., and Neinska, Y. *Micropor. Mesopor. Mater.* 24 (1998): 143–51.
259. Neinska, Y., Mihalyi, R. M., Mavrodinova, V., Minchev, Ch., and Beyer, H. K. *Phys. Chem. Chem. Phys.* 1 (1999): 5761–65.
260. Zhou, X., Xu, Z., Zhang, T., and Lin, L. *J. Molec. Catal. A: Chem.* 122 (1997): 125–29.
261. Jänchen, J., Vorbeck, G., Stach, H., Parlitz, B., and Van Hooff, J. H. C. *Stud. Surf. Sci. Catal.* 94 (1995): 108–15.

14 EPA Consent Decree Implementation

Jeffrey A. Sexton

CONTENTS

14.1 INTRODUCTION

The Clean Air Act (CAA), passed in 1970, created a national program to control the damaging effects of air pollution. The CAA Amendments of 1990 went further to ensure that the air Americans breathe is safe. The CAA protects and enhances the quality of the nation's air by regulating stationary and mobile sources of air emissions.

The CAA requires major stationary sources to install pollution control equipment and to meet specific emissions limitations. In addition, the 1990 CAA amendments required major stationary sources to obtain operating permits. Examples of stationary sources include manufacturers, processors, refiners, and utilities.

For fluidized catalytic cracking (FCC) units in the United States, the CAA created several regulatory requirements and emission standards including but not limited to the following:

- New Source Review Regulation (40 CFR 52.21)
- National Emission Standards for Hazardous Air Pollutants for Petroleum Refineries (40 CFR 63 Subpart CC)
- New Source Performance Standards for Petroleum Refineries (40 CFR 60 Subpart J & Ja)
- State Implementation Plans (SIPs)

EPA's civil enforcement programs are designed to assure compliance with these and other federal environmental laws. Civil enforcement includes the investigations and cases brought to address the most significant violations, and includes EPA administrative actions and judicial cases referred to the Department of Justice. EPA works closely with states that share responsibility for implementing federal programs, as well as with tribes and federal agencies. Civil enforcement actions serve a number of important goals, such as returning violators to compliance, eliminating or preventing environmental harm, deterring others from misconduct, and preserving a level playing field for responsible companies that work hard to abide by the law. The Agency emphasizes those actions that reduce the most significant risks to human health or the environment, and consults extensively with states and other stakeholders in determining risk-based priorities. For over two decades, EPA's enforcement programs have made a measurable contribution to reducing the amount of pollution that goes into the air we breathe or the water we drink, and by encouraging safer handling of hazardous waste and toxic materials.

14.2   EPA REFINERY INITIATIVE

The EPA had long conducted inspections and taken enforcement actions in the refining industry. They began to focus significant attention on refinery compliance concerns in 1996 when refineries became an enforcement priority due to the high rate of noncompliance and pollutant releases. The EPA reported in 1996 that the refining sector had the highest inspection-to-enforcement ratio of the 29 industry sectors ranked by the EPA. In 1996, the EPA ranked the refining industry #1 for releases of volatile organic compounds (VOCs), #2 for SO_x, #3 for NO_x, #4 for particulate matter (PM), and #5 for CO out of 496 total sectors [1]. This was largely due to the complexity and difficulty interpreting the multiple regulatory requirements for the refining industry. The refining industry designation as an enforcement priority meant that industry received special emphasis from the EPA.

The EPA began working with regional offices to explore ways to address compliance issues with marginal success. In 2000, the EPA shifted the refining program focus to pursue voluntary global settlements with refining companies resulting in consent decrees.

This was a significant turning point of the EPA's strategy. The "Refining Initiative" had the expressed goal to have 80% of the refining industry enter into voluntary consent decrees by 2005. The results as of December 2009 include:

- Since March 2000, the Agency has entered into 24 settlements with U.S. companies responsible for nearly 88% of the nation's petroleum refining capacity or more than 14,843,000 barrels per day.
- These settlements cover 99 refineries in 29 states and on full implementation will result in annual emissions reductions of more than 87,000 tons of nitrogen oxides and more than 250,000 tons of sulfur dioxide.
- Negotiations are continuing with other refiners representing an additional 7% of domestic refining capacity and investigations are underway on others.

EPA's investigations focused on the four most significant CAA compliance challenges for this industry and the emissions units that are the source of most of its pollution:

- New Source Review/Prevention of Significant Deterioration
- New Source Performance Standards
- Leak Detection and Repair Requirements
- Benzene National Emissions Standards for Hazardous Air Pollutants

Settling companies have agreed to invest more than $5 billion in control technologies, pay civil penalties of more than $73 million, and perform supplemental environmental projects valued at approximately $67 million.

The EPA's settlements were unique for each facility. In general, they all required the following:

- Significant reductions of nitrogen oxide
- Significant reductions of sulfur dioxide
- Additional emission reductions of benzene, volatile organic compounds, and particulate matter

EPA has reached innovative, multi-issue, multi-facility settlement negotiations with the following major petroleum refining companies:

- BP Exploration and Oil, Inc.
- Chevron USA Inc.
- CHS Inc. (Cenex)
- CITGO
- Coastal Eagle Point Oil Company (CEPOC)
- Conoco, Inc. (preconsolidation refineries only)
- ConocoPhillips
- Ergon Refining Inc.
- ExxonMobil Corporation
- Frontier
- Giant
- Holly Refining
- Hunt Refining
- Koch Industries
- Lion Oil
- Marathon Ashland Petroleum LLC
- Motiva Enterprises LLC/Equilon Enterprises/Deer Park Refining (Shell)
- Navajo Refining Company and Montana Refining Company
- Sinclair Oil Co.
- Sunoco, Inc.
- Total Petrochemicals U.S.A.
- Valero Eagle Refining Company
- Valero (Premcor)
- Wyoming Refining

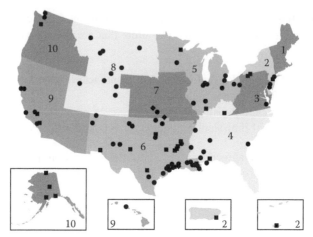

- ● Petroleum refineries under global consent decree
- ■ Petroleum refineries not under global consent decree
- ◆ Petroleum refineries under nonglobal consent decree

FIGURE 14.1   Consent decree status (December 2009).

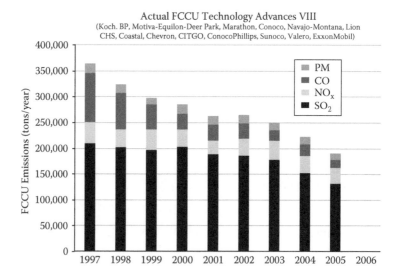

FIGURE 14.2   Overall emissions from consenters.

The status of EPA Consent Decree implementation has been documented previously [2–5]. Figure 14.1 highlights refineries that are covered by consent decrees and those that are not by regions.

The graph in Figure 14.2 shows emission reduction trends for all refineries under consent decrees through 2005.

Marathon Petroleum Company, LLC, a subsidiary of Marathon Oil Company, is the fourth largest U.S.-based integrated oil and gas company with seven refineries

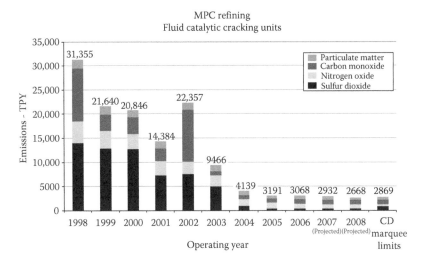

FIGURE 14.3 Overall emissions from Marathon.

located across the Midwestern United States. Marathon entered into a Consent Decree with the U.S. EPA on May 11, 2001. One of the stated objectives in Marathon's Consent Decree was to decrease the level of emissions from each of their seven FCC units. Catalytic additives were required to be used in six of the FCCUs. The level of control was to be similar to levels that could be achieved with hardware modifications with final limits determined through an 18-month demonstration of performance. Although every consent decree is unique, this approach was typical for several of the earlier consenters.

Marathon completed all additive demonstrations and reached agreement on final limits with the EPA by the end of 2006. Figure 14.3 documents the success of the process in lowering emissions for Marathon.

14.3 CONSENT DECREE NEGOTIATIONS

The negotiation of a Consent Decree for a given company and refinery is a complex process that is in principle driven by the strength and severity of the CAA violations alleged against a company and the company's desire to avoid litigation. The EPA negotiated with refining companies and offered an incentive in the form of relief from past liabilities in order to persuade the industry to sign consent decrees. The EPA worked with the industry's desire to obtain a level of certainty regarding regulatory risks with the joint desire of both parties to reduce emissions. As a result, the final consent decrees included controls expected to reduce emissions as well as requirements to go "beyond compliance" with regulations.

Relative to the FCCU, the EPA has targeted goals for SO_2 (25 ppm) and NO_x (20 ppm) long-term emissions limits. In some agreements, the government has also targeted reductions in CO and PM emissions. The refiner's goals are to achieve the emission reductions in a cost-effective manner, while minimizing any

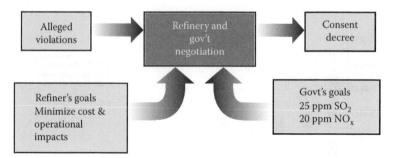

FIGURE 14.4 Consent Decree negotiation overview.

operational impacts of required environmental projects. This process is shown in
Figure 14.4.

The unique aspect of the process is that the EPA did not apply a one-size-fits-all
approach. The refiner and the EPA were able to select technology applications that
were cost-effective and resulted in the greatest reduction in emissions. The exact
emissions limits from FCC units, and the methods by which to achieve them, were
all subject to negotiation. This is why the various consent decrees from each settlement are all slightly different. Earlier consenters were allowed use of new technology and catalyst additives with limits set by demonstration. Later consenters were
typically given hard limits and had to determine the best way to achieve them. The
limits were generally lower since they could take advantage of the lessons learned
from previous consenters and technology applications.

Marathon's strategy was to leverage future clean fuels projects to minimize capital investment and maximize emission reductions. Units with feed hydrotreating
capacity typically used catalyst additives. Units with no hydroprocessing typically
relied upon other technology options in conjunction with catalyst additives. Since
each application was unique, implementation was required over a period of time.
Figure 14.5 shows the timeline when significant events occur for the Marathon
CD.

14.4 CONSENT DECREE IMPLEMENTATION

Consent decrees may specify hardware or additive solutions for individual applications. When a refiner agrees to implement a hardware solution, emissions limits are
typically specified in the Consent Decree. This requires the refiner to design and
implement an appropriately sized unit to meet these limits. With FCC additive solutions or hybrid solutions combining hardware and additives (such as a hydrotreater
and SO_x reduction additive), final emissions limits are not generally defined in the
Consent Decree. Instead, a testing and demonstration program is defined to determine the performance of the additive(s) in the FCC unit at optimized concentrations.
This may also be the case for some hardware solutions. The process to determine the
optimized additive rate and process conditions is also identified. A baseline period
and model is often used to determine additive effectiveness. A series of kick-out factors based upon additive performance are evaluated to determine the optimized level

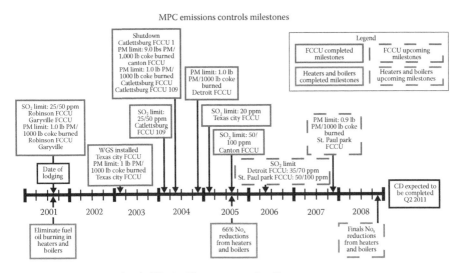

FIGURE 14.5 Marathon's CD significant events timeline.

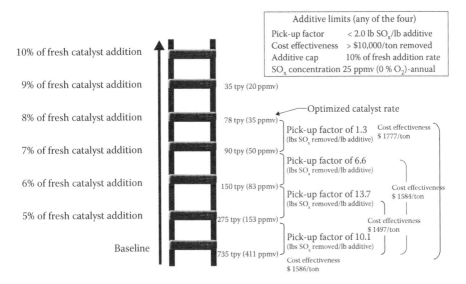

FIGURE 14.6 Typical catalyst additive kick-out ladder.

for demonstration. This ladder approach for a catalyst additive trial is demonstrated in Figure 14.6.

Upon completion of the demonstration, the data are reviewed to determine the final permitted limit. This rigorous data analysis typically bases the final permitted limits on the 95th percentile for the long-term and 99th percentile for the short-term from the demonstration period data. This process is illustrated in Figure 14.7.

Over time the EPA has expanded the specificity of the Consent Decrees as they relate to FCC additive testing and demonstration. The Consent Decrees now generally require:

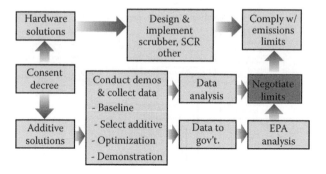

FIGURE 14.7   Implementation of Consent Decrees for FCC units.

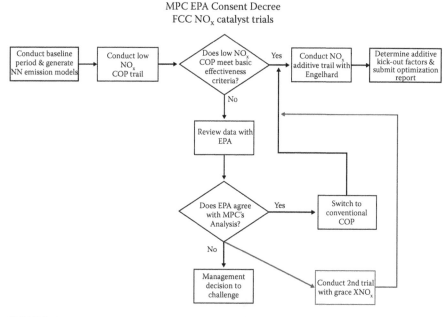

FIGURE 14.8   Marathon's low NO$_x$ COP additive trial protocol.

- EPA approval of additives used, testing protocols, and emissions models
- Comparative performance testing of multiple additives to determine the best performer
- Specific requirements for data collection and reporting

The increased structure and rigor of the Consent Decrees increases the complexity of executing FCC additive testing and demonstrations. A flowchart showing the various phases of the process identified for the Marathon CD is show in Figures 14.8 through 14.10.

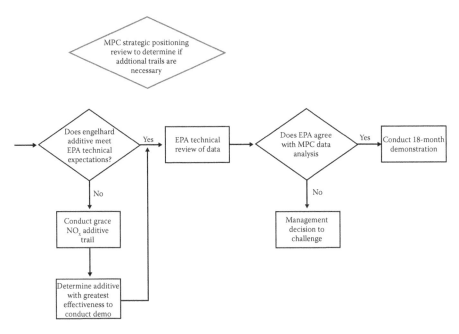

FIGURE 14.9 Marathon's NO_x additive trial protocol.

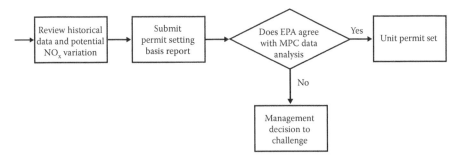

FIGURE 14.10 Marathon's NO_x additive permit setting protocol.

14.5 FCC TECHNICAL TEAM

The majority of refining companies in the United States have voluntarily entered into Consent Decrees with the EPA to reduce environmental emissions from the FCC unit. Although each consent decree is unique, there are common elements amenable to efficiency improvement through industry experience and practice. The consenting companies and the EPA formed a joint FCC Technical Team. The objective of this team is to facilitate implementation of Consent Decrees safely, on-time, and cooperatively while ensuring compliance of Consent Decree limits through optimization and reliability improvements. The goals of this team are:

- Facilitate implementation of and compliance with Consent Decrees
- Reduce learning curves and share lessons learned

- Minimize redundant efforts to improve efficiency
- Minimize emissions through process optimization and reliability improvements to ensure compliance with limits
- Coordinate interpretation and implementation of common elements of each consent decree to minimize interpretative differences
- Identify common issues and where possible common positions for discussion with the EPA to maximize efficiency of all parties
- Share consent decree implementation information including successes, failures, and insights
- Share best practices and lessons learned for implementation of technologies required and employed by the consent decree
- Share best practices and lessons learned for process optimization and reliability of technologies required and employed by the consent decree to ensure long-term compliance of limits
- Share technical information to benchmark catalyst additive performance, control technology performance, and emission limits
- Facilitate improved communications with the EPA

Members of this team jointly meet with the EPA to facilitate these goals. This working group has been a success in improving communications with all parties to implement the consent decrees and meet the stated objectives.

14.6   TECHNOLOGY APPLICATIONS

Several emission control technologies are considered during the refinery specific consent decree negotiations. Figure 14.11 is a schematic of typical options.

The FCC Technical Team summarized all the requirements of the various refinery specific consent decrees. All stated technology applications were considered. Figure 14.12 is a summary of stated technology applications across all consent decrees.

Consent decrees created a demand for technical solutions. Several refiners were encouraged through the CD negotiations to consider new technology. This demand

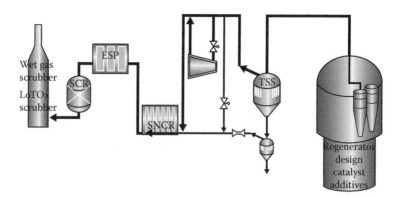

FIGURE 14.11   Typical FCC emission control technology options.

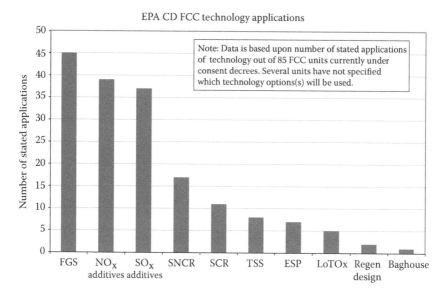

FIGURE 14.12   FCC emission control technologies stated in Consent Decrees.

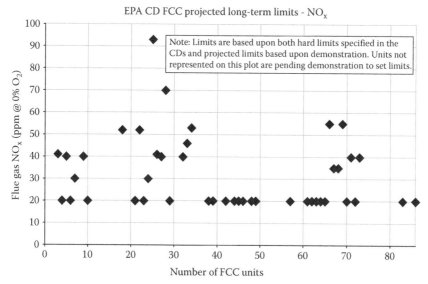

FIGURE 14.13   Long-term NO$_x$ limits set by EPA Consent Decree.

accelerated introduction of new additives and hardware that have been successful at reducing emissions.

14.7   CONSENT DECREE LIMIT SETTING BENCHMARK

Several FCC units have completed technology demonstrations and set permit limits. Figures 14.13 through 14.16 show a benchmark of all FCC units under Consent Decrees with the final limits. (Note all units have not set a limit.)

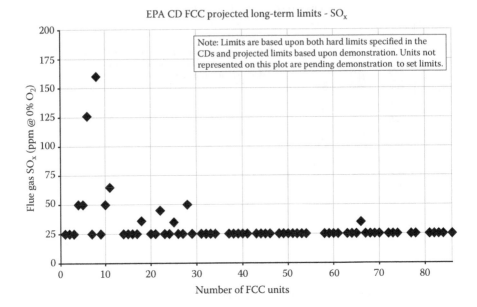

FIGURE 14.14   Long-term SO$_x$ limits set by EPA Consent Decree.

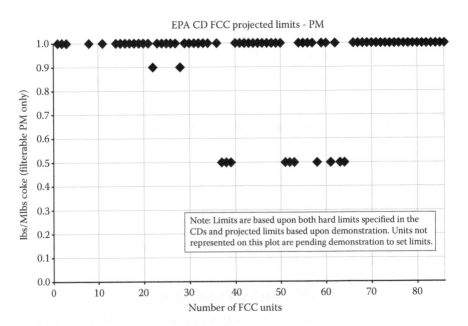

FIGURE 14.15   PM limits set by EPA Consent Decree.

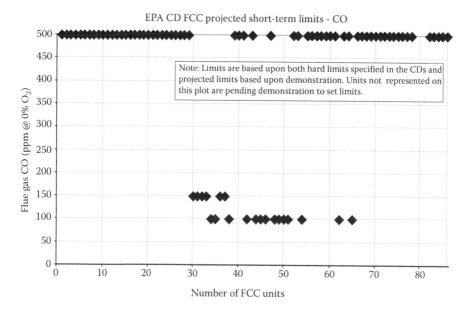

FIGURE 14.16   CO limits set by EPA Consent Decree.

FCC emission reduction technologies for regulated pollutants (SO_x, NO_x, CO, and PM) through consent decrees implementation will be further discussed in Chapters 15 through 18 in this book.

14.8   SUMMARY AND CONCLUSION

Historically, the EPA had largely approached enforcement on a facility-by-facility and issue-by-issue basis. The "Refining Initiative" program represented a radical departure from this practice. There are few industries as complex as the petroleum refining industry and there are few regulatory programs as complex as the CAA. Notwithstanding this complexity, the EPA and the refining industry successfully embraced the global consent decree as a mechanism to secure permanent, consistent compliance with the CAA on a company-wide basis. This was done in a cost-effective manner without a one-size-fits-all approach resulting in a significant reduction in pollutants.

ACKNOWLEDGMENTS

The following individuals and companies are acknowledged for their contributions: Patrick Foley (Senior Environmental Engineer, U.S. EPA), for his input and permission to use information and data. EPA FCC Technical Team Members, for team participation and willingness to share lessons learned in Consent Decree implementation.

REFERENCES

1. U.S.E.P.A. Office of Inspector General, Evaluation Report #2004-P-00021. *EPA Needs to Improve Tracking of National Petroleum Refinery Compliance Program Progress and Impacts.* June 22, 2004.
2. Sexton, J. *EPA Consent Decree FCC Technical Team: Activities & Experience To Date.* Grace Davidson North American Technical Refining Seminar, Washington, DC, June 23, 2005.
3. Sexton, J., Foley, P., and Joyal, C. *U.S. Refining Industry Implementation of EPA Consent Decrees for FCC Units.* Albemarle SCOPE Symposium, Athens, Greece, June 19, 2007.
4. Sexton, J., and Evans, M. *U.S. Refining Industry Implementation of EPA Consent Decrees for FCC Units.* Intercat FCC Seminar, Amsterdam, The Netherlands, October 2, 2007.
5. Sexton, J., Foley, P., and Joyal, C. *EPA Consent Decrees: Progress on FCC Implementation & Future Challenges.* NPRA Annual Meeting, San Antonio, Texas, March 20, 2007.

15 FCC Emission Reduction Technologies through Consent Decree Implementation: Heat Balance Effects on Emissions

Jeffrey A. Sexton

CONTENTS

15.1 INTRODUCTION

The fluid catalytic cracking (FCC) is a very dynamic unit that is typically the major conversion process in a refinery. Proper modeling and understanding of unit capabilities represents a tremendous opportunity to improve the overall unit operation and minimize unit emissions. The combustion chemistry in the FCC regenerator that produces environmental pollutants is extremely complex as numerous interactions and reactions occur between the various chemical species.

15.2 FCC REGENERATOR OPERATION

15.2.1 HEAT BALANCE

The FCC unit (FCCU) operates in heat balance, meaning that the total energy coming into the FCC must equal the energy going out. This has been documented in literature previously [1]. The primary source of energy in the FCCU is coke that forms on the catalyst surface. Combustion of coke in the FCC regenerator supplies the energy to run the cracking process. Coke yield for the cracking of vacuum gas oil is typically about 5 pounds formed per 100 pounds of feed processed. When cracking residual feeds, coke yield can often rise to 10 pounds of coke formed per 100 pounds of feed processed.

The primary consumption of energy in the FCCU is the process heat required to achieve the following:

- Heat and vaporize the feed (40–50% of total heat duty)
- Heat of reaction (15–30% of total heat duty)
- Heat the incoming combustion air (15–25% of total heat duty)
- Heat the incoming dispersion and stripping steam (2–8% of total heat duty)
- Account for unit heat loss (2–5% of total heat duty)

Total unit heat duty will typically be in the range of 500–1000 BTU per pound of feed to the unit. This set of process heat requirements establishes the amount of heat that must be supplied by combustion of coke. Because of the process control schemes that are normally employed in FCCUs, the unit operation will automatically adjust itself so that the energy produced via coke combustion equals the heat requirements of the process. If the balance is shifted by changes to the feed quality or operating conditions, shifts in catalyst circulation rate and regenerator temperature will occur until a new equilibrium set of conditions is established.

As indicated above, the heat produced in the FCC regenerator is ultimately balanced against the requirements of the reaction side. In addition to providing catalytic reaction sites, FCC catalyst also absorbs the heat of combustion and carries it to the riser to provide the heat required to vaporize and crack the feed. As catalyst circulation increases, the amount of heat transfer surface available in the feed vaporization zone increases proportionately.

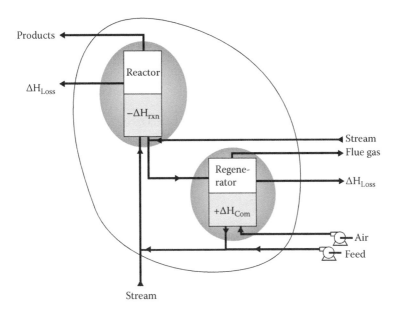

FIGURE 15.1   FCC heat balance envelope.

Understanding FCC heat balance is the key to understanding the interaction between FCC process variables, unit performance, and environmental emissions. Figure 15.1 represents the heat balance envelope of the FCC process.

15.2.2   CARBON BURNING CHEMISTRY

The operating characteristics of the FCC regenerator are dictated by the constraints of the heat balance. Considering the overall energy balance, the burning of coke in the regenerator provides all the energy required to satisfy the heat balance. The combustion of coke in the regenerator is considered a first-order reaction with respect to coke concentration and oxygen partial pressure:

$$dC/dt = -k*C*P_{O_2},$$

where: k = rate constant (hr-1 atm-1),
   C = carbon on catalyst (wt%), and
   P_{O_2} = oxygen partial pressure (atm).

The rate constant will be temperature dependent and can be represented by an Arrhenius equation:

$$dC/dt = -ko*e^{\wedge}(-E/RT)*C*P_{O_2},$$

where: E = activation energy,
   R = ideal gas constant, and
   T = regenerator temperature.

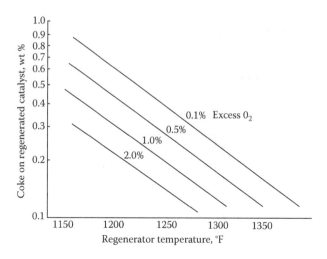

FIGURE 15.2   Catalyst regeneration versus excess oxygen and temperature.

The coke burning rate is a function of temperature, oxygen partial pressure, carbon content of catalyst, and residence time. A typical relationship between these variables is shown in Figure 15.2.

The impact of temperature on the rate of combustion is exponential. The rate increases by a factor of 2.4 going from 1200 to 1300°F. However, the rate increases by factor of 7.2 going from 1200 to 1400°F. The impact of carbon concentration on catalyst is also nonlinear. The relative amount of residence time required to decrease carbon concentration by 0.1% increases by a factor of 10 from an initial concentration of 1.0–0.15 wt%. The impact of oxygen partial pressure is linear. The unit feed rate will also influence coke burning kinetics. As feed is increased, the coke production will increase requiring more air for combustion. Since the bed level is constant, the air residence time in the bed will decrease causing the O_2 concentration in the dilute phase to increase. This will lead to afterburn, which is defined as the combustion of CO to CO_2 in the dilute phase or in the cyclones of the regenerator.

Afterburn is generally the result of oxygen breakthrough on one side of the catalyst bed reacting with the CO and/or hydrocarbon combustibles escaping from another part of the regenerator. These oxidation reactions occur either above the regenerator bed and/or across the regenerator cyclones. In some cases, this afterburning continues in the flue gas line. In this scenario, this breakthrough of oxygen, CO, and hydrocarbon combustibles is largely from uneven contact (nonuniform) of spent catalyst with air [2].

The oxidation of carbon on the catalyst surface proceeds through formation of solid surface oxides that decompose to CO and CO_2 as primary products. Previous studies have shown the CO_2/CO ratio at the catalyst surface is a function of temperature (Arthur's Ratio) and is typically ~1.0 for FCC catalyst and conditions [3]. However, the CO exiting the burn site can be further oxidized to CO_2 at a rate dependent on temperature, CO, O_2, H_2O, active metals on the catalyst, and even the presence of the catalyst itself. Also, transition metal oxides have been found to increase the coke burning rate. In the presence of this metal function, CO oxidation with O_2 to CO_2 occurs readily. In the absence of this function, the oxidation occurs slowly through the dense bed. The oxidation reaction is then accelerated in the dilute phase due to the catalyzing effect of fine decoked metals containing catalysts and the presence of water vapor.

The heat released comes from the reaction of carbon and hydrogen to form CO, CO_2, and H_2O with the following heats of combustion:

$$H_2 + \tfrac{1}{2} O_2 \rightarrow H_2O = 121{,}000 \text{ kJ/kg } H_2,$$
$$C + O_2 \rightarrow CO_2 = 32{,}700 \text{ kJ/kg } C,$$
$$C + \tfrac{1}{2} O_2 \rightarrow CO = 9200 \text{ kJ/kg } C.$$

The heat release from CO_2 combustion is about three times greater than the heat release from carbon to CO, so it is important that this combustion occur in the dense bed of catalyst. Without the catalyst bed to absorb this heat of combustion, the flue gas temperature increases very rapidly in the dilute phase of the regenerator.

In a typical fluid bed regenerator the dense phase catalyst particles are well mixed due to fluidization. Air entering at the bottom is considered to move in a plug flow fashion up the regenerator. As the combustion reactions progress to remove coke from the catalyst, oxygen is consumed and the combustion products (CO and CO_2) change in concentration along the regenerator axis. Changes in O_2 and CO concentration are of particular interest since NO and NO_2 are reduced in the presence of CO. The magnitude of change is expected to be different with different regenerator configurations and cocurrent versus countercurrent flow. In addition to regenerator configuration and mode of operation (partial vs. complete combustion), these compositions will also change with the flue CO/CO_2 ratio (in partial combustion), excess oxygen (in complete combustion), feed type, and other parameters that affect the coke on the catalyst.

The expected concentration profiles for CO and O_2 with regenerator elevation in a well mixed bed are depicted in Figure 15.3 with data from the Marathon FCC pilot plant.

An increase in excess oxygen would result in less reduction potential in the reducing zone. The relative concentrations of CO and O_2 largely dictate the emission chemistry. Thus, the above profile underscores the fact that the emissions cannot be readily assessed from changes in oxygen concentration at the entrance (air grid) or exit (excess O_2 in flue) of the regenerator; rather it is important to understand the axial variations and their impact on emissions. Industry experience has shown emissions chemistry in particular will be impacted by specific unit designs and bed hydrodynamics.

15.2.3 Catalyst Regeneration Variables

In order to establish an optimized FCC operation, it is often possible to influence factors impacting the heat balance. If coke combustion produces an amount of heat that causes regenerator temperature to rise above a preferred level, refiners may choose to reduce the feed temperature or lower the heat of combustion of the coke by reducing

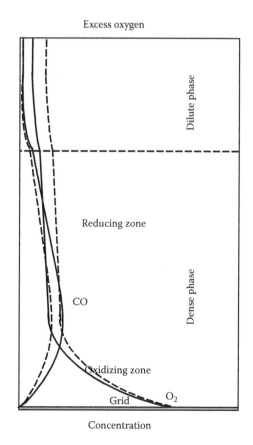

Excess oxygen

Concentration

FIGURE 15.3   Regenerator CO and O_2 profiles.

combustion of CO to CO_2. Another possible move would be to increase the stripping steam rate in order to minimize the hydrogen-in-coke carried into the regenerator. In extreme cases such as residual oil cracking, it may be necessary to install catalyst cooling devices to maintain the regenerator temperature within a practical range. On the other hand, if regenerator temperature is below a preferred range, refiners can increase the feed temperature with the use of a fired heater or can add components to the FCC feed that will result in increased coke production. Figure 15.4 is a graphical representation of catalyst movement during regeneration.

15.2.3.1   Temperature

Since the coke burning kinetics are first-order, increasing or decreasing the temperature will have a significant effect on the combustion rate. This can be significant when the coke or air distribution in the regenerator is less than ideal. The activation energy for CO combustion is considerably higher than for carbon combustion, so CO combustion is usually the rate-limiting step in the process. At temperatures below 1250°F, the thermally driven reaction rate of CO to CO_2 is very slow. To

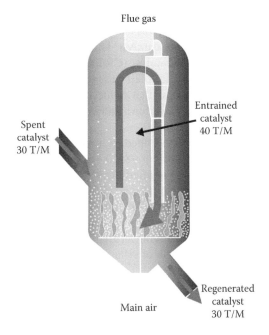

Flue gas

Entrained
catalyst
40 T/M

Spent
catalyst
30 T/M

Main air

Regenerated
catalyst
30 T/M

FIGURE 15.4   Catalyst regeneration movement.

help speed up the process, a combustion promoter (COP) is typically added to the
unit to increase the CO combustion rate. Without the catalyst bed to absorb this
heat of combustion, the flue gas temperature increases very rapidly and is called
afterburning. This represents an energy loss from the regenerator and a decrease
in the overall thermal efficiency of the regenerator. The use of promoters causes
the CO to burn in the catalyst bed so afterburning is controlled to acceptable levels. The following are typical actions used to improve coke burning kinetics via
temperature:

- Maximize feed preheat and increase delta coke
- Increase catalyst activity via higher catalyst additions and increase delta
  coke
- Decrease feed nozzle atomizing steam and increase delta coke
- Decrease stripping steam and increase delta coke
- Increase regenerator pressure
- Recycle slurry to increase delta coke
- Preheat air to regenerator
- Use torch oil

15.2.3.2   Spent Catalyst Distribution
Spent catalyst being introduced into the regenerator should be distributed as evenly
as possible across the catalyst bed. It is the nature of a fluidized bed that mixing

vertically occurs about 10 times faster than horizontally. An example of this was shown on a commercial unit that utilizes a horizontal distributor across the diameter of the regenerator to distribute catalyst. During the course of a 5 year TAR cycle, the carbon on regenerated catalyst (CRC) would increase from 0.05–0.10 wt% at start of run conditions to 0.15–0.20 wt% at end of run conditions. This was due to erosion of fluidization lances that reduced the spent catalyst distribution efficiency. This resulted in higher afterburn and increased use of Pt COP (factor of four times) to control unit operation.

Another commercial unit was revamped to improve spent catalyst distribution. This particular unit has a shallow bed of ~8–10' deep with a bed L/D ratio of ~0.4. The revamp resulted in a reduction of afterburn from ~190 to ~100°F and decrease in COP addition from ~45 to ~30 lb/d (Figure 15.5).

Another factor influencing afterburn is stripper efficiency. Hydrocarbon from the reactor stripper due to poor stripping could potentially flash off the spent catalyst and combust in the dilute phase generating an afterburn condition. Several units have seen stripper problems result in higher afterburn.

15.2.3.3 Air Distribution

The bed is fluidized with air from a distributor near the vessel tangent line. It is important that the air distribution be uniform, but biased slightly toward the area where the spent catalyst enters. The air jet size should be small enough to avoid forming large bubbles, and the jet exit velocity should be controlled to avoid catalyst attrition, jet erosion, or excessive jet penetration. One unit experienced an air grid failure following a unit upset. Upon inspection, one of four distributor quadrants had broken off creating an area of poor air distribution. The result was increased afterburn and higher CRC. Use of COP increased by a factor of four to maintain unit operations until the unit was shut down for repairs. Figures 15.6 and 15.7 show the bed temperature profiles and CRC trends from this incident.

15.2.3.4 Bed Depth

The depth of the catalyst bed is also an important variable to optimize coke burning. A deeper bed provides more time for the oxygen in the air to react with CO forming in the bed before the bubbles break free. This helps minimize afterburning. However, a deeper bed reduces the distance from the top of the bed to the cyclone inlets, increasing the catalyst load to the cyclones, causing increased cyclone erosion and catalyst losses from the regenerator. The regenerator operation must be a balance between minimizing afterburning and minimizing cyclone catalyst loading. With the proper CO combustion promoter usage, and good coke and air distribution, a bed depth of 10–15 feet is typically satisfactory. Increasing or decreasing this level will have an impact on afterburn and CRC. One unit increased the bed L/D from 0.5 to 0.6 and saw a decrease in CRC and afterburn. Figure 15.8 illustrates a physical layout of this modification.

Since catalyst is continuously added to the process to maintain activity, the bed level will typically increase with time with catalyst withdrawn as necessary to maintain a desire bed level. Figure 15.9 shows a typical response.

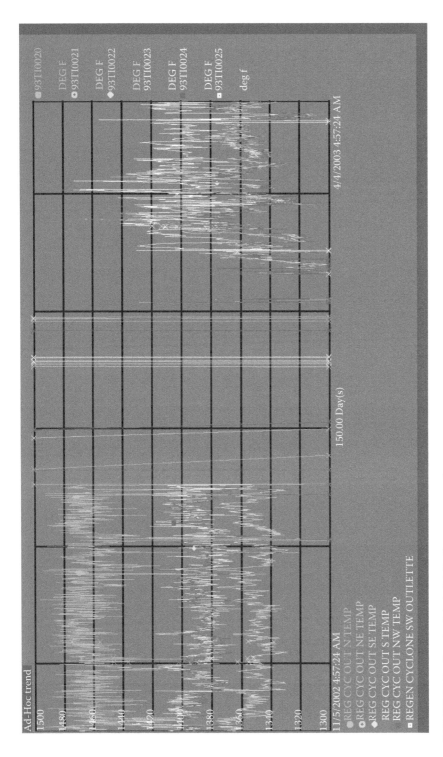

FIGURE 15.5 Spent catalyst distribution revamp performance.

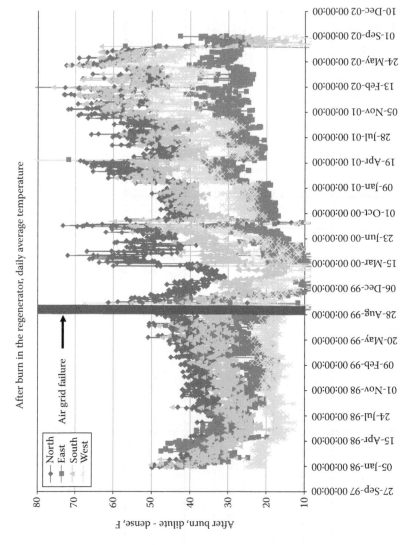

FIGURE 15.6   Regenerator temperature profile with damaged air grid.

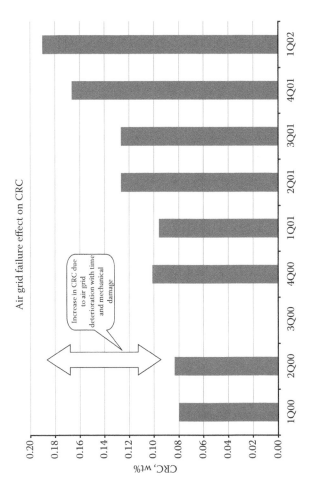

FIGURE 15.7 Regenerator CRC trends with damaged air grid.

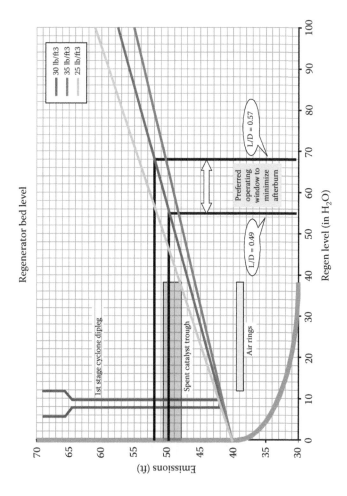

FIGURE 15.8   Bed level variation with overall elevation.

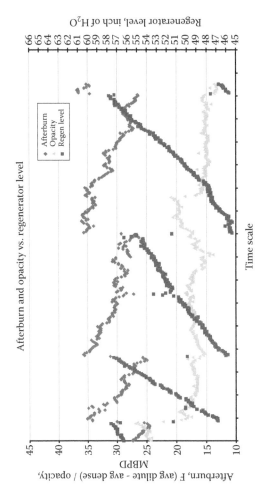

FIGURE 15.9 Impact of bed level on opacity and afterburn.

15.3   REGENERATOR DESIGN

Bubbling bed regenerators have been used to regenerate spent catalyst since the first
FCCU started up in 1942 and they are still widely used today. The FCC process
requires large volumes of fluidized catalyst to circulate back and forth between the
reactor and the regenerator. The historical way to do this is to arrange the regenerator so the catalyst enters the top of the bed and leaves at the bottom in a countercurrent flow to air. Some designs, such as the Exxon Flexicracker, introduce the spent
catalyst at the bottom of the bed with a catalyst overflow well in a cocurrent flow.
Following the introduction of CO promoter in the 1970s, many bubbling bed regenerators switched from partial combustion mode to complete combustion mode to
capture yield benefits.

Another type of design is the high-efficiency combustor regenerator [4]. The lower
vessel of the combustor is designed to completely burn the coke to CO_2 without CO
promoter and with low excess oxygen. This is accomplished by mixing and symmetrical design. The combustor is designed to operate at about twice the velocity of
a typical bubbling bed regenerator. The higher velocity results in a lower density than
in a bubbling bed so resistance to horizontal mixing is reduced. The lower part of the
combustor is a highly backmixed fast fluidized bed. A portion of the hot regenerated
catalyst is recirculated to the lower combustor to heat the incoming spent catalyst
and to control the combustor density to the desired level. As the catalyst enters the
combustor riser, the velocity is further increased and the two-phase mixture exits
through symmetrical downturned arms. A typical high efficiency regenerator in service today uses less than 1 mol-% excess oxygen and results in less than 20°F afterburn with no CO promoter.

The regenerator design has proven to be an important factor influencing NO_x
emissions from an FCC regenerator. The high-efficiency combustor and certain
countercurrent bubbling bed designs have demonstrated very low emissions. This
has been achieved through designs that offer good catalyst and air distribution and
operate with low excess oxygen and minimal CO promoter.

15.4   CO PROMOTER (COP) FUNCTION

In the 1970s, Mobil Oil discovered that certain Group VIII metals (platinum) can
be used at low concentrations (~1 ppm) to catalyze the combustion of CO to CO_2
without undesirable dehydrogenation reactions. Use of platinum as a COP has
become a common practice throughout the industry. This has allowed units to push
capacity beyond design limits without having to sacrifice higher CRC or afterburn.
Optimization of regeneration kinetics through use of COP has increased unit flexibility, improved operability and safety, reduced CRC, improved unit conversion and
selectivity, enhanced mechanical integrity and equipment life, and increased unit
utilization.

Platinum was historically used as an additive to increase the rate of CO combustion in the catalyst bed. Lack of Pt in the circulating catalyst inventory could
reduce overall combustion kinetics. Most units would operate with ~1 ppm Pt on
E-cat. Others have to operate much higher due to inherent design problems. Some

units with deep fluidized beds (L/D > 0.5), spent catalyst lift (Model III), or high efficiency combustors may not need COP unless pushed well beyond design limits.

FCC units that use antimony to passivate the deleterious impacts of nickel poisoning can also passivate the platinum (Pt) in the CO promoter. Antimony use can lead to an increase in afterburn or higher amounts of promoter.

The following are typical benefits and justification for use of COP and control of afterburn.

15.4.1 MECHANICAL INTEGRITY

It is a common practice for commercial FCCU not-to-exceed (NTE) limits for safe operation. These limits are based upon a time at temperature relationship and metal fatigue leading to an overstress condition set by applicable vessel codes of the regenerator cyclone support system. The exact limit will depend upon the cyclone vendor and the specific mechanical design. Most units have set a maximum temperature of ~1450°F for safe operation. It is also common for most units to be equipped with an automatic safety shutdown system. If the regenerator exceeds this limit for a certain amount of time, the unit will automatically trip off-line. Every occasion an FCCU is shutdown results in an economic penalty. The duration of the shutdown can range from a few hours to several days if the shutdown creates a problem elsewhere in the unit. COP is used to control afterburn and provide a comfortable operating margin away from this limit. Most units have established a maximum of < 1400°F as a target to ensure any unit upsets will not result in exceeding this maximum temperature.

15.4.2 EQUIPMENT LIFE

Regenerator cyclones have a typical life of 15–30 years depending upon erosion and mechanical fatigue. The base metal of the cyclones will deteriorate with time leading to graphitization. Once this happens, the metal cannot be welded upon and hence cannot be repaired during a normal unit TAR. This phenomenon is dependent upon time and temperature. COP is used to minimize the temperature to extend the cyclone life.

15.4.3 CATALYST DEACTIVATION

FCC catalyst is subject to hydrothermal deactivation. This occurs when the Al atom in the zeolitic cage is removed in the presence of water vapor and temperature. The result is a loss of activity and unit conversion. The effect of temperature on this process is nonlinear. The deactivation rate increases exponentially with temperature. Units that experience high afterburn have attributed high rates of catalyst deactivation on the higher dilute phase temperatures. This phenomenon is more apparent on units with high combustion air superficial velocities. The high velocity not only increases afterburn, but also increases catalyst entrainment to the cyclones and dilute area. COP is used to decrease afterburn and minimize catalyst deactivation.

15.4.4   EFFECTIVE CATALYST ACTIVITY (CARBON ON REGENERATED CATALYST)

Carbon on regenerated catalyst has a direct impact on unit conversion. Reducing the coke on catalyst will restore the catalytic activity. The impact of CRC on unit performance can be dramatic. At constant processing conditions, reducing the CRC will increase unit conversion. The improvement in conversion will depend upon the starting CRC and feed nitrogen content. Nitrogen is a temporary catalyst poison and is preferentially last to be removed from the catalyst during the regeneration process. Units with coker feedstocks will see a stronger effect with CRC. A typical effect from decreasing CRC by 0.1 wt% would be an increase in conversion of ~1%. A nominal 25 kbpd FCCU using typical economics would result in $1 MM/yr increased profitability from the lower CRC. COP is used to ensure optimum catalyst regeneration conditions to minimize CRC.

15.4.5   UNIT CAPACITY VERSUS EMISSIONS

Every FCCU will have its own relationship between excess oxygen, CRC, and CO emissions. The majority of FCCU also operate at or near an air blower or coke limit. For these units, minimizing excess oxygen will result in increased capacity and profitability. The trade-off is CRC and CO emissions. COP is used to accelerate CO combustion and allow the unit to operate at maximum profitability at low excess oxygen and stay within allowable CO emission limits. Figure 15.10 shows a typical relationship between excess O_2, CO, and NO_x.

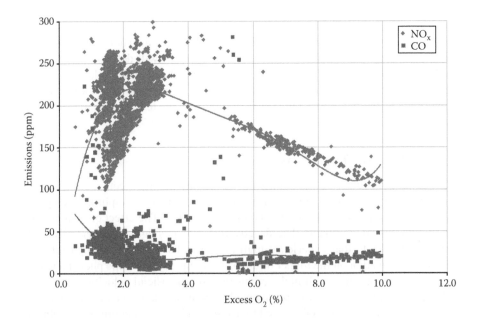

FIGURE 15.10   Relationship between excess O_2, CO, and NO_x.

15.4.6 UNIT OPERABILITY

Some units have design challenges that make unit operability a key operating parameter. Several commercial units with bubbling bed regenerators have high afterburn due to their mechanical configuration. Unit operating changes such as feed quality or process variables can result in dramatic changes in the regenerator operation. Pressure-balanced controlled units will also see changes in feed preheat result in dramatic coke burn shifts that will reduce excess oxygen and increase CO emissions. In all of these occasions, COP is added to improve CO combustion, minimize regenerator temperature swings, reduce CO emissions, and provide for stable unit operations.

15.5 OXYGEN ENRICHMENT

Oxygen enrichment is a resource to provide supplemental coke burn capacity. Several units may operate at a coke burn constraint due to mechanical equipment (air blower), cyclone velocity, or regenerator vessel size. Oxygen enrichment of regenerator air has been one of the attractive approaches to overcome the various limitations by removing the nitrogen component in the combustion air. However, a comprehensive understanding of the regenerator emission chemistry is needed to predict the impact on flue gas emissions (NO_x/SO_x).

Marathon studied the impact of oxygen enrichment on regeneration conditions using a circulating riser pilot plant at our research center in Catlettsburg, Kentucky. The ranges of variables were selected to cover the current and future potential operating conditions with oxygen enrichment under complete CO combustion mode of operation. The effects of oxygen enrichment (up to 40% O_2) were explored in combination with those of other operating parameters such as excess oxygen and addition of an SO_x transfer agent.

The key observation from this study was that under complete CO combustion conditions, NO_x emission on an lb NO_2 equivalent/lb coke basis was significantly reduced by O_2 enrichment. This result is consistent with the expectation that oxygen enrichment of regenerator gases will increase the rate of oxidation of the reduced nitrogen species near the air grid, and at the same time promote NO_x reduction reactions in the upper portion of the regenerator dense bed by decreasing the diluent N_2 and thereby increasing the concentration of reducing species that react with and destroy NO_x. An increase in CO concentration in the dense bed with oxygen enrichment was confirmed by the probe analysis along the regenerator axis. The combination of these effects increases the formation of elemental nitrogen and decreases NO_x emission.

It should be pointed out that often NO_x emissions are reported in terms of concentration, as ppmv (parts per million, volume). The concentration may actually go up with oxygen enrichment due to the reduced nitrogen dilution of the flue gas. The results of this program were jointly published by Moore and Menon [5]. The expected concentration profiles for CO and O_2 as functions of oxygen enrichment and excess oxygen are shown in Figure 15.11.

The concentration of N_2 will be the remainder in the profile. Under oxygen-enriched operating conditions, the reduction potential is increased in a portion of the

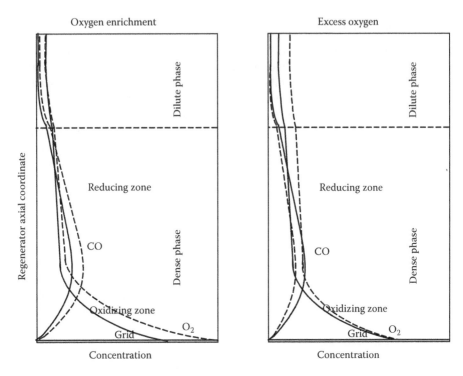

FIGURE 15.11   Effect of oxygen enrichment and excess oxygen on CO and O_2 profiles.

regenerator even in a complete combustion operating mode. An increase in excess oxygen would result in less reduction potential in the reducing zone. Temperature is also impacted by use of oxygen enrichment. Bed temperature will increase ~7°F for every 1% increase in oxygen in the air to the regenerator. This is set by the heat balance since less nitrogen in air results in a lower heat demand. The relative concentrations of CO and O_2 along with temperature will largely dictate the emission chemistry.

It is known that an increase in excess oxygen leads to increased NO_x in the flue gas. This observation may lead to the misconception that oxygen enrichment will increase NO_x. However, effects of oxygen enrichment and excess oxygen are contrary to each other. In fact, increased oxygen concentration at the air grid increases the reducing potential in portions of the regenerator, whereas increased excess oxygen lowers the reducing potential, thereby decreasing the rate of NO_x destruction.

15.6   UNINTENDED CONSEQUENCES

The FCC is a very dynamic unit that is typically the major conversion process in a refinery. Proper modeling and understanding of unit capabilities represents a tremendous opportunity to improve the overall unit operation and minimize unit emissions. The combustion chemistry in the FCC regenerator that produces environmental pollutants is extremely complex as numerous interactions and reactions occur between the various chemical species. Ultimately, all of the combustion chemistry revolves around the competition for, and the availability of, oxygen in the regenerator. It has been shown that catalysts, additives, or process conditions that reduce one undesirable flue gas component can inadvertently lead to an increase in the emissions of another. An example of one of the interactions that occurs in the regenerator is the interaction between CO and NO_x. Laboratory research studies have shown that even in the oxidizing environment of the regenerator, CO can act as a reductant, reducing NO_x to N_2. Thus, the elimination of CO by the complete oxidation of CO to CO_2 can result in an increase in NO_x emissions from the FCC [6]. It is important to be aware that when attempting to minimize the emissions of these undesired chemical species in the flue gas, care must be taken in the design and use of different emissions control technologies. Otherwise unintended consequences and undesirable results could be attained.

15.7   SUMMARY AND CONCLUSION

The heart of the FCC process is the heat balance. The FCCU operates in heat balance requiring the total energy coming into the FCC to equal the energy going out. The primary source of energy in the FCCU is coke formed on the catalyst surface. Combustion of coke in the FCC regenerator supplies the energy to run the cracking process. Catalyst regeneration, combustion kinetics, catalyst distribution, air distribution, oxygen enrichment, regenerator design, and mechanical integrity are all critical factors influenced by the heat balance and have a direct impact on emissions. Changes in any of these parameters to minimize one pollutant can have an unintended consequence of increasing another. Understanding the role of the heat balance in FCCU emissions is critical to success in controlling them.

REFERENCES

1. Upson, L., Dalin, I., and Wichers, R. *Heat Balance—The Key to Cat Cracking*. 3rd Katalistiks FCC Symposium, Amsterdam, The Netherlands, 1982.
2. Sedebeighi, R., and Sexton, J. *Principles and Troubleshooting of Regenerator Afterburn.* 2004 NPRA Cat Cracker Symposium, Houston, Texas, 2004.
3. Arthur, J. R. *Transactions of the Faraday Society* 47 (1951): 164.
4. Rosser, F., Schnaith, M., and Walker, P. *Integrated View to Understanding the FCC NO_x Puzzle*. AICHe Symposium, Austin, Texas, 2004.
5. Menon, R., Limbach, K., Tamhankar, S., Ganguly, S., Bakse, V., and Moore, H. Oxygen Enrichment to Reduce NOx Emissions. *PTQ*, Q1, 2001.
6. Sexton, J., and Fisher, R. Marathon Refineries Employ New FCC CO Combustion Promoter. *Oil & Gas Journal,* April 27, 2009.

16 FCC Emission Reduction Technologies through Consent Decree Implementation: FCC SO$_x$ Emissions and Controls

Jeffrey A. Sexton

CONTENTS

16.1   INTRODUCTION

The Clean Air Act requires EPA to set national air quality standards for sulfur dioxide (SO_2) and five other pollutants considered harmful to public health and the environment. EPA's National Ambient Air Quality Standard for SO_2 is designed to protect against exposure to the entire group of sulfur oxides (SO_x). SO_2 is the component of greatest concern and is used as the indicator for the larger group of gaseous SO_x. Other gaseous SO_x (e.g., SO_3) are found in the atmosphere at concentrations much lower than SO_2.

Sulfur dioxide (SO_2), a colorless, reactive gas, is produced during the burning of sulfur-containing fuels. It belongs to a family of gases called sulfur oxides (SO_x). Major sources include power plants, industrial boilers, petroleum refineries, smelters, iron, and steel mills. Generally, the highest concentrations of SO_2 are found near large fuel combustion sources.

SO_x can react with other compounds in the atmosphere to form small particles. These small particles penetrate deeply into sensitive parts of the lungs and can cause or worsen respiratory disease and aggravate existing heart disease. EPA's NAAQS for particulate matter (PM) are designed to provide protection against these health effects.

Acid deposition or, acid rain, occurs when SO_2 and oxides of nitrogen (NO_x) react with water, oxygen, and oxidants to form acidic compounds. It is deposited in dry form (gas, particles) or wet form (rain, snow, fog), and can be carried by wind hundreds of miles across state and national borders. Acid rain harms lakes and streams, damages trees, crops, historic buildings, and monuments.

Everywhere in the United States meets the current SO_2 NAAQS. Annual average ambient SO_2 concentrations, as measured at area-wide monitors, have decreased by more than 70% since 1980. Currently, the annual average SO_2 concentrations range from approximately 1–6 parts per billion [1].

Fluid catalytic cracking (FCC) sulfur emissions in the form of SO_x (SO_2 and SO_3) from the regenerator vary significantly depending on the feed sulfur content and the FCC unit (FCCU) design. In the FCCU reactor, 70–95% of the incoming feed sulfur is transferred to the acid gas and product side in the form of H_2S. The remaining 5–30% of the incoming feed sulfur is attached to the coke and is oxidized into SO_x that is emitted with the regenerator flue gas. The sulfur distribution is dependent on the sulfur species contained in the feed, and in particular the amount of thiophenic sulfur. SO_2 can range from 200 to 3000 parts per million dry volume basis (ppmdv), whereas SO_3 typically varies from an insignificant value to a maximum of 10% of the SO_2 content.

Several technology solutions have been developed to reduce FCCU SO_x emissions including catalyst additives, wet gas scrubbers, and regenerative scrubbers. Table 16.1 summarizes various options to reduce SO_x emissions.

For many refineries, switching to a low sulfur feed is both expensive and unrealistic. Feed hydrotreating is the most capital intensive of the available solutions.

TABLE 16.1

Options for Reducing SO$_x$ Emissions from FCC Units

Potential Solution	Advantages	Disadvantages
1. Processing of low sulfur feedstocks	Low SO$_x$ emissions	Low sulfur feeds are generally high in cost and limited in availability
2. Feed hydrotreating	Low SO$_x$ emissions Improved FCC yields	Very high capital and operating costs
3. Flue gas scrubbing	Low SO$_x$ emissions Low particulate emissions	Very high capital and operating costs
4. Catalytic SO$_x$ control using additives	Ultra-low SO$_x$ emissions for some FCCUs No capital required Generally lowest operating costs	At high levels in catalyst inventory, additives can dilute catalyst activity and increase opacity
5. Combinations of the above	Can optimize the efficiency of each solution to achieve desired SO$_x$ reduction	Must monitor multiple solution systems

Flue gas scrubbing also has high capital and operating costs associated with it, but also offers PM control. An additive-only solution is almost always the least expensive option when including both capital and operating expenses for SO$_x$ reduction depending on the feed sulfur. SO$_x$ reduction additives can also be effectively used in conjunction with one of the other solution options. For example, several refiners with FCC feed hydrotreaters use a SO$_x$ reduction additive to trim the SO$_x$ emissions to the required 25 ppm level rather than to hydrotreat the FCC feed more severely or to install a flue gas scrubber. The optimum choice for a given unit is often site specific.

16.2 SO$_X$ FORMATION CHEMISTRY

Feed quality is the most significant factor affecting SO$_x$ emissions from an FCCU. The sulfur content and the particular sulfur species present in the feed strongly determine the extent of potential SO$_x$ emissions [2]. Typically, about 10% of the sulfur in feed goes to SO$_x$, but it can vary from 5 to 30 wt% (Figure 16.1).

16.3 CATALYST ADDITIVES CONTROL TECHNOLOGY

16.3.1 SO$_x$ Reduction Additive Mechanism

SO$_x$ additive chemistry has been described previously in literature [3]. SO$_x$ reduction additives remove SO$_x$ from the regenerator flue gas and release the sulfur as H$_2$S in the FCC reactor. In a full burn regenerator, the amount of SO$_2$ removed is directly proportional to the amount of additive used. Normal additive levels in the catalyst inventory range from 1–10%, with up to 20% being used in some units. Typical SO$_x$ removal rates have historically been in the 20–60% range. With the introduction of new super additives, rates in excess of 95% are commonly being achieved [4].

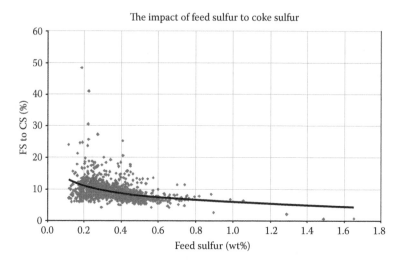

FIGURE 16.1   Impact of feed sulfur on coke sulfur with hydrotreated feed.

SO_x reduction catalysts are basically two component systems. The first component catalyzes the oxidation of SO_2 to SO_3 in the regenerator:

$$SO_2 + \tfrac{1}{2} O_2 \Rightarrow SO_3.$$

The second component, commonly referred to as the pick-up agent, removes the SO_3 from the regenerator as a metal sulfate and releases it as H_2S in the reactor or stripper. Both components of the additive must work together for maximum SO_x removal.

Magnesium-based materials are used as the pick-up agent in currently available additives. In the FCC regenerator, the additive reacts with SO_3 to form magnesium sulfate:

$$MgO + SO_3 \Rightarrow MgSO_4.$$

Once the additive has picked up SO_3, the sulfate circulates with the catalyst to the reactor. In the reducing environment of the reactor, the hydrogen sulfide is released and the additive reverts back to its original state:

$$MgSO_4 + 4H_2 \Rightarrow MgO + H_2S + 3H_2O.$$

The H_2S formed exits the FCCU in the dry gas and is removed downstream in the sulfur recovery unit. The increase in H_2S production, 5–20%, can typically be managed within a refinery's operations (Figure 16.2). A different mechanism of SO_x uptake has been presented by Magnabosco [5].

16.3.2   SO$_x$ REDUCTION ADDITIVE DEVELOPMENT

SO_x reduction catalysts have been under development since the late 1970s. The initial SO_x reduction catalysts were alumina based. While these were shown to be effective

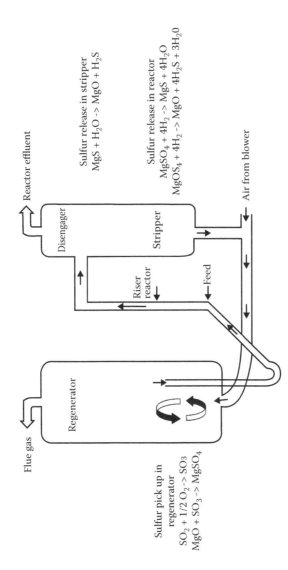

FIGURE 16.2 Schematic diagram of SO_x reduction chemistry in a typical FCC unit. (With permission from Intercat, Inc.)

when promoted with rare earth elements, particularly cerium, they were very susceptible to deactivation. Although this additive removed SO_x, its effectiveness was reduced by its limited sorption capability and short life.

The first technology to offer an alternative was a magnesium aluminate spinel-based technology that has been further advanced and is still offered today. The spinel technology was originally developed by ARCO in the 1970s [6], with advancements in the use of magnesium for SO_3 pickup (Figure 16.3).

Realizing the importance and effectiveness of the magnesium species in the sorption of SO_x, in the late 1980s, Akzo Nobel patented the use of hydrotalcite and related compounds for use in an FCC to reduce SO_x emissions [8,9,10]. Hydrotalcite contains more active Mg species than spinel. Hydrotalcite-based compounds typically contain 3–4 moles of Mg per mole of Al, while spinels contain only 1 mole of Mg per 2 moles of Al (Figure 16.4).

Early hydrotalcite technology required it to be supported or otherwise bound, resulting in less than optimal performance. In 1997, INTERCAT developed a self-supporting hydrotalcite that overcame these previous technology barriers [7]. With increased commercial utilization, there has been a continuing evolution in the performance of these products. The level of contained MgO in SO_x additives has been increased to improve effectiveness. This has been achieved without any degradation in physical properties. Some of the newer products (typically named Super versions) contains over 55% MgO. These products have been effective in achieving ultra low SO_x emissions without requiring high percentage levels in the unit inventory. The

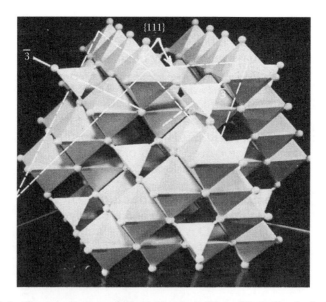

FIGURE 16.3 Crystal structure of $MgAl_2O_4$ spinel. (From Vierheilig, A. A., *Process for Making, and Use of, Anionic Clay Materials,* U.S. Patent 6,479,421, 2002. With permission from Intercat, Inc.)

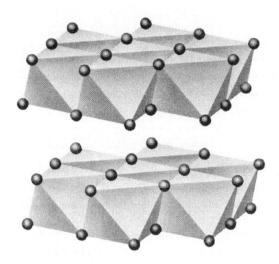

FIGURE 16.4   Crystal structure of hydrotalcite $Mg_6Al_2(OH)_{18}.4.5\ H_2O$. (From Vierheilig, A. A., *Process for Making, and Use of, Anionic Clay Materials,* U.S. Patent 6,479,421, 2002. With permission from Intercat, Inc.)

science and technology of SO_x reductions in FCCUs has been reviewed in detail by Magnabosco [5].

16.3.3   FACTORS AFFECTING SO_x REDUCTION ADDITIVE EFFICIENCY

A number of factors can affect the efficiency of SO_x reduction additives in the FCCU. Some of these factors are related to the operation of the FCCU, and some are directly related to the composition of the SO_x additive itself. The following are significant factors that affect additive performance:

- Flue gas excess oxygen content. Oxygen is required to drive the SO_2 to SO_3 reaction that must take place in the regenerator before the SO_x additive can pick up the SO_3 and transport it to the reactor. In many FCCUs, the availability of oxygen is the rate-limiting step in the process. However, increasing O_2 concentration above a certain level has no impact on improving SO_x additive efficiency.
- The FCC catalyst may affect SO_x emissions. The active alumina in FCC catalysts plays a limited role as a pick-up agent for SO_3 (similar to MgO). However, the fresh catalyst lacks the oxidants that enhance the effectiveness of SO_x reduction additives.
- The presence of CO promoter catalyzes the oxidation of SO_2 to SO_3 and therefore enhances the SO_x removal process. Higher concentrations of SO_3 are also produced in the presence of excess oxygen, so SO_x reduction additives tend to be more effective in full combustion regenerators.

- Increasing catalyst circulation rate increases the availability of fresh metal oxides for SO_3 pick-up and hence reduces SO_x emissions.
- Lower regenerator temperatures tend to favor SO_3 formation, while good air distribution and mixing in the regenerator enhances SO_3 pick-up.
- Large regenerator inventories will reduce the efficiency of an additive. Inefficient strippers increase the amount of sulfur going to the regenerator, and hence increase the SO_x emissions.

The compositional factors affecting the efficiency of SO_x additives are:

- The additive must have a high degree of physical integrity so that it is able to withstand the severe hydrothermal environment of the FCCU. Factors such as attrition resistance, apparent bulk density, and particle size distribution are critical in retaining the additive in the unit.
- MgO content of the additive is a parameter affecting the SO_x reduction efficiency. However, for an additive to be effective and stable, the MgO must be integral to the structure of the additive.
- The active sites of the additive must be accessible to the reactant SO_x molecules.
- The additive formulation must be optimized to carry out the oxidation/ reduction reactions in the regenerator and reactor. Increasing the concentration of the active MgO requires a corresponding increase in the content of cerium and vanadium.

16.3.4 PARTIAL BURN REGENERATORS CONSIDERATION

Recently, the focus on SO_x additive development has been on improving performance in partial burn FCCUs. It has been apparent for some time that oxygen availability is the factor limiting SO_x reduction additive performance in partial burn units. Oxygen availability is critical for the SO_x additive to oxidize SO_2 to SO_3 in order to capture the SO_3 generated in the regenerator, and release it as H_2S in the riser and reactor. In partial burn units, it is well known that O_2 availability is extremely limited. Recent detailed flue gas analyses have shown that in deep partial burn regenerators only about 30% of the sulfur species are present in an oxidized form (Table 16.2). The majority of the sulfur species are in reduced forms such as CS_2 and H_2S.

TABLE 16.2
Sulfur Oxide Species Present in FCC Flue Gas

Flue Gas Analysis (Mol/hr)	CO Boiler Inlet	CO Boiler Outlet	%S in CO Boiler Inlet in Oxidized Form
SO_2	669	2566	
SO_3	179	57	
Total SO_x	849	2623	32%

This discovery has led to increased research into the oxidation function of these additives and the development and commercialization of the first SO_x reduction additive specifically designed for partial burn units, Lo-SO_x PB. This additive contains an oxidation package specifically designed to improve additive performance in oxygen deficient environments. However, even with a specially formulated additive, there is a limit to how much SO_x can be removed by additives in partial burn regenerators. This limit is heavily unit specific, and can only be determined by commercial testing. In general, partial-burn regenerators that are poorly mixed can achieve higher levels of reduction than well-mixed, true countercurrent flow regenerators.

16.4   WET GAS SCRUBBERS CONTROL TECHNOLOGY

The control of PM and SO_2 emissions with wet scrubbing systems is becoming commonplace in industry. The FCCU application presents the additional requirement that in order to match the reliability of the FCCU, the air pollution control equipment must also operate on line for 3–5 years without interruption. It must be able to tolerate significant fluctuations in operating conditions, withstand the severe abrasion from catalyst fines, and maintain operation through system upsets. The robust design of the wet scrubbing system must tolerate all operating conditions without requiring a shutdown. It is paramount that the operability of the air pollution control system is no less than that of the FCCU process.

SO_3 control is also an increasingly important issue for refinery FCC operators. Regional haze, $PM_{2.5}$, hazardous air pollutants, and visible stack emissions all in some way relate to SO_3. The present level of control achievable in wet scrubbers is limited. However, use of wet ESPs integral to the scrubber has been used commercially.

16.4.1   A BASIC WET SCRUBBING SYSTEM

For many years, refiners worldwide have chosen to use wet scrubbing systems to reduce both particulate and SO_2 emissions. With the proper wet scrubbing technology, both particulate and sulfur emissions are removed simultaneously and very efficiently. This technology is well proven in providing flexibility to handle added capacity that may result from FCCU expansions or to increase reduction efficiency as regulatory pressures increase and in providing uninterrupted operation and performance exceeding that of the FCCU itself. Each refiner's specific reasons for choosing wet scrubbing differ, but these have generally been related to environmental compliance as well as relative costs, reliability, and flexibility of wet scrubbing compared to other emission control options.

Several licensors currently offer FCC wet scrubbing technology. For the general purpose of this chapter, the discussion will focus on application of the Belco technology (BELCO, EDV, and LABSORB are registered trademarks of Belco Technologies Corporation). This system has been described previously in literature [11].

Figure 16.5 represents a graphic of the Belco EDV Wet Scrubbing System. This is one of the technologies available to the refining industry for reducing particulate and SO_2 emissions. The EDV Wet Scrubbing System is the technology with the most installed units in oil refineries. The schematic in Figure 16.5 represents only one of

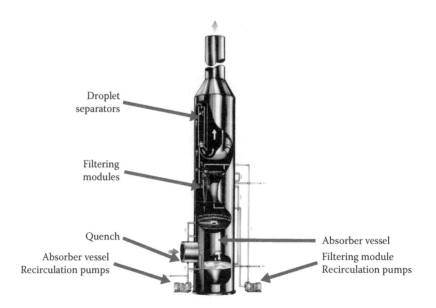

Droplet separators

Filtering modules

Quench

Absorber vessel
Recirculation pumps

Absorber vessel

Filtering module
Recirculation pumps

FIGURE 16.5 EDV 5000 model wet scrubbing system. (With permission from Belco Technologies.)

the EDV configurations sold by Belco Technologies Corporation. Other configurations are used depending on the specifics for each project.

The system treats hot flue gas containing particulates such as FCC catalyst fines and SO_2 from the flue gas and discharges cleaned gas to the atmosphere through an integral stack. At the scrubber inlet, FCCU flue gas is quenched and saturated by means of multiple water sprays in the spray tower's horizontal quench section. Normally the flue gas enters the wet scrubber after passing through a heat recovery device (boiler tubes or flue gas cooler, etc.). However, the system can be designed to also accept the flue gas directly from the flue gas source at its normal exit temperature. An example of this would be in cases where a CO boiler on an FCCU application requires to be bypassed. In that case the flue gas from the FCCU can be diverted directly to the EDV Wet Scrubbing System without any concerns and without having to make any adjustments to the operation. This not only results in a more reliable and simpler operation (from the standpoint of the plant operation), but also allows the plant to continuously reduce emissions even during bypass and upset conditions.

The EDV Wet Scrubbing System utilized proprietary nozzles to produce high-density water curtains through which the gas must pass. Each nozzle sprays water droplets that move in a cross-flow pattern relative to the flue gas. These cover the entire gas stream and uniformly flush the vessel's surfaces clean. The spray nozzles are nonclogging and are designed to handle highly concentrated slurries.

SO_2 absorption and particulate removal begins at the quench section and continues as the flue gas rises up through the main spray tower where the gas is again contacted with high-density water curtains produced by additional spray nozzles. The spray tower itself is an open tower with multiple levels of the BELCO spray nozzles. Since it is an open tower, there is nothing to clog or plug in the event of a process upset. In fact, this design has handled numerous process upsets and reversals without any concern.

The scrubbing liquid is controlled to a neutral pH with reagent addition to drive SO_2 absorption. Caustic soda (NaOH) is typically used as the alkaline reagent. However, other alkalis, such as soda ash, magnesium hydroxide, and lime have also been utilized with excellent results in terms of performance and reliability. For FCCU applications, however, where a 5–7 year continuous operation is required, the use of lime as a reagent is not recommended. Multiple levels of spray nozzles provide sufficient stages of gas/liquid contact to remove both particulate and SO_2. An illustration of the spray tower and the spray nozzles is provided in Figure 16.6.

Makeup water is added to the system, replacing water lost to evaporation in the quench zone and also for water purged from the system. Captured pollutants, including suspended catalyst fines and dissolved sulfites/sulfates ($NaHSO_3$, Na_2SO_3, and Na_2SO_4 in cases when sodium based reagents are used) resulting from SO_2 removal is purged from the spray tower recycle loop.

In order to remove very fine particulate, flue gas leaving the spray tower is distributed to a bank of parallel filtering modules. Within each module, the flue gas first accelerates (compresses) and then decelerates (expands). This action causes water to condense from the flue gas. The water uniformly washes the module's walls. More importantly, water condenses on the fine particulate and acid mist (mostly H_2SO_4 from condensation of SO_3 in the saturated flue gas) present in the flue gas, increasing both their size and mass. Some agglomeration also takes place.

A proprietary F nozzle, located at the exit of the filtering module and spraying countercurrent to gas flow, provides the mechanism for the collection of the fine

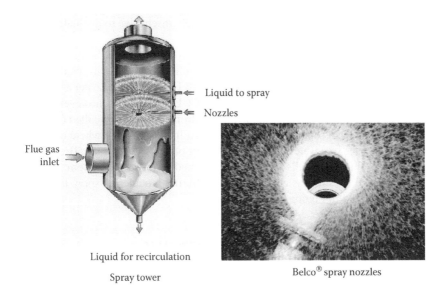

Spray tower   Belco® spray nozzles

FIGURE 16.6 Spray tower and BELCO spray nozzles. (With permission from Belco Technologies.)

Filtering module   Cyclolab™ droplet separator

FIGURE 16.7 Filtering module for fine particulate control and droplet separators. (With permission from Belco Technologies.)

particulate and mist, which has been enlarged and agglomerated. This device has the unique advantage of being able to remove fine particulate and acid mist with an extremely low pressure drop and no internal components that can wear and be the cause of unscheduled shutdowns. It is also relatively insensitive to fluctuations in gas flow. This device is illustrated in Figure 16.7.

Prior to being discharged to the atmosphere through a stack, the flue gas enters the system's droplet separators. These separate/collect free water droplets allowing the flue gas to exit the stack free of water droplets. For boiler and heater applications Chevron type droplet separators are used. For FCCU applications, however, the EDV system uses large tubes with fixed spin vanes as droplet separators (called Cyclolabs). The gas entering each separator passes through a fixed spin vane where centrifugal acceleration causes free water droplets to impinge on the separator's walls. Collected water droplets flush the walls uniformly clean and drain to the bottom. Collected water is recycled for flue gas cleaning in the filtering modules or spray tower. This device is illustrated in Figure 16.6.

16.4.2 TYPICAL WET SCRUBBING SYSTEM PERFORMANCE

In order to illustrate the typical performance of this system, the EDV system installed at a U.S. refinery is utilized. At this refinery, an EDV wet scrubbing system was installed to bring a new small RCC in compliance with New Source Performance Standards (NSPS) for particulate and SO_2 emissions.

Emission testing was performed in January 1998, to verify the emissions performance of the system. Testing was performed at both the inlet to the wet scrubbing system and at the stack. All emission guarantees were met although the actual conditions vary significantly from the original design conditions provided by the owner.

First, the testing at the inlet to the EDV wet scrubbing system demonstrated that the system was operating at higher than design values for gas flow and considerably higher than design SO_2 loading while having a lower than design loading for particulate. The flue gas flow rate was approximately 20% over design on a mass basis. SO_2 was approximately 3.1 times the design value on a mass basis. However, the particulate was approximately 50% of the design value on a mass basis. A summary of the average inlet test values, compared to the system design values, is presented in Table 16.3.

The performance of the system was excellent. Stack SO_2 was only a small fraction of the design outlet value. The mass outlet SO_2 emissions were only 12% of the design values while the tested removal efficiency was 99.92% versus a required design efficiency of 97.90%. Particulate emissions were also very low. The mass emission rate was approximately 24% of the design value while the tested removal efficiency was 92.24% versus the required design removal efficiency of 83.70%. A summary and comparison of this data is in Table 16.4.

TABLE 16.3
EDV System Inlet Values

Item	Tested Inlet Value	Design Inlet Value
Flue gas flow	141,781 kg/hr	118,769 kg/hr
	227,679 Am³/hr	181,329 Am³/hr
Flue gas temperature	236°C	274°C
Particulate loading	146 mg/Nm³	407 mg/Nm³
	17.2 kg/hr	34.5 kg/hr
SO_2 loading	1313 ppm	626 ppm
	440 kg/hr	141 kg/hr

TABLE 16.4
EDV System Outlet Emissions Values

Item	Tested Stack Value	Design Stack Value
Particulate emissions	10.8 mg/Nm³	66.4 mg/Nm³
	1.34 kg/hr	5.62 kg/hr
	92.24% removal efficiency	83.70% removal efficiency
SO_2 emissions	1 ppm	13 ppm
	0.36 kg/hr	2.97 kg/hr
	99.92% removal efficiency	97.90% removal efficiency

From an operations and maintenance perspective the system has also been excellent. Over the first several months of operation, the RCC experienced multiple process upsets that resulted in as much as 20–30% of the catalyst inventory being carried out of the regenerator and into the wet scrubbing system. The wet scrubber readily handled all of these process upsets. The operation of the scrubber was not interrupted. The system continued to operate.

From an operator's perspective, the scrubbing system required very little attention. Normally, the only attention required is a routine walk-by of the equipment to ensure that everything is in normal operation. Since 1998 the refinery took the RCC out of operation and moved the EDV Wet Scrubbing System over to their SRU and applied it as a tail gas cleaning device. The system is presently operating at very high efficiency as a tail gas unit to rave reviews of the refinery client.

16.4.3 Discharge of Scrubber Purge

Captured pollutants, including suspended catalyst fines and dissolved sulfites/sulfates resulting from SO_x and NO_x removal are purged from the spray tower recycle loop. The liquid that is purged from the scrubber is typically processed in a purge treatment unit (PTU). The purge treatment system removes the suspended solids and converts the sodium sulfite to sodium sulfate to reduce the chemical oxygen demand (COD) so that the effluent can be safely discharged from the refinery.

In order to remove the suspended solids, the purge treatment system contains a clarifier to separate the suspended solids and a filter press or dewatering bins to concentrate the solids into a filter cake, which is cohesive and can be readily disposed. The scrubber purge enters the clarifier from a deaeration tank. The solids settle out in the clarifier and are removed from the clarifier in the underflow. The underflow from the clarifier is sent to a filter press or dewatering bins where the excess water is removed. The solids are sent to disposal while the water is returned to the clarifier. The effluent is then sent to the oxidation towers.

The oxidation system consists of towers where air is forced into the effluent to oxidize the sodium sulfite to sodium sulfate. Effluent from the oxidation towers, which is now cleaned of catalyst (suspended solids) and has a low COD level, can be processed in the refinery wastewater system or possibly directly discharged from the refinery. A typical purge treatment system that employs a filter press is illustrated in Figure 16.8.

Total dissolved solids (TDS) levels in purge water have posed water quality issues, especially at refineries that discharge into small or impacted bodies of water. These TDS issues have lead to significant permit delays and disposal cost issues.

16.4.4 Reagent Options

Historically, most wet scrubbing systems on FCCUs have utilized caustic (NaOH) as the reagent. Caustic is readily available in refineries, is easy to handle, and has no solid reaction by-product. These systems have proven to be very effective and reliable, with continuous operation in excess of 5 years while handling all upset conditions that can occur.

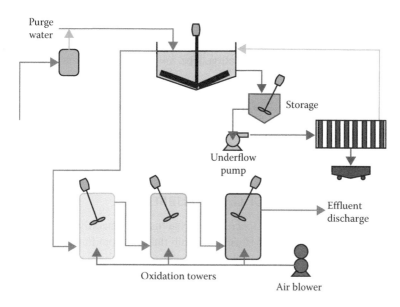

FIGURE 16.8 Typical purge treatment system. (With permission from Belco Technologies.)

With the escalating cost of caustic and the need to reduce the total liquid effluent from the system, some refiners are using soda ash (Na_2CO_3) as a reagent. The primary difference between soda ash and caustic is that soda ash is delivered as a bulk solid and mixed into a liquid on site. However, it has an advantage of having no chlorides. High concentrations of chlorides attack the 316L stainless steel material used in the wet scrubber, so the level of chlorides must be controlled. With no chlorides from the soda ash, the dissolved solids concentration in the wet scrubber can be increased, thus reducing the amount of liquid that must be purged.

Operating costs also vary greatly. A caustic system has the highest operating cost, due to the reagent cost. A soda ash scrubber has a lower operating cost, primarily due to lower reagent cost. A soda ash system with a crystallizer has a cost near that of a caustic system, mostly due to steam needs and additional power requirements. However, this option has the added benefit of no liquid effluent discharge.

16.4.5 SCRUBBER APPLICATIONS

The economics of the technologies described is very much dependant on the combination of the processes used and the amount of pollutants in the flue gas stream that is being treated. However, it is essential to recognize that all costs associated with the different technologies can be optimized by careful planning. Facilities that do not require PM reduction, NO_x reduction, SO_3 reduction, or elimination of the scrubber purge should consider if any of these may be required in the future. The costs associated with adding additional control technology options in the future can be minimized if it is properly addressed when planning for a present installation.

16.4.6 OTHER DESIGNS

There are several different wet gas scrubber designs that have been applied to FCCUs. All are unique in design and application. The following is a summary of licensors with FCC applications:

- Belco
- GEA Bischoff
- ExxonMobil (Hamon)
- MECS

16.5 REGENERATIVE WET GAS SCRUBBERS

The nonregenerable wet gas SO_2 scrubbing systems are generally known as throw-away systems. The reagent that is used to scrub the SO_2 from flue gas is consumed in the absorption process and discharged as the alkali sulfate salt in the process wastewater or landfill streams. NaOH, purchased as 20 or 50 wt% solutions, dry soda ash (Na_2CO_3), or dry lime (CaO) are reacted with SO_2 in the FCCU regenerator offgas and oxidized to form either sodium sulfate or calcium sulfate. On a dry basis, 1 ton of SO_2 produces 2.1 tons of calcium sulfate (gypsum or $CaSO_4$) or 2.2 tons of sodium sulfate (Na_2SO_4). Sodium sulfate is discharged from the scrubbing system, dissolved in water, and is added to the refinery wastewater effluent stream, while calcium sulfate is filtered from the scrubber effluent and discharged as a wet or dry cake and disposed of in a landfill.

The discharge of aqueous sodium sulfate or gypsum slurry from the scrubber may not be permissible under existing environmental permits, or may not be allowed as part of the overall repermitting process that accompanies the installation of the scrubbing system itself. If effluent discharge limits prevent the use of nonregenerable processes, then the application of a regenerable process may not only be preferred, but mandatory under revised permit requirements.

Regional landlocked operators, for example, may have a need to eliminate the liquid purge from scrubbing systems. Regenerable process options, which absorb SO_2 in an aqueous medium and that regenerate it from the solution, include the CANSOLV SO_2 Scrubbing System and the Belco LABSORB and Dual Alkali regenerative wet scrubbing technology.

Both of these processes direct the SO_2 absorbed from the FCCU flue gas to the refinery SRU, where it is converted to elemental sulfur and added to the marketable sulfur that is generated by the SRU from H_2S. Alternately, the SO_2 can be converted to sulfuric acid in a dedicated sulfuric acid plant, or in combination with an existing refinery spent acid regeneration unit. When the SO_2 is directed to the SRU, 1 ton of SO_2 captured in the scrubber is converted to 0.5 tons of marketable elemental sulfur and less than 0.1 ton of sodium sulfate waste is generated per ton of SO_2 absorbed. In an acid plant, 1 ton of SO_2 generates 1.5 tons of 98% sulfuric acid. Steam is also generated from the conversion of SO_2 in both the SRU and the acid plant, which moderates somewhat the steam consumption rate of the solvent regenerator for both the LABSORB and CANSOLV systems.

The CANSOLV SO_2 scrubbing process uses an amine to absorb SO_2 from the flue gas and regenerate it in a solvent regenerator. The process is very similar to amine processes used throughout the refining industry to remove H_2S from refinery gas streams. Operational, maintenance, and technical service requirements for most of the CANSOLV system are similar to those associated with the H_2S/amine processes. Particular differences exist, however, because flue gas contains significant amounts of both particulates and SO_3 that must be managed.

Particulate management in the FCCU is critical because particulate emission rates can vary in response to upsets in the FCCU. For example, failure of the regenerator cyclones can lead to an order of magnitude increase in steady state particulate emission rates and pressure reversal upset incidents can result in massive, short-term particulate emission rates that must be accommodated by the SO_2 scrubbing system.

High efficiency particulate removal systems, such as the Belco particulate scrubber and high efficiency, venturi type, particulate removal systems are capable of removing most of the particulates exiting with the FCCU regenerator offgas. But during FCCU upsets, the increased particulate load leaving the FCCU cannot be entirely contained in the gas prescrubbing system. Solvent filtration systems in the regenerative scrubbing section also need to be able to handle upset conditions in the FCCU.

16.5.1 Belco LABSORB System

A LABSORB regenerative system can be used to nearly eliminate the production of liquid or solid waste stream from the reduction of SO_2. This regenerative process utilizes the EDV Wet Scrubbing System to remove the SO_2 from the flue gas with a regenerable, nonorganic buffer. This buffer is sent to a regeneration plant where the buffer is regenerated and the SO_2 is extracted from the buffer as a concentrated stream of SO_2. The concentrated SO_2 stream can be sent to the sulfur recovery unit for recovery as elemental sulfur. Or it can be sent to an acid plant where valuable sulfuric acid can be produced. This approach has a major side benefit in the reduction of operating costs since the buffer that removes the SO_2 is reused.

The SO_2 rich buffer is pumped from the EDV absorber vessel to the regeneration plant. Before entering the regeneration process, it is heated in a series of heat exchangers. The first heat exchanger utilizes the heat from the regenerated buffer being returned to the absorber vessel, while the second heat exchanger utilizes steam. After being heated, the buffer is sent to a double loop evaporation circuit. These circuits use a heat exchanger, separator, and condenser to separate water and SO_2 from the buffer. Buffer, which is free of SO_2, is sent to a mixing tank, while the evaporated water and SO_2 are sent to a stripper.

In the stripper/condenser, the gas is cooled by counterflowing condensate from the condenser. The temperature of the SO_2 rich gas that leaves the condenser is used to control the amount of cooling medium that must be sent to the condenser. Condensate from the stripper is returned to the buffer mix tank. The SO_2 rich gas, containing at least 90% SO_2 with the remainder being water, is ready for transport to a process unit. In the refinery, this normally will be the SRU or acid plant, where it will be converted to elemental sulfur or sulfuric acid. The SO_2 can also be compressed and transported if necessary. Figure 16.9 shows a typical Belco LABSORB System.

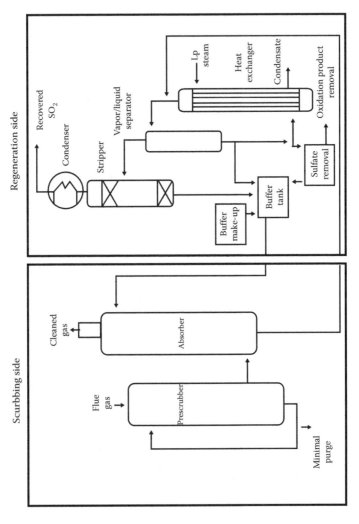

FIGURE 16.9   Belco LABSORB regenerative scrubber process diagram. (With permission from Belco Technologies.)

16.5.2 BELCO DUAL ALKALI REGENERATION

The dual alkali regeneration process takes the effluent streams from the EDV Wet Scrubbing System, and treats it in a series of process operations (the regeneration process) to produce a potentially marketable gypsum by-product. In so doing, the effluent stream from the scrubber is regenerated and sent back to the scrubber for reuse. The regeneration process train is designed to ensure overall system reliability. In addition, key process components have been spared in the train to further ensure complete operations over a sustained operating period. The following process description discusses the various portions of the regeneration process.

Lime Reactor: The effluent from the EDV system is pumped to an agitated reactor vessel where the EDV effluent is reacted with lime to form calcium sulfite and active sodium species via the following reactions:

$$2NaHSO_3 + Ca(OH)_2 \text{ ------} \rightarrow CaSO_3 + Na_2SO_3 + H_2O,$$

$$Na_2SO_3 + Ca(OH)_2 \text{ -----} \rightarrow CaSO_3 + 2NaOH.$$

In addition, some of the sulfite gets oxidized in the EDV to form sulfate, and the sulfate material also reacts with lime as shown below to form calcium sulfate:

$$Na_2SO_4 + Ca(OH)_2 \text{ -----} \rightarrow CaSO_4 + 2NaOH.$$

The lime addition rate is determined by the system pH and will be controlled. The lime will be slaked on-site prior to being fed to the reaction vessel. A small quantity of lime grits will be produced and will need to be disposed.

Thickener: The lime reactor products flow by gravity to a thickener where the calcium solids settle. The thickener overflow, which is a clear liquor, is separated and is pumped to a clarifier to remove any dissolved calcium sulfate. The thickener underflow is pumped to a rotary vacuum filter to further concentrate the calcium sulfite solids.

Clarifier: In the clarifier, the thickener overflow is treated to precipitate any dissolved calcium sulfate. The solids from the clarifier are returned to the thickener where it is settled and discharged with the thickener underflow. The clear liquor from the clarifier is returned to the EDV as make-up for SO_2 and particulates removal.

Rotary Drum Vacuum Filter: The underflow from the thickener is the feed to this filter. The filter concentrates the solids (predominately calcium sulfite), which then leave the filter in a cake form and are transferred to a cake repulp tank where they can be reslurried. The liquid generated from the filtering process (filtrate) is collected and pumped to the lime reactor vessel.

Cake Repulp Tank: The cake repulp tank is an agitated vessel whose function is to reslurry the calcium sulfite solids so that they can be oxidized. The solids are mixed with recycled filtrate and fresh make-up water to produce a slurry that is then oxidized.

Oxidizer Column: The oxidizer column is where the calcium sulfite slurry is reacted with air to form the by-product calcium sulfate. Air compressors are used to supply the oxidation air. In addition, the proper pH is maintained to sustain the oxidation reaction. The product leaving the oxidizer column is basically by-product gypsum that will contain any ash/catalyst/coke fines that enter the regeneration system.

Hydroseparator: The effluent from the oxidizer column next flows to a hydroseparator in which any lime grit, catalyst, or coke fines are physically removed. The solids are sent to a horizontal vacuum belt filter, and the overhead liquor from this separator will be further treated to reclaim most of the process liquor and to produce a filter cake for disposal.

Waste Clarifier: This clarifier will treat the overhead liquor from the hydroseparator to concentrate any solids and to reclaim process liquor for reuse in the process. The overflow from this clarifier will be returned to the process and the underflow will be filtered to produce a filter cake for disposal. The filtrate from the filter press will be returned to the process.

Horizontal Belt Vacuum Filter: The final piece of equipment is the horizontal belt filter, in which the underflow from the hydroseparator (basically gypsum diluted with process liquor) will be filtered to produce marketable gypsum of ~12% moisture. The filter cake will be washed to reduce the amount of dissolved solids and will be conveyed to a product gypsum storage area. Filtrate from this filter will be disposed of as an aqueous purge stream.

16.5.3 CANSOLV

CANSOLV is an amine based regenerative scrubbing technology. It was originally developed by Union Carbide and acquired by Shell Global Solutions in 2008. The technology uses a unique amine to absorb SO_x. The process chemistry is outlined in Figure 16.10.

The process has been commercialized on 13 applications. However, only one is on an FCC flue gas. The process flowscheme is shown in Figure 16.11.

The process diagram illustrates a typical flowscheme for the CANSOLV SO_2 Scrubbing System. In it, gas is shown to leave the FCCU battery limits at a hot and dry condition. While a CO boiler and ESP are shown, the FCCU system may

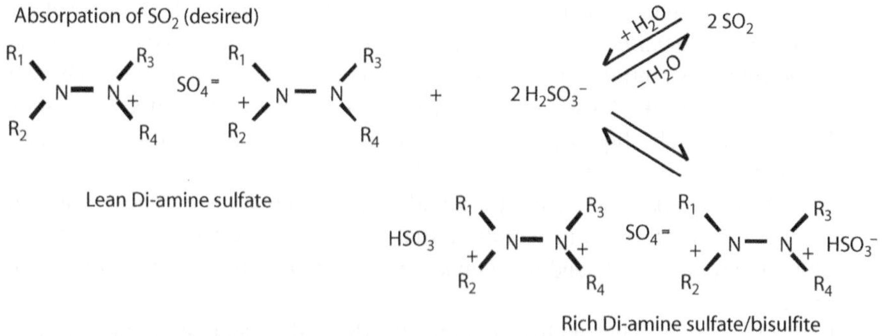

FIGURE 16.10 CANSOLV process chemistry. (With permission from CANSOLV Technologies.)

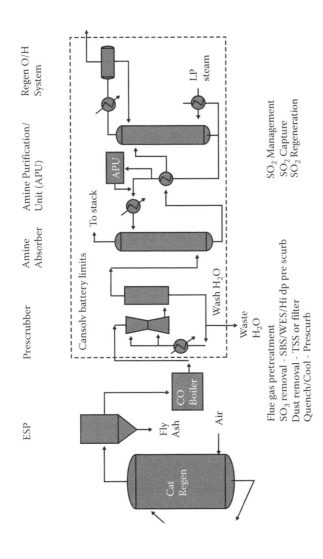

FIGURE 16.11 CANSOLV regenerable SO_2 scrubbing system process diagram. (With permission from CANSOLV Technologies.)

alternatively incorporate a full burn regenerator and no CO boiler, a third stage or fourth stage particulate separator, an electrostatic precipitator, or a waste heat recovery system to clean and cool the gas prior to its discharge to the scrubbing system. Regardless of the upstream equipment selected, gas must be completely oxidized to eliminate H_2S and hydrocarbons from the FCCU regenerated gas prior to its introduction to the SO_2 scrubbing system, or else elemental sulfur will be produced in the prescrubbing system if both H_2S and SO_2 are present and plugging can result.

16.5.3.1 Gas Conditioning

Hot gas, containing SO_2, water, nitrogen, oxygen, CO_2, particulates, and some SO_3 must first be cooled upstream of the CANSOLV system. Gases first enter a particulate removal and quench device and then flow to the CANSOLV SO_2 absorber. Before the gas can be contacted with the CANSOLV amine absorbent, it must first be quenched to saturation conditions and then subcooled to temperatures that are appropriate for the necessary amount of SO_2 removal. As the flue gas contacts the prescrubber water, SO_3 and some portion of the particulates in the gas are absorbed and the pH of the recirculating liquid drops to a steady state value of less than 1 as the acid concentrates. The prescrubber recirculating water stream must be partially purged to remove contaminants. A makeup stream of water is fed to the prescrubber to replace the water lost due to evaporation and to satisfy the purge requirements of the prescrubber loop. Waste water that is discharged from the prescrubber must be stripped of SO_2 and then neutralized prior to final disposal in the refinery wastewater treatment system.

16.5.3.2 SO_2 Absorber

SO_2 is absorbed from the feed gas by contact with the CANSOLV absorbent in the SO_2 multistage, countercurrent absorption tower. Trays or random packing may be used as mass transfer internals for the absorber, but structured packing is preferred, because it minimizes pressure drop and maximizes the SO_2 removal efficiency for a given height of packing. Cool, lean amine is fed to the top of the tower. As the absorbent flows down the column countercurrently to the flow of feed gas, SO_2 is absorbed into the amine absorbent, which then exits the column as rich amine from the bottom of the tower. The absorbent is pumped through a lean/rich amine heat exchanger and on to the SO_2 regeneration tower. The SO_2 content of the treated gas is largely determined by the quality of the lean amine. If low SO_2 levels in the treated gas are required, the lean amine must be well regenerated to a low SO_2 content. The absorber temperature and the lean amine temperature must also be carefully selected to ensure that target SO_2 specifications are met. A polishing NaOH scrubber can also be used to polish gas from the absorber, when regulations call for an extremely low level of SO_2 in the treated FCCU regenerated off gas.

Losses of amine up the stack are undetectable in the CANSOLV SO_2 Scrubbing System. Losses from evaporation do not occur because the amine is in salt form in solution, and losses from entrainment are avoided by using trough type distributors instead of spray nozzles to distribute the amine over the packing. In extreme cases where upsets in gas flow are anticipated, a demister may also be used.

16.5.3.3   Absorbent Regeneration

The regeneration section of the CANSOLV SO_2 Scrubbing System has four main components: a lean-rich heat exchanger, the regeneration column, a reboiler, and a condenser. Depending on the required delivery pressure, a blower or compressor might also be necessary to transfer SO_2 to the receiving sulfur recovery unit or sulfuric acid plant. The rich SO_2 laden absorbent from the absorption tower is fed to the regeneration tower via the lean/rich heat exchangers, where sensible heat is recovered from the hot, lean amine leaving the regenerator tower. The regeneration tower is packed with structured packing in order to achieve high mass transfer efficiency and a low pressure drop across the tower. Low pressure steam is used in the reboiler to generate stripping steam that removes SO_2 from the amine solution. SO_2 that is stripped from the liquid is carried overhead by the stripping steam and cooled in the overhead condensers where most of the steam condenses. The SO_2 is transferred offsite to the by-product conversion unit. The lean amine leaves the reboiler and flows through the lean/rich exchangers and then to the absorption tower, via the lean amine cooler. A slipstream of the lean absorbent is directed through the lean absorbent mechanical filters and then to the CANSOLV amine purification unit (APU).

16.5.3.4   Amine Purification Unit

The absorbent in the CANSOLV SO_2 Scrubbing System accumulates nonregenerable salts [also called Heat Stable Salts (HSS)] and dust that are removed from the gas over time. These contaminants must be removed from the absorbent continuously to avoid excessive build-up. An APU incorporates both an ion exchange unit (IX) for the removal of HSS and a filtration unit for the removal of dust.

The biggest operating concern is handling the HSS and dealing with caustic carryover from the prescrubber.

16.6   SUMMARY AND CONCLUSION

SO_x is a colorless, reactive gas produced during the burning of sulfur-containing fuels. It is a known pollutant that can react with other compounds to form small particles that are a respiratory health danger and can also form acid rain, which is an environmental danger. Several methods to control SO_x emissions from FCCUs are available including catalyst additives, wet gas scrubbers, and regenerative scrubbers. All have been able to demonstrate compliance with environmental regulations and consent decree requirements.

ACKNOWLEDGMENTS

The following individuals and companies are acknowledged for their input and permission to use information and data: Nic Confuorto, Business Manager, Belco Technologies; Martin Evans, Director of Technical Service, Intercat Inc.; Rick Birnbaum, Sales Manager Oil & Gas, CANSOLV Technologies.

REFERENCES

1. EPA Web site, http://www.epa.gov
2. Huling, G. P., et al., Feed Sulfur Distribution in FCC Product. *Oil & Gas Journal*, May 19, 1975.
3. Evans, M. SOx REDUCTION. *Hydrocarbon Engineering*, January 2003.
4. Maholland, M. K., Aru, G. W., and Clark, P. A. SOx 25. *Hydrocarbon Engineering* 9, no. 9 (2004).
5. Magnabosco, L. Principles of SOx Reduction Technology in FCCUs. *Fluid Catalytic Cracking VII, Studies in Surface Science and Catalysis,* Vol. 166, 254–305. Amsterdam: Elsevier, 2007.
6. Yoo, J. and Jaecker, J., Catalyst and Process for Conversion of Hydrocarbons, U.S. Patent 4,469,589, 1984.
7. Vierheilig, A. A. *Process for Making, and Use of, Anionic Clay Materials*. U.S. Patent 6,479,421, 2002.
8. Van Broekhoven, E., Catalyst composition and absorbent which contain an anionic clay, U.S. Patent 4,866,019, 1989.
9. Van Broekhoven, E., Cracking process employing a catalyst composition and absorbent which contain an anionic clay, U.S. Patent 4,946,581, 1990.
10. Van Broekhoven, E., Process for removing sulfur oxides with an absorbent which contain an anionic clay, U.S. Patent 4,952,382, 1990.
11. Sexton, J., Confuorto, N., Sarresh, N., and Barrasso, M. *LoTOx™ Technology Demonstration at Marathon's Texas City Refinery*. 2004 NPRA Annual Meeting, San Antonio, Texas, March 22, 2004.

17 FCC Emission Reduction Technologies through Consent Decree Implementation: FCC NO$_x$ Emissions and Controls

Jeffrey A. Sexton

CONTENTS

17.1 INTRODUCTION

The Clean Air Act requires EPA to set national ambient air quality standards for criteria pollutants. Currently, nitrogen oxides and five other major pollutants are listed as criteria pollutants. The sum of nitric oxide (NO) and NO_2 is commonly called nitrogen oxides or NO_x. Other oxides of nitrogen including nitrous acid and nitric acid are part of the nitrogen oxide family. While EPA's National Ambient Air Quality Standard (NAAQS) covers this entire family, NO_2 is the component of greatest interest and the indicator for the larger group of nitrogen oxides. NO_x is also a precursor for ozone formation. There is a NAAQS for ozone as well. NO_x is generally regulated as a precursor to ozone rather than for ambient NO_2 reasons.

All areas in the United States presently meet the current (1971) NO_2 NAAQS, with annual NO_2 concentrations measured at area-wide monitors well below the level of the standard (53 ppb). Annual average ambient NO_2 concentrations, as measured at area-wide monitors, have decreased by more than 40% since 1980. Currently, the annual average NO_2 concentrations range from approximately 10 to 20 ppb [1]. However, many areas in the U.S. do not meet the ozone NAAQS and is why NO_x continues to be a pollutant of concern.

EPA expects NO_2 concentrations will continue to decrease in the future as a result of a number of mobile source regulations that are taking effect. Tier 2 standards for light-duty vehicles began phasing in during 2004, and new NO_x standards for heavy-duty engines are phasing in between 2007 and 2010 model years.

NO_x reacts with ammonia, moisture, and other compounds to form small particles. These small particles penetrate deeply into sensitive parts of the lungs and can cause or worsen respiratory disease and aggravate existing heart disease. NO_x has also been identified as the primary cause for formation of ground level ozone (smog) that is formed when NO_x reacts with VOCs in the presence of heat and sunlight. For this reason, regulatory agencies have increased attention on reducing NO_x emissions from stationary sources. In 1997, as part of the revision to the Clean Air Act, the EPA issued a stricter ozone standard of 0.08 ppm averaged over an 8 hour period, as opposed to the older standard of 0.12 ppm averaged over a 1 hour period. According to the EPA, motor vehicles account for 49% of the NO_x emissions, utilities contribute about 27%, and industrial and commercial factories account for about 19% of the emissions [2].

While petroleum refining represents only 5% of the total emissions, these emissions are concentrated in small areas and generally in or near metropolitan areas not meeting the ozone NAAQS. Many refineries are located in so called nonattainment areas and can significantly contribute to local concentration of NO_x and the concomitant ozone. For many refineries, the NO_x emission limits from the Fluid Catalytic

TABLE 17.1

Options for Reducing NO_x Emissions from FCC Units

Potential Solution	Advantages	Disadvantages
Regenerator design	Inherent with process	Controllability
Feed Hydrotreating	Effectiveness	Cost
Catalyst additives and promoters	No capital cost	Effectiveness
SNCR	Low cost	Effectiveness
SCR	Effectiveness	Cost and operability
$LoTO_x$	Effectiveness	Operating costs

Cracking (FCC) units have been established through consent decree between the EPA and the refinery or by the application of new source review (NSR) provisions of the Clean Air Act when making refinery modifications that result in a significant emission increase. NSR standards require facilities to apply best available control technology (BACT) in ozone attainment areas and the lowest achievable emissions rate (LAER) in ozone nonattainment areas. In addition, in mid-2009, the U.S. EPA revised the Standards of Performance for Petroleum Refineries (40 CFR 60 Subpart Ja) to include NO_x standards for process heaters and FCCUs.

Since the FCCU is one of the largest single sources of emissions, in terms of tons emitted per year, the FCCU is a primary area of focus for NO_x reduction. Uncontrolled NO_x emissions from the FCCU regenerator can vary greatly and depends on many variables, such as the feed to the FCCU, regenerator design (partial or full burn), and the design of the secondary combustion device (CO boiler) if applicable. Uncontrolled NO_x emissions can range from 50 to 400 ppm, although most facilities see uncontrolled levels in the range of 75 to 150 ppm.

There are several ways to control NO_x produced by an FCC regenerator. Several methods have been described previously [3]. In situ methods include equipment design, operating conditions, and additives, which include nonplatinum based CO promoters and NO_x reducing additives. Feed hydrotreating can reduce total feed nitrogen 20–90% depending on severity and consequently reduce the NO_x potential of the feed. Flue gas treating methods include selective catalytic reduction (SCR) systems, selective noncatalytic reduction systems, and ozone injection systems. Table 17.1 contains a summary of NO_x reducing technology options.

17.2 NO_x FORMATION CHEMISTRY

By definition, NO_x includes NO plus NO_2. Other nitrogen bearing oxides such as N_2O are also present in FCC unit flue gas, but are not considered to be NO_x and are not regulated. It is generally understood that the predominant NO_x species inside an FCC regenerator is NO that is further oxidized to NO_2 upon release to the atmosphere. NO_x in the regenerator may be formed by two mechanisms, thermal NO_x produced from the reaction of molecular nitrogen with oxygen and fuel NO_x produced from the oxidation of nitrogen-containing coke species deposited on the catalyst during feedstock conversion inside the reactor.

Simple thermodynamic calculations show that thermal NO_x is not a significant contributor to total NO_x. Thermal NO_x occurs at very low rates below 1600°F. Consequently thermal NO_x is not a concern for FCC unit regenerators, most of which operate below 1400°F. However, thermal NO_x is often a significant fraction for partial-burn regenerators because the CO boilers utilized in partial-burn operation typically operate in excess of 1600°F.

The chemistry of coke-bound nitrogen conversion to NO_x is very complex. It is known that nitrogen is converted to reduce gaseous nitrogen compounds such as hydrogen cyanide (HCN) and ammonia (NH_3). These reduced species are then oxidized to NO_x. The selectivity of competing reaction pathways to form NO_x and N_2 is strongly influenced by the regeneration conditions. In the FCCU regenerator, there are two reductants available to reduce NO_x to molecular nitrogen, CO, and coke. In a full combustion regenerator, the objective is to reduce NO_x after it is already formed. A reducing environment (i.e., partial combustion) inhibits the formation of NO_x and favors formation of the intermediate products and N_2. These intermediates are subsequently converted to NO_x and N_2 in the downstream CO boiler.

In addition to the pathways involving nitrogen species there are several other competing reactions. Each of these competing reactions can affect the nitrogen pathway by either occupying a catalytic site or by affecting the concentration of a reactant. The most critical one is the CO promotion pathway. The presence of a platinum additive promotes oxidation of CO to CO_2. A decrease in the CO concentration in the regenerator reduces the rate of the NO + CO reaction, causing the NO_x emission from the regenerator to increase. Platinum is also a known NH_3 oxidation catalyst. It likely increases the rate of NH_3 oxidation to NO_x under FCC regeneration conditions. An increase in NO_x emissions when using a platinum based CO combustion promoter has been observed in many commercial units.

A schematic of the key aspects of the nitrogen chemistry in the regenerator is shown in Figure 17.1.

Production of NO_x intermediates can be monitored with the degree of partial combustion. HCN appears to be minimized when CO is less than 2%. This window

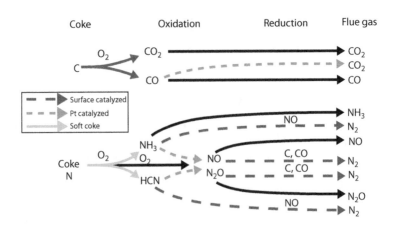

FIGURE 17.1 FCC NO_x chemistry.

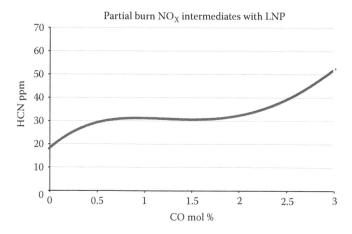

FIGURE 17.2   HCN vs. NO_x in partial burn regeneration.

TABLE 17.2
FCC Feed Nitrogen Distribution to Products

Feed Type	Unit 1 HT VGO	Unit 3 VGO + Sweet Reside	Unit 2 Partially HT + Coker GO
Feed Total Nitrogen, ppm	469	696	986
Sour water (%)	12	8.1	12.1
Gasoline (%)	5.8	1	3.8
LCO (%)	5.7	10.1	26.3
Slurry (%)	8.7	7.8	13.2
Coke (%)	67.8	73.1	44.6
Coke N_2–NO_x (%)	5	5	2.5

of operation typically results in the lowest NO_x emissions. Figure 17.2 shows HCN as it varies with CO concentration in the regenerator flue gas. These data were collected on a commercial FCC regenerator with an FTIR analyzer.

17.3   FEED NITROGEN DISTRIBUTION

The majority of nitrogen in the feed is produced in the coke. Nitrogen species typically attach to the FCC catalyst and are deposited as coke. However, the majority of the nitrogen in coke is liberated as N_2 with only a small amount resulting in NO_x. Table 17.2 is a survey showing FCC feed nitrogen distribution to products.

17.4   CATALYST ADDITIVES

17.4.1 NO_x REDUCING CATALYST ADDITIVES

The modern catalytic converter installed on most automobiles is a washcoat consisting of precious metal oxides, supported on a ceramic monolith. After passage of the

Clean Air Act in 1970, the EPA mandated the control of CO, hydrocarbons, and NO_x emissions from automobiles. This is particularly difficult to accomplish because CO and hydrocarbons must be oxidized while NO_x must be reduced. Chemically speaking, the oxidation and reduction processes are diametric opposites. The discovery of unique chemical substances that store and release oxygen, has allowed the catalytic converter to meet this challenge.

The same technology used in a catalytic converter was the starting point for use of additives to reduce NO_x in FCC units. The ability of these metal oxide materials to store and release oxygen affects the oxidation and reduction of coke nitrogen in the regenerator.

Several vendors offer NO_x reducing catalyst additives. All have seen commercial success in reducing NO_x. However, the results have not been consistent. The additives do not function like SO_x reducing additives that absorb the desired pollutant and transfer it to the reactor. Current generation NO_x additives affect the availability of nitrogen species to be oxidized and reduced, which is highly dependent on the bed hydrodynamics. As such, performance depends on the application.

Several commercial trials have demonstrated large NO_x reductions (>75%) when used in conjunction with conventional Pt-based COPs [4]. However, when used with low NO_x promoters (LNP), incremental reductions have been marginal. The EPA Consent Decrees required use of NO_x reducing additives as well as low NO_x promoters. Several refiners that initially used both additives as part of the Consent Decree requirements have subsequently removed the NO_x reducing additives after setting limits due to the success of LNP in reducing NO_x. Development of new NO_x reducing additives to meet this challenge remains a topic of research for the industry.

17.4.2 Low NO_x Promoters

For nearly 35 years, the preferred method of controlling afterburn and CO emissions has been the use of a platinum-containing combustion promoter to catalyze the oxidation of CO to CO_2 in the regenerator dense bed. Historically these combustion promoters have contained low concentrations of highly dispersed platinum (Pt) on an inert alumina support. Platinum (Pt) is an excellent combustion promoter that, unlike many other oxidation catalysts, can be used in low enough concentrations to catalyze CO oxidation without also catalyzing undesirable dehydrogenation reactions in the FCC reactor. However, it has been discovered that while Pt is an effective oxidation catalyst for the conversion of CO to CO_2, it also catalyzes the oxidation of nitrogen intermediate species (e.g., NH_3 and HCN) found in the regenerator. This results in increased NO_x emissions and runs counter to the desire to minimize all of these environmental pollutant emissions.

In recent years, a shift has been made away from Pt-containing CO combustion promoters to promoters containing elements other than Pt. This was driven by the requirements of various EPA Consent Decrees for several units to conduct trials of these new promoters. The desire is to use elements that are still effective in oxidizing CO to CO_2 and preventing afterburn, but that do not generate NO_x in the process. Many of the non-Pt combustion promoters that are currently available have substituted Palladium (Pd) for Pt as one of the active components. However, commercial performance results show that the performance of these materials varies widely, and that additives available from different suppliers are not all equivalent.

In developing a non-Pt CO combustion promoter, initial products substituted a different oxidizing agent for Pt, leaving all other additive properties unchanged. While many of these new additives have been successful in reducing NO_x emissions, most are not as effective as conventional CO combustion promoters, requiring from two to five times more additive to achieve the same degree of combustion effectiveness and afterburn control as with a Pt promoter.

Recent second-generation promoters have been proven much more effective [5]. This has been done by incorporating modifications to the base support as well as to the catalytically active components. The effectiveness of a CO promoter in the FCCU depends on several physical and chemical factors. The support must be a highly attrition resistant material, have a high particle density, and have a minimum amount of particles smaller than 40 microns. These properties are necessary to minimize first cycle losses of the promoter and give high retention in the unit inventory. The support should also have a moderately high surface area and a high pore volume to provide ready access of the reactants to the catalytically active sites. The catalytically active components must be uniformly dispersed on the support for maximum effectiveness. The chemical composition of the additive should not be easily poisoned by other components in the FCCU, nor should it generate undesirable side reactions.

INTERCAT's product COP-NP is one of the improved low NO_x promoters. Marathon conducted trials of COP-NP on four of their FCC units in a phased manner, successfully replacing first generation non-Pt promoters. The results found COP-NP to be two to three times more active than other vendor's first generation non-Pt promoters. This CO combustion promotion improvement comes without an increase in NO_x emissions as is seen with Pt-containing promoters. Overall, the improved combustion promotion effectiveness of COP-NP has resulted in use of significantly less additive for afterburn and CO emissions control, and has led to significant cost savings for each of Marathon's FCC units. The data are summarized in Table 17.3.

TABLE 17.3
Second Generation Low NO_x Promoter Performance

Refinery	Afterburn Reduction	CO Reduction	Additive Addition Reduction	Approximate Cost Savings
Garyville, LA	48%	47%	50%	2.8 cpb
Robinson, IL	92%	N/A	67%	2.9 cpb
Detroit, MI	19%	54%	38%	2.1 cpb
St. Paul, MN	1%	4%	56%	3.3 cpb

Source: With permission from Intercat, Inc.

17.5 SELECTIVE NONCATALYTIC REDUCTION

The selective Noncatalytic reduction (SNCR) process is a postcombustion NO_x reduction technology. NO_x is reduced through the controlled injection of a reagent, either ammonia or urea, into the combustion products of boiler, heater, or FCC regenerator. This process is typically applied on partial burn applications with a CO boiler (COB).

17.5.1 SNCR PROCESS DESCRIPTION

A basic process flow diagram for the SNCR TDN process is shown in Figure 17.3. Anhydrous or aqueous ammonia is vaporized and mixed with a carrier gas of air or steam for transport to injection/distribution modules. The injection distribution modules distribute ammonia and/or hydrogen reagent and carrier gas to proprietary spray nozzles or injection lances. Reagent flow control can be controlled and trimmed by outlet NO_x signals or ammonia slip measurements.

At the proper temperature, the injected reagent reacts selectively in the presence of oxygen to reduce the oxides of nitrogen (NO_x) primarily to molecular nitrogen (N_2) and water (H_2O). Figure 17.4 shows the results of pilot-scale tests of the use of urea and ammonia in the SNCR performance.

The SNCR process is capable of high levels of NO_x reduction under ideal conditions. The reaction between the reagent and NO_x occurs within a specific range of temperature. At about 1800°F (982°C) the reactions essentially go to completion in a short residence time. At 1600°F (871°C), a much longer residence time is required to achieve similar NO_x reduction. Because the reaction occurs slower at low temperatures, the potential for unreacted ammonia (ammonia slip) increases as temperature is reduced. At lower temperatures, use of an additional reagent (eSNCR) in the flue

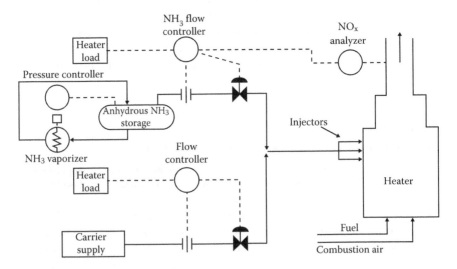

FIGURE 17.3 Basic process flow diagram for SNCR application. (With permission from Hamon.)

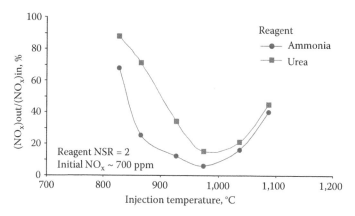

FIGURE 17.4 Impact of gas temperature on SNCR performance.

gas stream will extend the operating window for efficient NO_x reduction. Above 2000°F, SNCR becomes impractical because the equilibrium shifts in favor of NH_3 oxidation (creation of NO_x) rather than reduction. This results in an optimal temperature window of 1600–2000°F for ammonia injection and may be extended < 1400°F with the addition of hydrogen (eSNCR).

17.5.2 SNCR PROCESS CHEMISTRY

The overall chemical reactions for reducing NO_x with ammonia are:

$$4NO + 4NH_3 + O_2 \rightarrow 4N_2 + 6H_2O,$$

$$6NO_2 + 8NH_3 \rightarrow 7N_2 + 12H_2O.$$

Enhanced selective noncatalytic reduction (eSNCR) is the same process with an additional reagent, such as hydrogen, to extend the temperature window for efficient NO_x reduction. Kinetic modeling suggests that the hydrogen generates OH radicals, which accelerates the performance of NH_3 agents by increasing the production rate of NH_2 radicals:

$$NH_3 + OH \rightarrow NH_2 + H_2O,$$

$$NH_2 + NO \rightarrow N_2 + H_2O.$$

This effect becomes especially pronounced at low temperatures where the generation of active centers by the N–H–O system slows down and the hydrogen accelerates the radical initiation process. The injection of hydrogen can reduce the NO_x reduction temperature window enough so it could be applied on some conventional flue gas streams without COBs. Others have documented use of sodium salts as another additive that can generate OH radicals and has the same effect as hydrogen addition. However, these additives have not been commercialized. Figure 17.5 shows the overall reactions for SNCR.

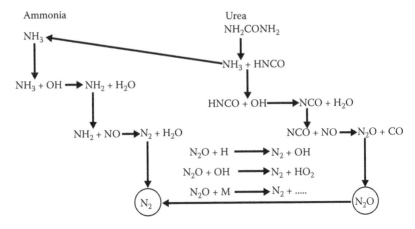

FIGURE 17.5 Overall SNCR reactions.

TABLE 17.4
Parameters Affecting SNCR Process Performance

Item #	Parameter	Furnace Characteristics	Control
1	Temperature	X	
2	Residence time	X	
3	Initial NO_x level	X	
4	Flue gas composition (N_2, CO_2, H_2O, O_2, CO, NO, SO_2, SO_3)	X	
5	Reagent stoichiometry (NH_3/NO_x ratio, H_2 injection rate for eSNCR)		X
6	Reagent distribution/mixing		X

17.5.3 SNCR DESIGN CONSIDERATIONS

In practical combustion systems, such as CO boilers, the flue gas experiences spatial and temporal variations. Constituent concentration, streamline residence time, and temperature are critical to determining an efficient process design. Computational fluid dynamics (CFD) modeling and chemical kinetic modeling are used to achieve accurate design assessments and NO_x reduction predictions based on these parameters. The critical parameters affecting SNCR and eSNCR design are listed in Table 17.4.

17.5.4 SNCR COMMERCIAL EXPERIENCE AND APPLICATION

A number of vendors offer SNCR technology based on either ammonia or urea. Exxon Mobil Thermal DeNO$_x$ (TDN) technology is a common SNCR process applied to FCC units. The technology is licensed exclusively to Hamon Research-Cottrell Inc., and has been utilized to achieve postcombustion NO_x reduction in CO furnaces, thermal oxidizers, overhead regenerators, and power boilers. Thermal

DeNO$_x$ technology is a relatively low cost method to achieve postcombustion NO$_x$ reduction. In the TDN process, combustion products are treated with ammonia and (optional) hydrogen. The TDN process can use either anhydrous or aqueous ammonia as a reagent source. GE Energy is another vendor with a number of applications of eSNCR technology in refineries in the United States and worldwide. Fuel-Tech licenses SNCR technology utilizing urea as the primary reagent. Most EPA Consent Decree applications have achieved 5–30% reduction with others in industry achieving up to 70% depending on process conditions.

The key to maximizing SNCR efficiency is uniform injection of ammonia through the ammonia injection grid (AIG). The reagent(s) must be distributed properly throughout the flow-field to achieve the correct stoichiometry in the correct temperature zone, with sufficient residence time along all flow streamlines. Improper control of reagent distribution will cause undertreatment, overtreatment, or less than optimal temperatures for good process performance. Optimal reagent distribution is critical for maintaining maximum NO$_x$ reduction while minimizing ammonia slip. For units with a wet gas scrubber downstream, the unreacted ammonia can be captured within the scrubber and allow higher DeNO$_x$ operation at higher slip rates.

A potential drawback of this technology is the potential formation of ammonium sulfate salts and their resulting fouling. This can be mitigated through either use of high pressure soot blowers or on-line water washing in the economizer. These salts will exist as small particulates that will increase opacity.

The components of an injection system are different for each type of reagent, anhydrous ammonia, aqueous ammonia, or urea. They are described as follows:

Anhydrous Ammonia: Anhydrous ammonia is injected into the flue gas as a vapor. Anhydrous Ammonia is stored at approximately 265 psig with an electric heater that cycles to maintain pressure in the tank. The anhydrous ammonia is mixed with dilution air supplied by fans. The ammonia concentration is typically 3–8% by volume, which is half of the explosion range of 16–25%. The NH$_3$/air mixture is then injected into the flue gas through an AIG. Anhydrous ammonia is the most economical reagent to use, however, it is more difficult to transport, handle, and store than other reagents.

Aqueous Ammonia: Aqueous ammonia is also injected into the flue gas as a vapor, but since it is typically 19–29% ammonia, the water must be vaporized by use of hot air. Ambient air is pumped through an electric heater into a vaporizing vessel. Aqueous ammonia is sprayed into the top of the vessel where it contacts the heated air. The NH$_3$/air mixture then proceeds to the AIG.

Urea: Urea is injected as a liquid into the flue gas stream, so a heater for vaporization is not needed. Since urea will crystallize at temperatures below 70°F, the entire system must be heat traced and insulated. A reagent circulation module consisting of a high head, high-flow delivery system is designed to supply filtered reagent to the injection zone-metering module. The injection zone metering module is used to precisely meter and independently control the concentration of the reagent to each zone of injection.

17.6   SELECTIVE CATALYTIC REDUCTION (SCR)

Selective Catalytic Reduction (SCR) process is very similar to SNCR with the exception that a catalyst is used to accelerate the reactions at lower temperatures allowing it to be applied to both full and partial-burn regenerators. An SCR system consists of a catalyst bed installed in the flue gas line of a combustion system. Ammonia is injected into the flue gas with air in the presence of a catalyst. The catalyst is typically oxide forms of vanadium and tungsten. The ammonia selectively reacts with NO_x to form molecular nitrogen and water via an exothermic reaction that has achieved > 90% reduction in NO_x when applied to an FCCU.

17.6.1   SCR PROCESS

The ideal temperature range for an SCR system is 600–750°F, which gives the smallest catalyst volume for a given NO_x reduction. If the temperature is outside of this range, additional catalyst volume will be required to maintain the same NO_x reduction. The temperature can also vary depending on the concentration of SO_3 in the flue gas stream. Higher concentrations of SO_3 require the temperature to be raised to keep ammonium bisulfate and ammonium sulfate from precipitating and plugging the catalyst. High SO_3 concentrations also lower the amount of NO_x reduction at a given temperature. Controlling the temperature is typically accomplished through integration with the waste heat boiler (WHB). Either boiler feed water flow control or a flue gas bypass can be used to regulate inlet temperature. Steam production from the flue gas may decrease with application of SCR due to the desired temperature for SCR being higher than desired for maximum energy recovery. Figure 17.6 shows a typical SCR process flow diagram.

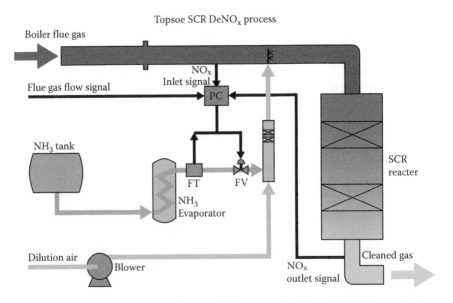

FIGURE 17.6   SCR process flow diagram. (With permission from Haldor-Topsoe, Inc.)

The same reagents used for SNCR can also be used for SCR service including anhydrous ammonia (NH_3), aqueous ammonia (NH_4OH), or urea ($NH_2–CO–NH_2$). The amount of ammonia reagent required to remove NO_x is slightly above a 1:1 stoichiometric ratio of NH_3 to NO_x accounting for maldistribution and allowable NH_3 slip. Therefore, if each mole of injected urea or ammonia reduces NO_x to the theoretical maximum amount, utilization is 100%. One hundred percent chemical utilization is approached in SCR systems, but in SNCR systems the value ranges from 30 to 60%.

Ammonia is adsorbed on the surface of an SCR catalyst in a diffusion limited laminar flow regime. The ammonia combines with vanadium pentoxide V_2O_5, a catalytic metal impregnated on the surface of the catalyst, to form a Bronsted acid site. NO_x reduction takes place on this acid site to form nitrogen and water. The spent V^{4+}-OH site is restored to V^{5+}-OH via oxidation to repeat the catalytic cycle. Once the vanadium site can no longer revert back into the {+5} oxidative state, then that site is no longer active for NO_x reduction. Figure 17.7 shows the catalytic cycle for the SCR reactions.

17.6.2 SCR Catalyst

The catalyst must be designed to handle the abrasive environment where catalyst fines are always present in the flue gas yet still perform with a low pressure drop typically below 5 inches of water column. It must also maintain activity continuously for a 5 year cycle, yet be selective enough to limit undesirable reactions like SO_2 oxidation. The catalyst must also be able to withstand periodic blasts of steam or pressurized air coming from the soot blower system found in many of the newer FCCU SCR units.

The SCR catalyst designed for FCCU regenerator flue gas service is a homogenous monolith, typically made from 1 mm thick material. Some catalysts are extruded clays that receive a wash coat of titanium dioxide before impregnation of the vanadium and tungsten metals. Another type involves painted plates of expanded metal

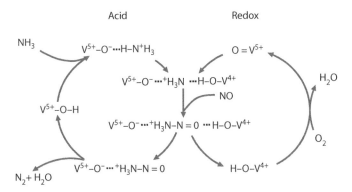

FIGURE 17.7 Catalytic cycle for SCR reaction over vanadium/tungsten catalyst. (With permission from Haldor-Topsoe, Inc.)

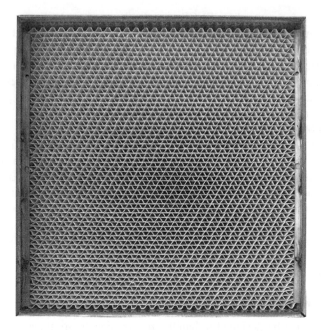

FIGURE 17.8 Inlet surface of the SCR catalyst. (With permission from Haldor-Topsoe, Inc.)

mesh. The paint is titanium dioxide and the catalytic metals are also vanadium and tungsten. The SCR catalyst shown in Figure 17.8 is made from a corrugated reinforced fiberglass matrix for mechanical strength and vibration tolerance. It also is a titania/vanadia/tungsten catalyst.

With the honeycomb system, the catalyst pitch determines the solids handling capabilities. Pitch is the distance from centerline to centerline of one gas path in the honeycomb and typically varies from 2 to 9 mm with higher pitches being used for heavier dust applications. Typical FCC SCR catalyst would have approximately a 5 mm pitch. The following causes can lead to catalyst deactivation:

Fouling: Salt formation can build-up on the catalyst surface effectively limiting accessibility. Ammonium bisulfate can form at low temperatures. This foulant can be removed by increasing temperature and is considered a temporary poison.

Mechanical Failures: Cracks or debonding of the catalyst from the substrate material can occur from thermal stresses as well as dynamic forces on the modules. The catalyst must be carefully handled to prevent premature fracturing. Each requires a warm-up and cool-down rate.

Thermal Degradation: Catalyst sintering can occur at flue gas temperatures > 800°F. This will result in the pore distribution shifting to larger pores. The loss of small pores will generally not have a large effect on activity since diffusion is not a critical parameter. The majority of conversion occurs on the exterior surface of the catalyst.

Erosion: Excessive catalyst fines loading can lead to erosion of the catalyst from the substrate.

Plugging: Catalyst fines can potentially deposit on the catalyst system and reduce accessibility. An ESP or TSS is often installed before the SCR to reduce catalyst loading. However, recent designs have moved the SCR before the PM removal device to minimize collection of very fine particulates that can lead to pressure drop increases.

17.6.3   SCR Technology Evolution

Selective Catalytic Reduction (SCR) has been commercially used since the mid 1980s on fired equipment with the first application on a boiler in 1976. The first SCR unit installed on a fluid catalytic cracking unit was at Saibu Oil Company in Yamaguchi, Japan in April 1986. Since then, nearly two dozen FCC units have installed SCR units to remove NO_x from the flue gas and more are slated to be built in the future. Vendors and catalyst suppliers of this technology include Haldor-Topsoe, Mitsubishi Power Systems, Hitachi, Technip, BASF, and Cormetech.

The following is a list of SCR units installed on FCC units:

Saibu Oil, Yamaguchi, Japan	Showa Yokkaichi Oil, Yokkaichi, Japan
Nippon Petroleum Refining, Negishi, Japan	KOA Osaka, Japan
CPC Taiwan, PRC	Kyokutou Petrochemical, Chiba, Japan
Idemitsu Petrochemical, Hokkaido, Japan	Scanraf, Gothenburg, Sweden
Tamoil, Switzerland	Shell, Hamburg, Germany
Shell Pernis, Rotterdam, The Netherlands	ExxonMobil, Torrance, CA
ExxonMobil, Beaumont, TX	BP, Carson, CA
BP, Whiting, IN	ConocoPhillips, Wood River, IL
Shell Oil Company, Deer Park, TX	CITGO Petroleum, Lemont, IL
Chevron, El Segundo, CA	SUNOCO, Philadelphia, PA
Sinclair Oil Company, Tulsa, OK	VALERO, Benicia, CA
ExxonMobil, Billings, MT	SUNOCO, Toledo, OH

The majority of these units have a Third Stage Separator (TSS) or electrostatic precipitator (ESP) located before the SCR catalyst bed to protect against upsets in the FCC regenerator. The catalyst can easily be designed to handle the normal dust loading (60 mg–700 mg/Nm^3), which is much lower than the typical coal fired boiler (5 g–9 g/Nm^3). To handle a FCC upset (>15 g/Nm^3) without a PM removal device, the catalyst volume would need to increase.

The first U.S. FCC unit that installed a SCR was Mobil (now ExxonMobil) in Torrance, CA in April 2000. It was significant relative to the previous experience on other FCCU SCR units because this unit was designed to run continuously for a 5 year run between scheduled turnarounds. The Japanese units typically took a maintenance and inspection outage every year.

17.6.4 SCR DESIGN CONSIDERATIONS

The FCC unit application for SCR comes with unique challenges. These include:

- Two-phase flow with catalyst fines entrained in the flue gas
- High NO_x reduction in a high inlet NO_x environment
- Low SO_2 oxidation in a high inlet SO_x environment
- Low pressure drop requirement
- Low NH_3 slip requirement
- Guaranteed continuous performance for 4–5 years

Along with NO_x reduction requirements, refiners must also contend with other emission reductions, specifically CO, SO_x, and particulate matter. Thus, there is an optimal amount of catalytic activity to achieve NO_x reduction over several years in a dusty application, yet not too active to oxidize SO_2 to SO_3 above accepTable limits.

There are two design options when considering an SCR unit for the FCCU: upstream or downstream of an ESP or TSS. If the SCR unit is placed upstream of an ESP or TSS, then the refiner has to incorporate soot blowers for catalyst fines removal from the catalyst surface and use a wider pitch catalyst to handle the higher levels of catalyst particulates. The wider pitch catalyst contains more void volume and thus will directionally increase the catalyst bed dimensions since the NO_x reduction is based on the total amount of surface area, not just catalyst volume.

If the SCR is placed downstream of an ESP or TSS, the design can take advantage of a cleaner flue gas. This would allow for smaller catalyst volumes using finer pitch catalyst and thus smaller SCR reactors. Problems occur when the ESP or TSS collection efficiency no longer removes the particulates from the flue gas. Not only does the SCR catalyst bed foul, requiring increased run frequency on the soot blowers, the stack opacity will also increase.

Several units with a PM collection device located upstream of the SCR have seen increased pressure drop from fine particulates accumulating on the catalyst bed. Soot blowers have been partially successful in this application. When the SCR is applied to a CO boiler with limited pressure drop, the SCR has typically been located upstream of PM removal to avoid this problem. Some refiners have chosen to install a spare SCR reactor to provide redundancy due to pressure drop concerns. Others have used a bypass where local regulations allow.

Several refiners have recently dismantled older ESPs and built wet gas scrubbers with wet ESPs to remove SO_3 at the stack. In these applications, the SCR unit can be designed to handle the catalyst fines as well as full range equilibrium catalyst during an upset condition.

Similar to SNCR, the distribution of ammonia and design of the AIG is critical. The ability to tune this flow and distribution during operation can improve performance. Another important consideration is use of an on-line NH_3 slip analyzer.

A tunable IR laser spectroscopy analyzer has been demonstrated on several applications to accurately control slip.

17.6.5   SCR CHEMISTRY

The main chemical equations associated with SCR are as follows:

$$4NO + 4NH_3 + O_2 \rightarrow 4N_2 + 6H_2O \qquad \Delta H_0 = -1{,}627.7 \text{ kJ/mol,}$$

$$NO + NO_2 + 2NH_3 \rightarrow 2N_2 + 3H_2O \qquad \Delta H_0 = -757.9 \text{ kJ/ mol,}$$

$$6NO_2 + 8NH_3 \rightarrow 7N_2 + 12H_2O \qquad \Delta H_0 = -3{,}067.1 \text{ kJ/ mol.}$$

NO (nitrogen monoxide) is the primary NO_x component in the flue gas meaning that the first equation above is the more significant one. Stoichiometry reveals that one mole of ammonia is required to reduce one mole of NO_x and convert it to nitrogen and water. Reaction rates are indicative of the Arrhenius equation that describes temperature dependent reactions.

There are undesirable side reactions associated with SCR. They are presented below:

Oxidation of SO_2 to SO_3: $\qquad\qquad SO_2 + \frac{1}{2}O_2 \rightarrow SO_3,$

Ammonia oxidation to NO_x: $\qquad\qquad 2NH_3 + 3\frac{1}{2}O_2 \rightarrow 2NO_2 + 3H_2O,$

Sulfuric acid formation: $\qquad\qquad\qquad SO_3 + H_2O \rightarrow H_2SO_4,$

Ammonium bisulfate formation: $\qquad\quad NH_3 + H_2SO_4 \rightarrow NH_4HSO_4.$

The oxidation of SO_2 to SO_3 is undesirable for several reasons. SO_3 will result in a blue plume off the stack and increase opacity. SO_3 will also continue to react to form sulfuric acid and ammonium bisulfate. Sulfuric acid will corrode downstream equipment like economizers and ductwork as well as leave the stack as an acid mist. SO_3 will also reactant with NH_3 to form ammonium bisulfate salt (ABS), which is corrosive and tacky. ABS will sublime from a gas to its solid form at and below its dew point.

SO_2 oxidation is kinetically driven. It increases with temperature and catalyst activity. SCR catalysts are reduction catalysts. However, they will also oxidize SO_2 to SO_3 depending on the catalyst design and operating conditions. A platinum based CO promoter is also a highly efficient oxidizing catalyst used to convert CO to CO_2. Unfortunately, it will also oxidize SO_2 to SO_3. Thus, it becomes significantly important to identify the presence of oxidizing promoters/additives in the circulating inventory to better manage the SCR.

Ammonia oxidation begins to occur at high temperatures above 900°F. Incremental NO_x is produced through ammonia oxidation. As the FCCU is typically

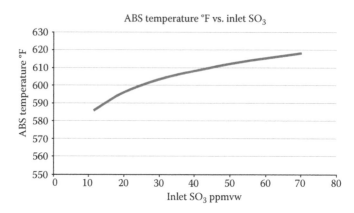

FIGURE 17.9 ABS formation temperature versus SO_3 concentration. (With permission from Haldor-Topsoe, Inc.)

a high inlet NO_x SCR application, the creation of more NO_x through ammonia oxidation is undesirable. Therefore, the refiner is challenged to accurately inject an equi-molar amount of ammonia relative to NO_x.

Ammonia slip on large units like an FCCU translates to wasted chemical expense with no benefit to the refiner. Excess ammonia in a high temperature SCR will oxidize to create NO_x, which is counter to the intended purpose of the unit. Excess ammonia in a low temperature SCR will tend to form ammonium bisulfate (ABS) per Figure 17.9.

Ammonium bisulfate is a corrosive salt that sublimes from a vapor to a solid at temperatures below its dew point. The salt will foul downstream equipment resulting in higher pressure drops across economizers and preheaters. Ammonium bisulfate can be returned to its vapor state if the temperature is increased above its ABS dew point.

ABS formation on a FCCU SCR is a major concern because it is a sticky foulant. Once formed, it traps catalyst fines to its surface. Once these particles are no longer moving within the flue gas stream, it is very difficult to reentrain them even with the use of soot blowers.

Figures 17.10 and 17.11 illustrate the color and severity of ABS formation and fouling. This is an SCR on a refinery heater. The ABS precipitated on the edge of the access door probably due to cooler temperatures created because of a seal leak. The interior side and top edges of the door showed ABS corrosion once the maintenance personnel performed their inspection. If the seal problem went uncorrected, eventually ABS would have destroyed the door requiring it to be replaced.

17.6.6   SCR Commercial Experience and Application

The design of a FCCU SCR comes with unique challenges for the determination of the specific catalyst type and required volume to achieve NO_x reduction. These include:

FIGURE 17.10   ABS corrosion on SCR access door. (With permission from Haldor-Topsoe, Inc.)

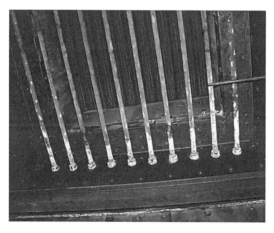

FIGURE 17.11   ABS corrosion on NH$_3$ injection lances. (With permission from Haldor-Topsoe, Inc.)

- Two-phase flow as catalyst fines are entrained in the flue gas
- Continuous operation targeting a 4–5 year run life
- Low pressure drop in a dusty operating environment

Once the design parameters are identified, sizing of the SCR catalyst follows. Typically, a multilayer SCR reactor is considered with the flue gas directed in a vertical down flow orientation. Some U.S. refiners have installed bypasses around the

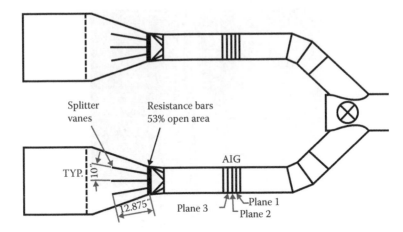

FIGURE 17.12   Split flow SCR design. (With permission from Haldor-Topsoe, Inc.)

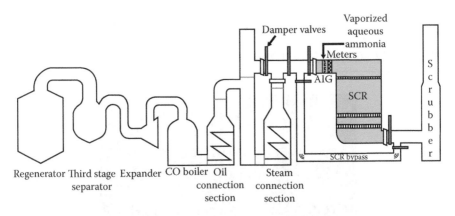

FIGURE 17.13   FCCU regenerator flow scheme with SCR reactor and bypass line. (With permission from Haldor-Topsoe, Inc.)

SCR reactor or installed parallel reactors in a wish-bone design. This was required to meet reliability requirements for a 5-year cycle and provide redundancy for any pressure drop concerns. Both designs are shown in Figures 17.12 and 17.13.

The design data for a U.S. FCCU SCR is provided in Table 17.5.

After deciding on a catalyst and determining the required volume configured in two identical layers, CFD can be used to further develop the design. Root mean square maldistributions for flue gas flow, NH_3: NO_x, and temperature are quantified and corrected within acceptable tolerances, +15%, +10% RMS, and +20°F, respectively. Turning vanes, static mixers, and adequate mixing time enable the even distribution of flow and NH_3 prior to entering the SCR catalyst.

CFD analysis allows design engineers the opportunity to quickly determine the viability of different configurations at low expense with today's advanced computer technology and modeling software. Since the FCCU flue gas is a two-phase flow

TABLE 17.5
Specific Design Data for SCR

Flue gas flow rate, pounds per hour	1,198,137
Temperature, °F	615
Pressure, inches of WC	30
NO_x, ppmvdc @ 0% reference O_2	315
SO_x, ppmvdc @ 0% reference O_2	1555
H_2O, vol. %	7.85
O_2, vol. %	2.77
CO_2, vol. %	13.64
N_2, vol. %	76.84
Ar, vol. %	0.90
Particulate, pounds per hour	1000
Flue gas flow maldistribution, ± % RMS	15
NH_3 to NO_x maldistribution, ± % RMS	5
Temperature maldistribution, °F	25

stream, CFD is used to identify the location and quantity of guide vanes to achieve the ideal flow pattern. Figure 17.14 is a CFD model output for this application.

The model results show the velocity profile of a FCCU flue gas traveling vertically upward then making a 90° turn leading to the SCR. Notice the uniformity of the stream velocity as it travels upward. As soon as the stream encounters the 90° turn, the velocities stratify with nearly stagnant flow at the corner, and very high velocities at the far wall. This occurs because the denser catalyst particles are carried out further in the duct by their momentum relative to the lighter gas molecules.

Based on these results, it becomes necessary to install guide vanes to prevent flue gas flow stagnation at the corner as well as erosion of the far wall. The goal of this work is to deliver a well-mixed homogeneous flue gas stream to the inlet of the SCR catalyst. Angular entry of the flue gas into the SCR catalyst is avoided to protect the catalyst from erosion.

Construction of a plexiglass scale model for cold flow studies typically occurs after completion of the CFD analysis. Smoke entrained in an air stream is used to empirically confirm the SCR design. Tuning efficiencies of the guide vanes and mixing ability of the static mixers are some of the design qualities confirmed by the cold low model. Figure 17.15 is a picture of an actual model used for commercial scale-up.

Turndown scenarios using a two-phase stream are also investigated with the model. FCCU catalyst fines are represented by salt granules or silica beads. The model is used to evaluate the rate of particulate accumulation due to turndown conditions. An SCR unit should recover and return the settled catalyst that collects on straight run ducting and horizontal transitions back into the flowing stream.

After the CFD analysis and scale model studies are complete, the SCR unit design is done. Scale up and fabrication of the unit follows. The SCR catalyst modules and other equipment are produced. SCR catalyst for an FCCU is delivered as modules to reduce the number of crane lifts to load the reactor. The individual catalyst cubes

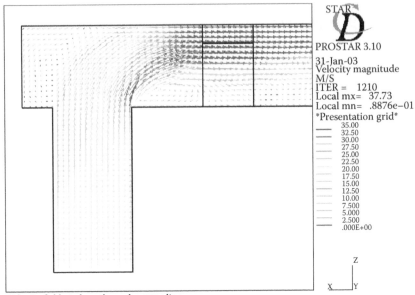

STAR
D
PROSTAR 3.10

31-Jan-03
Velocity magnitude
M/S
ITER =   1210
Local mx=   37.73
Local mn=   .8876e−01
Presentation grid

35.00
32.50
30.00
27.50
25.00
22.50
20.00
17.50
15.00
12.50
10.00
7.500
5.000
2.500
.000E+00

Z
X ⎿ Y

Velocity field at plane through centre line.
Shell, Deer park, existing and new T-piece
by Force Technology, Jan. 2003

FIGURE 17.14 CFD model results of a 90° elbow turn in the FCCU flue gas line. (With permission from Haldor-Topsoe, Inc.)

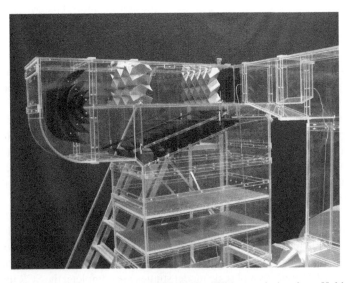

FIGURE 17.15 Plexiglass scale of the FCCU SCR. (With permission from Haldor-Topsoe, Inc.)

are bundled and welded onto a steel pedestal frame for strength and support. A protective top grid constructed of angle iron and wire mesh is included in the module design. A typical module is shown in Figure 17.16.

At elevation, an electric motor driven pallet mover picks the module up at the base frame and sets it in its intended place. The weight of an individual SCR catalyst module is approximately 2000 pounds and therefore requires safe handling. Loading the SCR reactor is labor intensive. The use of a motorized lifting tool increases the efficiency of the crew to complete their task as shown in Figure 17.17. Before the use of this machine, loading a layer of 36–40 modules required 2 days. The pallet mover reduced this time in half.

Figure 17.18 shows catalyst modules installed inside the SCR reactor. The modules are loaded at the same time the sealing gutters are installed. A small space is allotted between adjacent modules to slide in a narrow steel gutter that creates the seal between itself and the modules' pedestal frame at the base. Adjacent modules are not touching rake soot blowers. In this configuration, the distance from the rake soot blowers to the top grid of the catalyst modules is approximately 3 feet.

The ductwork shown in Figure 17.19 weighs 13 tons and contains guide vanes, static mixers, and the ammonia injection lances.

Figure 17.20 shows the static mixers inside the duct that create eddies in the flue gas. Ammonia is injected at this location and the guide vanes direct the flue gas downstream toward the SCR reactor.

An AIG designed for circular ducts is shown in Figure 17.21. It is made from carbon steel pipe and there are multiple planes for injection. This design offers a high degree of tuning as each injection lance can be controlled by a valve.

FIGURE 17.16 SCR catalyst module for high particulate flue gas service. (With permission from Haldor-Topsoe, Inc.)

FIGURE 17.17 Pallet mover used to place SCR modules. (With permission from Haldor-Topsoe, Inc.)

FIGURE 17.18 Catalyst modules loaded inside SCR reactor. (With permission from Haldor-Topsoe, Inc.)

FIGURE 17.19   Duct containing static mixers and ammonia injection lances for SCR. (With permission from Haldor-Topsoe, Inc.)

FIGURE 17.20   Static mixers inside the flue gas duct work. (With permission from Haldor-Topsoe, Inc.)

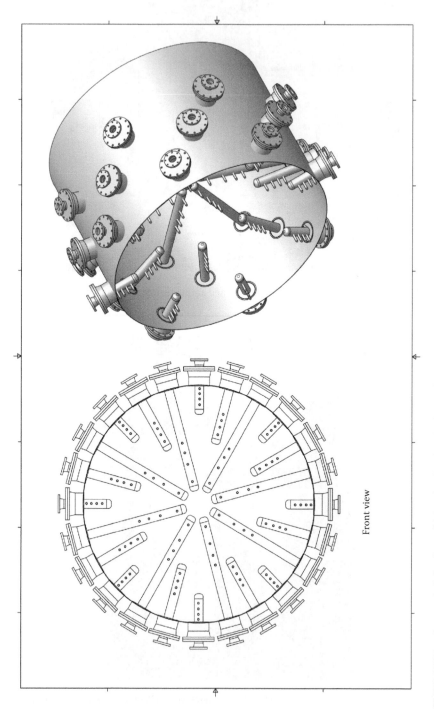

Front view

FIGURE 17.21 Ammonia injection grid for circular duct in the FCCU SCR. (With permission from Haldor-Topsoe, Inc.)

FIGURE 17.22   Assembly of the SCR reactor. (With permission from Haldor-Topsoe, Inc.)

Figure 17.22 shows the sections where the catalyst layers are located during assembly. The layers are spaced approximately 10 feet apart to accommodate the modules and soot blowers. The volume of catalyst required to achieve the high levels of NO_x reduction is substantial, typically around 150 cubic meters. This drives the catalyst bed design to be split into multiple layers.

The typical NO_x requirement for SCR units is to maintain a 20 ppmvdc @ 0% O_2 for a 365 day rolling average. Figure 17.23 shows commercial data with performance well below target. As the FCCU operating changes due to feedstock quality and market demands, so too does the NO_x entering the SCR. Even with a varied range of inlet NO_x concentrations, the SCR is able to adjust and maintain good performance.

Figure 17.24 shows the operating conditions of another commercial FCCU SCR. The inlet flue gas flows into the SCR reactor at 637°F and contains 208 ppm of NO and 20 ppm NO_2. After passing through the two layers of SCR catalyst, the outlet NO and NO_2 measures 7 ppm and 5 ppm, respectively.

17.7   LOTO$_x$

Ozone is the fourth strongest oxidant known to man and is more effective at sterilizing water than chlorine. Ozone's reaction with NO_x is a natural phenomenon that takes place in the upper atmosphere that depletes the ozone layer and creates acid

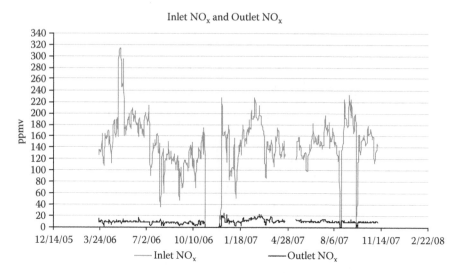

FIGURE 17.23 Inlet and outlet NO_x data for the U.S. FCCU SCR. (With permission from Haldor-Topsoe, Inc.)

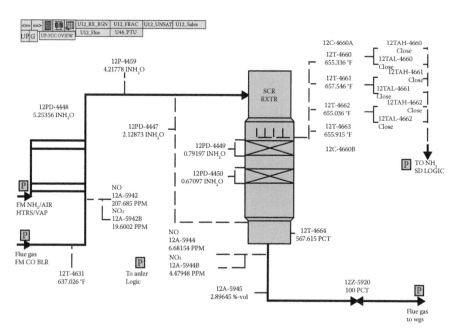

FIGURE 17.24 DCS display of the FCCU SCR. (With permission from Haldor-Topsoe, Inc.)

rain. As such, ozone is a chemical that can be used to reduce NO_x emissions in a controlled device.

Reduction of NO_x in FCC wet scrubbing units has been achieved by applying an ozone injection technology called $LoTO_x$. $LoTO_x$ is a technology approach sold by Belco Technologies Corporation (BELCO) under license from BOC. The injected ozone in the quench section of the BELCO EDV Wet Scrubbing System reacts with NO_x and ultimately form soluble nitrate salts. These salts are then removed by the remaining EDV wet scrubbing system.

SCRs have been applied on boilers and FCC units around the world for many years. However, some oil refiners continue to be concerned about potential SCR problems such as plugging (due to catalyst carry over or just due to normal dust), sintering (due to high temperature spikes), and poisoning (due to potential compounds in the flue gas). Any of these problems could result in potential downtime for the SCR and possibly the process. Further concerns are ammonia slip, conversion of SO_2 to SO_3 and formation of the sticky ammonium sulfate/sulfite particles that cold deposit within the catalyst or downstream ductwork and equipment. In addition, the flue gas temperature constraints of the SCR limit its usefulness in certain applications. An SCR reduces NO_x within a certain temperature range. If that range of temperature is not easily available then significant modifications would be required in the heat recovery scheme to fit an SCR system and potentially reduce energy efficiency. Additionally, if the heat recovery system is bypassed, the SCR will need to also be bypassed thus allowing NO_x to be emitted untreated.

The $LoTO_x$ process represents a high-efficiency alternative to SCR. When combined with the EDV Wet Scrubbing System, $LoTO_x{}^M$ can provide similar or higher NO_x reductions as an SCR, but without any of the associated concerns.

17.7.1 LoTO$_x$ Process Description

The name $LoTO_x$ was derived from Low (Lo) Temperature (T) Oxidation (Ox). In this patented process, ozone is injected into the flue gas stream to oxidize insoluble NO_x to highly soluble compounds that are subsequently removed in a wet scrubber. The ozone is produced on site and on demand by flowing oxygen through an ozone generator. Since $LoTO_x$ is a low temperature process, it does not require an increase in the flue gas temperature, as does SNCR or potentially SCR. Since the process does not use ammonia, deposition of ammonium sulfate/bisulfate on downstream heat transfer surfaces is avoided. And because it is applied as a gas in the open tower design EDV Wet Scrubbing System, there are no concerns of plugging. $LoTO_x$ can be employed at flue gas temperatures below 300°F and operates very efficiently at the natural flue gas saturation temperature.

Ozone is generated on demand from oxygen and ozone storage is not required. The amount of ozone produced is varied in response to the amount of NO_x at the system inlet and the emission level set point at the stack. Once injected, ozone rapidly reacts with the relatively insoluble NO and NO_2 to form soluble N_2O_5. This highly oxidized species of NO_x is very soluble and rapidly reacts with water to form nitric acid. The conversion of NO_x to nitric acid occurs as the N_2O_5 contacts the liquid sprays in the scrubber. The reaction is rapid and in the presence of an alkali, results in the irreversible capture of NO_x, allowing nearly complete removal from the flue gas. There are no known adverse effects of acid gases or particulate on the $LoTO_x$ system. Figure 17.25 depicts a typical $LoTO_x$ application.

The rapid reaction rate of ozone with NO_x makes the process highly selective for treatment of NO_x in the presence of other compounds such as CO and SO_2. This

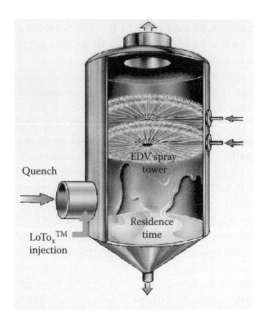

FIGURE 17.25 Simplified LoTO$_x$ schematic. (With permission from Belco Technologies.)

results in high conversion of NO$_x$ to N$_2$O$_5$, but virtually no reaction between ozone and CO or SO$_2$.

17.7.2 LoTO$_x$ Process Chemistry

The LoTO$_x$ process uses ozone to oxidize NO and NO$_2$ to N$_2$O$_5$, which is highly soluble. In contact with the wet scrubbing liquid, N$_2$O$_5$ is easily and quickly absorbed, converted to HNO$_3$, and then neutralized to NaNO$_3$. Many reactions occur, but for the sake of brevity, the LoTO$_x$ process can be summarized by the following reactions:

$$NO + O_3 \rightarrow NO_2 + O_2,$$

$$2NO_2 + O_3 \rightarrow N_2O_5 + O_2,$$

$$N_2O_5 + H_2O \rightarrow 2HNO_3,$$

$$HNO_3 + NaOH \rightarrow H_2O + NaNO_3.$$

N$_2$O$_5$ is an extremely soluble gas and reacts with water instantaneously. As a result, it is easily removed by the system even before the SO$_2$. N$_2$O$_5$ is estimated to be at least 100 times more soluble than SO$_2$.

It is important to understand that the ozone injected reacts with NO$_x$ and not with any other compounds that exist in the flue gas. Compounds of potential concern include CO and SO$_2$. While the reaction with CO to form CO$_2$ could offer positive environmental benefits, the ozone that would be consumed by this reaction would

TABLE 17.6
Reaction Rate Constants

Reaction	K (cm^3/molecule/sec @ 298°K)
$NO + O_3 \rightarrow NO_2 + O_2$	1.8×10^{-14}
$NO_2 + O_3 \rightarrow N_2O_5 + O_2$	3.5×10^{-17}
$CO + O_3 \rightarrow CO_2 + O_2$	1.1×10^{-21}
$SO_2 + O_3 \rightarrow SO_3 + O_2$	2.2×10^{-22}

cause the system operating costs to be very high. CO is much better controlled by more conventional methods (e.g., CO-promoting catalysts or CO boilers). SO_2 could also potentially react with ozone to form SO_3, fortunately, neither of these reactions occurs to a significant extent, and the ozone instead reacts preferentially with NO_x.

Table 17.6 shows reaction rate constants (from the *Journal of Physical and Chemical Reference Data*) for the reactions of interest. As can be seen, the reaction of NO to NO_2 is the quickest, with the reaction of NO_2 to N_2O_5 being slightly slower. The reaction of CO is four orders of magnitude slower than the NO_2 reaction and the reaction of SO_2 is five orders of magnitude slower. This ensures that neither of these reactions occurs. These relative reaction rates explain the lack of significant secondary reactions, as has been verified through extensive testing.

17.7.3 LoTO$_x$ COMMERCIAL EXPERIENCE AND APPLICATIONS

There are many commercial installations of EDV Wet Scrubbing Systems around the world. These have been reliably and efficiently removing particulate and SO_x for many years. More than 50 of them are applied to flue gas generated by the very demanding FCCU. Many more are applied to flue gas generated by heavy oil fired heaters, utility boilers, incinerators, and sulfur recovery units. Since its recent introduction, the installation list for the LoTO$_x$ technology lists 11 applications where the full system is used. Many other clients have purchased an EDV system prearranged to accept the LoTO$_x$ technology at a later time when environmental regulations may require it.

The performance of the EDV system and the LoTO$_x$ technology can be illustrated by examining some of these installations. The first installation is on a small coal-fired boiler in Ohio. Performance was extensively measured at this installation and Figure 17.26 illustrates the typical performance over a 7-hour time frame. As can be seen in the Figure, the flue gas flow rate varied widely from approximately 12,000 ACFM to 30,000 ACFM. The inlet NO_x was also quite variable, ranging from roughly 50 to 70 ppm. Even with these varying conditions, the outlet NO_x essentially remained steady at the set point of 5 ppm. This demonstrates the ability of the system to continuously maintain in excess of 90% NO_x reduction under widely varying conditions. Similar results were seen when the system was operated with different outlet NO_x set points.

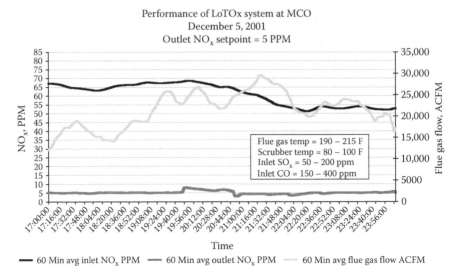

Performance of LoTOx system at MCO
December 5, 2001
Outlet NO$_x$ setpoint = 5 PPM

Flue gas temp = 190 – 215 F
Scrubber temp = 80 – 100 F
Inlet SO$_x$ = 50 – 200 ppm
Inlet CO = 150 – 400 ppm

Time

— 60 Min avg inlet NO$_x$ PPM   — 60 Min avg outlet NO$_x$ PPM   — 60 Min avg flue gas flow ACFM

FIGURE 17.26   Performance of LoTO$_x$ installation on boiler. (With permission from Belco Technologies.)

In the fall of 2002, Marathon conducted a review of several options to reduce nitrogen oxides (NO$_x$) from its 52,000 barrels per day FCCU in Texas City, Texas. The LoTO$_x$ technology was identified as a potential NO$_x$ control technology. Unfortunately, the LoTO$_x$ technology had never been used on an FCCU. Marathon needed more assurance that applying it to an FCCU would result in similar performance to that experienced in the other application. To that end, Marathon, BOC, and BELCO conducted a demonstration of BELCO's EDV Wet Scrubbing System with LoTO$_x$ technology. The demonstration was conducted October through November 2002 [6].

The demonstration achieved all objectives and proved that the EDV Wet Scrubbing System with LoTO$_x$ technology can easily be applied to FCCU applications for reducing both the NO$_x$ and SO$_x$ together. The demonstration proved that even when used on an FCCU flue gas, the combination system can reduce SO$_2$ greater than 99% and NO$_x$ greater than 90%. In addition, the EDV system's ability to reduce particulate emissions makes this a real all-pollutants reduction system. Figure 17.27 documents the results from the demonstration.

17.7.4 LoTO$_x$ System FCC Commercialization

Marathon proceeded to purchase and install LoTO$_x$ technology for application to their existing EDV wet scrubbing system at Texas City based upon the results of the technology demonstration. This system started commercial operation in February, 2007 [7]. The initial NO$_x$ reduction results from the commercial application have confirmed the results of the demonstration. Results from the first week of operation on the first LoTO$_x$ unit installed on an FCCU is shown in Figure 17.28. Subsequent testing has confirmed operation below 10 ppm. This outlet NO$_x$ is adjusted by varying the set point on the system controller.

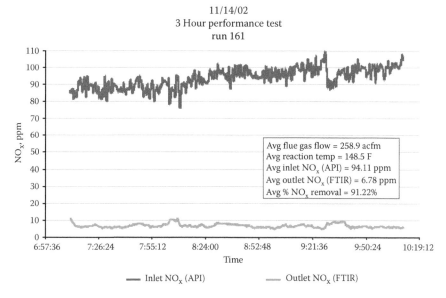

FIGURE 17.27 LoTO$_x$ demonstration on Marathon FCC unit. (With permission from Belco Technologies.)

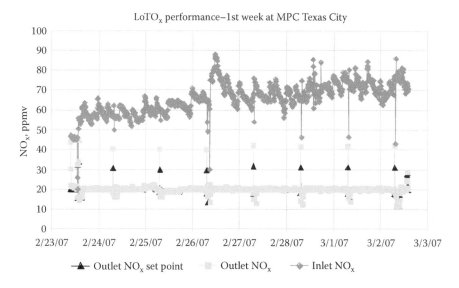

FIGURE 17.28 Initial results of commercial LoTO$_x$ operation at Marathon's Texas City Refinery. (With permission from Belco Technologies.)

Because the Marathon application was a retrofit to an existing EDV wet scrubbing system, a separate tower had to be built to provide the ozone injection and the required reaction zone. Figure 17.29 shows the actual unit layout. Although this approach was a more expensive approach, it was still significantly more cost effective than an SCR system based upon Marathon's analysis.

FIGURE 17.29   Commercial LoTO$_x$ unit layout at Marathon's Texas City Refinery. (With permission from Belco Technologies.)

Applying the LoTO$_x$ technology to new scrubbing systems is the most economical approach that is achieved all in an upflow scrubbing configuration. Most recent licensees of the EDV Wet Scrubbing Systems are electing to preinvest in the LoTO$_x$ technology by incorporating the minor scrubber modifications necessary for future addition of that technology to the scrubber. This approach will allow them to quickly add the ozone generation, controls, and injection systems to the scrubber at the required time.

Six commercial LoTO$_x$ are in operation on FCC units to date. Another 10 are in various stages of design and construction.

17.7.5   WASTEWATER PLANT CONSIDERATIONS WITH LoTO$_x$

The current NPDES permit limits for the refinery should be reviewed for any nitrate (NO$_3$) or total dissolved solids (TDS) limits when the LoTO$_x$ technology is considered. The increase in nitrate and TDS concentration of the final effluent should be compared with the permit to assure limits are not exceeded.

The potential exists for denitrification to occur in the secondary clarifier sludge bed in the refinery wastewater treatment plant under anoxic conditions. The biological reduction rate of nitrate to nitrogen in the sludge blanket is typically slow due to limited soluble Chemical Oxygen Demand (COD), the food source for denitrifying bacteria. However, if the denitrification rate is too high, the resulting nitrogen gas production could cause the sludge bed to rise, potentially causing excessive effluent of total suspended solids (TSS) and COD.

The denitrification rate in the clarifier can be controlled by return sludge recirculation rates and management of sludge solids inventory to maintain oxygenated sludge blankets. A simple alternative is to install a holding tank upstream of the aeration basins to serve as an anoxic reactor to accomplish the denitrification reaction in a controlled manner and release the nitrogen gas to prevent upsets in the main clarifiers. Current information does not provide limits for nitrate concentrations that will guarantee prevention of clarifier bed lifting. Should denitrification controls be required, Marathon has chosen to design anoxic reactors to achieve the 10 mg/L NO_3-N level on a monthly average and 25 mg/L NO_3-N for a daily maximum concentration.

17.8   OTHER NO_x CONTROL TECHNOLOGIES

There are several other technologies available to control NO_x from an FCC unit. The following is a list of applications:

- WGS +
- ESP + NH_3
- $CONO_x$
- Regenerator design

17.9   SUMMARY AND CONCLUSIONS

NO_x reacts with ammonia, moisture, and other compounds to form small particles promoting the formation of PM_{10} particulates and contributing to respiratory health concerns. NO_x has also been identified as the primary cause for formation of ground level ozone formed when NO_x reacts with volatile organic compounds (VOCs) in the presence of heat and sunlight. For this reason, NO_x is generally regulated as a precursor to ozone rather than for ambient NO_x reasons. The FCCU is one of the largest single sources of emissions in a refinery. As such, the FCCU is a primary area of focus for NO_x reduction. Several technology solutions have been developed to reduce FCCU NO_x emissions. In situ methods include equipment design, operating conditions, and additives. Backend control methods include SCR systems, SNCR systems, and ozone injection systems. All have been able to demonstrate compliance with environmental regulations and consent decree requirements either separately or in conjunction with multiple controls.

ACKNOWLEDGMENTS

The following individuals and companies are acknowledged for their input and permission to use information and data: Nic Confuorto, Business Manager, Belco Technologies; Neal Dahlberg, Vice-President, Hamon-Research Cottrell; Dennis Salbilla, Technical Manager, Haldor-Topsoe.

REFERENCES

1. *National Air Quality—2000 Status & Trends*. U.S. Environmental Protection Agency, September 2001.
2. *NO$_x$—How Nitrogen Oxides Affect the Way We Live and Breathe*. EPA, Office of Air Quality Planning and Standards, September 1998.
3. Sexton, J. *EPA FCC Consenters NO$_x$ Technology Profile*. NPRA Cat Cracker Seminar, Houston, Texas, August 20, 2008.
4. Sexton, J., Cantley, G., Wick, J., Kelkar, C. P., Stockwell, D., Winkler, S., and Tauster, S. *New Additive Technologies for Reduced FCC Emissions AM-02-56*. NPRA Annual Meeting, Salt Lake City, March 20, 2002.
5. Sexton, J., and Fisher, R. Marathon Refineries Employ New CO Promoter. *Oil & Gas Journal*, April, 27, 2009, 47–52.
6. Sexton, J., Confuorto, N., Sarresh, N., and Barrasso, M. *LoTO$_x$™ Technology Demonstration at Marathon's Texas City Refinery*. 2004 NPRA Annual Meeting, San Antonio, Texas, March 22, 2004.
7. Sexton, J., and Confuorto, N. *Wet Scrubbing Based NO$_x$ Control Using LoTO$_x$™ Technology—First Commercial FCC Start-Up Experience*. NPRA Environmental Conference, Austin, Texas, August, 2007.

18 FCC Emission Reduction Technologies through Consent Decree Implementation: FCC PM Emissions and Controls

Jeffrey A. Sexton

CONTENTS

18.1   INTRODUCTION

The Clean Air Act requires EPA to set National Ambient Air Quality Standards (NAAQS) for six criteria pollutants. Particle pollution (also known as particulate matter) is one of these. The Clean Air Act established two types of national air quality standards for particle pollution. Primary standards set limits to protect public health, including the health of sensitive populations such as asthmatics, children, and the elderly. Secondary standards set limits to protect public welfare, including protection against visibility impairment, damage to animals, crops, vegetation, and buildings. Particle pollution contains microscopic solids or liquid droplets that can enter human lungs and cause health problems. Numerous scientific studies have linked particle pollution exposure to a variety of health issues.

The nation's air quality standards for particulate matter were first established in 1971 and were not significantly revised until 1987 when EPA changed the indicator of the standards to regulate inhalable particles smaller than, or equal to, 10 micrometers in diameter (PM_{10}). Ten years later, after a lengthy review, EPA revised the PM standards, setting separate standards for fine particles ($PM_{2.5}$) based on their link to serious health problems. The 1997 standards retained a revised standard for PM_{10}, which was intended to regulate inhalable coarse particles that ranged from 2.5 to 10 micrometers in diameter. PM_{10} measurements, however, contain both fine and coarse particles.

EPA revised the air quality standards for particle pollution in 2006. The 2006 standards tightened the 24-hour $PM_{2.5}$ standard from 65 micrograms per cubic meter ($\mu g/m^3$) to 35 $\mu g/m^3$, and retained the annual $PM_{2.5}$ standard at 15 $\mu g/m^3$. The agency kept the existing 24-hour PM_{10} standard of 150 $\mu g/m^3$. The annual PM_{10} standard was revoked because available evidence does not suggest a link between long-term exposure to PM_{10} and health problems.

US EPA has provided the following definitions:

Particulate matter: All finely divided solid or liquid material, other than uncombined water, emitted to the ambient air as measured by applicable reference methods, or an equivalent or alternative method, or by a test method specified in an approved State implementation plan.

Primary particulate matter (PM): Particles that enter the atmosphere as a direct emission from a stack or an open source. It is comprised of two components: Filterable PM (FPM) and Condensable PM (CPM).

Secondary PM: Particles that form through chemical reactions in the ambient air well after dilution and condensation have occurred. Secondary PM is usually formed at some distance downwind from the source.

Filterable PM: Particles that are directly emitted by a source as a solid or liquid at stack or release conditions and captured on the filter of a stack test train.

Condensable PM: Material that is vapor phase at stack conditions, but which condenses and/or reacts upon cooling and dilution in the ambient air to form solid or liquid PM immediately after discharge from the stack.

Figure 18.1 shows the relative size of regulated PM.

Uncontrolled particulate (catalyst) emissions from the fluid catalytic cracking (FCC) will vary depending on the number of internal and external cyclone stages. Although cyclones are effective in collecting the greater constituent of catalyst recirculated in the FCCU regenerator, the attrition of catalyst causes a significant amount of finer catalyst particles to escape the cyclone system. Typically, emissions will range from 3 to 8 lbs per 1,000 lbs of coke burn-off.

New source performance standards (NSPS) for particulate matter from FCCUs will require refiners to reduce particulate matter in their stacks to below 80 mg/Nm^3 in most areas of the world. Even more stringent standards may follow in coming years including limits on the amount of less than 2.5 micron material. Options for reducing PM emissions from FCCUs are listed in Table 18.1.

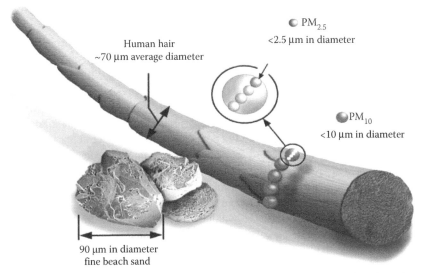

Image courtesy of EPA, Office of Research and Development

FIGURE 18.1 Relative size of regulated particulate matter (PM).

TABLE 18.1

Options for Reducing PM$_x$ Emissions from FCC Units

Potential Solution	Advantages	Disadvantages
Third stage separator (TSS)	Cost and operability	Efficiency
Electrostatic precipitator (ESP)	High efficiency	Safety
Wet gas scrubber (WGS)	High efficiency + SO$_x$ removal	Cost
WGS with wet ESP	High efficiency + SO$_3$ removal	Cost
Baghouse (BH)	High efficiency	Maintenance

18.2 PM TESTING METHODS

The pollutant is defined by the testing method. Several stack test methods exist to measure the different types and sizes of particulate matter as shown in Table 18.2.

These various test methods can generate a large disparity in the PM emission rate so it is critical to understand which PM fractions are being sought during testing. In addition, test conditions are important factors in obtaining the correct emission data. This is especially true when the stack pressure is high (above 20 psig). The test method is the same, but the preparations can be significantly different. For example, when a stack is to be tested at near atmospheric pressure, leak testing the probe under a vacuum is recommended. However, when the stack is at pressure a significantly larger pressure drop between the inside of the probe and the outside is possible. A leak in that part of the probe external to the stack and the metering valve could cause the sample collection to be much greater than intended, yet the volume of gas drawn by the pump is the same.

Sample ports are also a key issue. While the EPA accepts five pipe diameters before and two pipe diameters downstream of the sample port, experience has shown that the recommended eight pipe diameters before and two diameters after the port improves testing accuracy. The proper lengths are important to flow measurement, but they are also critical to obtaining representative dust samples. Turbulence in gas flow will result in mass emission test results that are not representative. The particulate matter will be maldistributed after an elbow and the heaviest particles will be biased to the outside wall. Even if appropriate gas rates are collected, the amount of dust may be biased to the outside wall but collected at too small a rate.

All samples should be collected using isokinetic sampling when determining mass emissions. This means the gas sample should be pulled through the sample probe at the same velocity as the velocity of the process gas. If not, the total particulate catch will not be representative of the gas stream.

The EPA Method 2 probe uses a standard S-type Pitot tube to determine the velocity pressure by measuring gas flow as a unidirectional vector. This method is typically 10–20% higher than the calculated flue gas rate from the FCC heat balance. The newly develop EPA Method 2F probe is a five-holed prism tip with a thermocouple. A centrally located tap measures the stagnation pressure, while two lateral taps measure the static pressure. The yaw angle is determined by rotating the probe until the difference between the two lateral holes is zero. This method closely matches the

TABLE 18.2
PM Testing Methods

Type of PM	Particulate Matter (PM) Sizes		
	Less than 2.5 microns	**Less than 10 microns**	**Total PM**
Filterable	U.S. EPA method 201A	U.S. EPA method 201A	U.S. EPA method 5[a]
Condensable[b]	U.S. EPA method 202 *or* OTM-28	U.S. EPA method 202 *or* OTM-28	U.S. EPA method 202, *or* OTM-28
Total	U.S. EPA methods 201A and OTM-28	U.S. EPA methods 201A and 202	U.S. methods 5 and 202

[a] U.S. EPA Test method 5 has several variations to measure total particulate matter including methods 5B, 5C, 5D, and 5F.
[b] OTM-28 has been listed as conditional test methods for condensable particulate matter.

calculated flow from the heat balance and gives a more representative flow value and hence PM measurement.

Selecting the right stack sampling company and understanding how they conduct the sample collection is very important in obtaining valid results.

The EPA has developed several different methods to measure filterable PM. These include methods 5, 5B, 5F, 17, and 201A. One significant difference is the temperature at which the sample is taken. Method 5 operates at 248°F. Method 5B operates at 320°F. As such, the colder sample train is subject to incremental condensable PM. Comparison sampling on two FCC units found this difference to be ~0.1 lb/Mlb coke between the two methods. Figure 18.2 shows the current method 5 sampling train.

The EPA has developed Method 202 to measure condensable PM (CPM). This method involves the analysis of the sample train impinger solution. As such, it is subject to measuring artificial PM that does not occur naturally at atmospheric conditions. The largest bias includes SO_2 that dissolves in the solution to form sulfites and bisulfites. Additionally, free ammonia can increase this contribution. The EPA recently updated this method to include a dry impinger train. The dry impinger method is now referenced as other test method 28 (OTM-28). Figure 18.3 shows the current dry impinger sampling train (OTM-28).

This method has resulted in lower measured PM since the artifact PM is minimized. Figure 18.4 shows a comparison of the two methods.

18.3 THIRD STAGE SEPARATORS (TSS)

Third stage separators (TSSs) have been used widely in FCC service since the 1960s. Traditionally, TSSs were used upstream of a power recovery turbine (PRT) to protect the turbine blades. These TSSs were originally designed to remove particulate matter (PM) greater than 10 microns in size. In most instances, the fines were removed from the flue gas, diverted around the PRT, reintroduced to the flue gas, and expelled via the stack. This configuration did not take advantage of the inherent flue gas upgrading associated with the TSS.

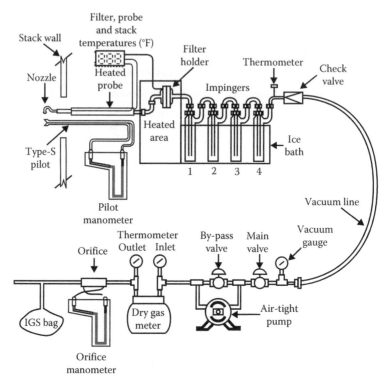

FIGURE 18.2   EPA method 5 sampling train.

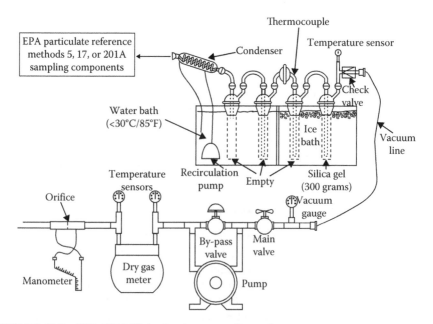

FIGURE 18.3   EPA OTM-28 dry impinger sampling train.

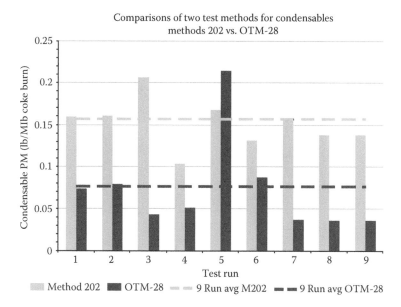

FIGURE 18.4 EPA OTM-28 dry impinger vs. old method 202 results.

Recently, refiners have begun to use TSSs to control particulate matter from FCCs for environmental reasons. MACT II regulations require that FCC particulate emissions be reduced to less than 1 lb particulate per 1000 lb coke burn (this limit is the same as the NSPS limit for particulate matter). There are four TSS technology companies licensing TSSs to meet the MACT II/NSPS limits. Figure 18.5 contains a sketch of a UOP TSS unit.

The flue gas passes through a number of small diameter high-efficiency cyclonic elements arranged in parallel and contained with the separator vessel. The UOP design uses an axial flow cyclone. After the catalyst particles are removed, the clean flue gas leaves the separator. A small stream of gas, called the underflow, exits the separator through the bottom of the TSS. In an environmental application, the underflow is diverted to a fourth stage separator (FSS) that is typically a barrier filter. The underflow rate is typically 2–5% of the total flue gas rate and is set by use of a critical flow nozzle.

The particulate removal efficiency of a TSS is difficult to calculate with a single theoretical relationship. The technology licensors have utilized pilot plants and cold flow modeling to improve their removal efficiencies to meet stricter environmental regulations. While there is no theoretical relationship that exactly matches removal efficiencies, the following efficiency relationship from Rosin, Rammler, and Intelmann [2] is often used to understand cyclone fundamentals:

$$D_{pmin} = [9\mu B_c/(\Pi N_{tc} V_c(\rho_s - \rho))],$$

where

D_{pmin} = critical particle diameter
μ = gas viscosity
B_c = inlet duct width

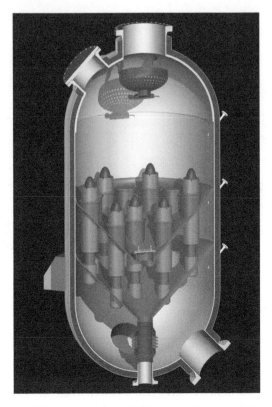

FIGURE 18.5   UOP third-stage separator. (With permission from UOP, LLC.)

N_{tc} = number of turns made by the gas in a cyclone
V_c = gas inlet velocity
ρ_s = particulate density
ρ = gas density

The above relationship shows that the opportunities for improved removal efficiency are by:

- Decreasing temperature that decreases viscosity (μ) and increases gas density (ρ).
- Decreasing inlet duct width (B_c) and increasing aspect ratio (inlet height/width).
- Increasing vortex height that increases the number of gas turns (N_{tc}).
- Increasing gas inlet velocity (V_c).
- Reducing dust reentrainment. This opportunity is not shown in the above equation, but is based on proprietary dust outlet designs.

The technology licensors have taken different approaches to improving dust removal efficiency, but the designs are all based on the fundamentals discussed above.

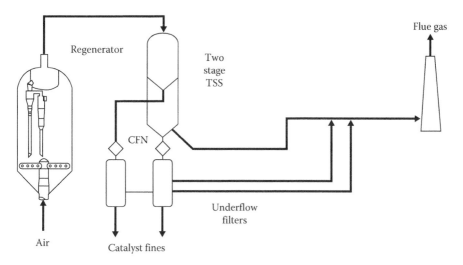

Flue gas

Regenerator

Two
stage
TSS

CFN

Underflow
filters

Air

Catalyst fines

FIGURE 18.6 UOP Two-stage TSS application. (With permission from UOP, LLC.)

The TSS has less potential for maximum emission control as the WGS and electrostatic precipitator (ESP). While all three technologies were designed to meet NSPS/MACT, only the WGS and ESP can meet more stringent requirements. TSS performance can typically achieve d50 grade efficiency down to 2 microns. Most reported performance values result in emissions of 0.4–0.8 lb/Mlb coke. Future $PM_{2.5}$ regulations are a concern for TSS applications. By definition, TSS units cannot effectively remove small particles due to the cyclonic operation.

A design consideration for an environmental application of TSS technology is location. In a PRT unit, the TSS is always located in a hot flue gas position. However, a cold flue gas position could be used for environmental applications. Depending upon the pressure control scheme, the TSS can also be located at high pressure before the flue gas slide valve and orifice chamber. This offers significant cost savings, lower material cost, and requires less plot space. On two EPA consent decree applications, this benefit was ~40% lower capital cost.

An option to improve TSS performance is a two-stage design [3]. This application uses two TSS stages in series that operate at different efficiencies to improve overall PM removal. This has been applied on one commercial unit with a 10% improvement in efficiency over a conventional arrangement. A schematic of this approach is shown in Figure 18.6.

Another option to improve TSS performance is use of a sintered metal filter. This technology has typically been applied only as a fourth stage application on the TSS underflow. Pall has commercialized this barrier filter on the entire FCC flue gas on one commercial unit.

18.4 ELECTROSTATIC PRECIPITATOR (ESP)

The ESP has been used to control particulate emissions in the flue gas following an FCC regenerator since the 1940s with well over 200 FCCU applications. The majority of the installed units have been supplied by Hamon Research-Cottrell Inc. and Buell Corporation, subsequently acquired by Hamon Research-Cottrell Inc. The ESP is designed to remove solid particulate matter only from the FCCU regenerator exhaust gas. ESPs of the twenty-first century have been pushed to extremely high levels of performance through recently enacted regional legislation (see SCQAMD rule 1105.1) requiring emissions of less than 0.005 gr/acf that result in efficiencies of +99.9% (typically <0.2 lb/1000 lbs of coke). The majority of today's regulations and EPA consent decrees require emissions ranging from 0.5 to 1.0 lb/1000 lbs of coke burned, and are easily achieved by ESPs. This makes the ESP an attractive choice in applications where the feed is deeply hydro-treated and only particulate remains to be removed.

18.4.1   ESP CONFIGURATIONS

In most cases, the ESP is installed after the FCC slide valve where the operating pressure is reduced to less than 0.5 psig. However, several units have been designed and installed prior to the slide valve with operating pressures up to 3 psig. In today's modern refinery, the ESP has been installed on both partial combustion (Figure 18.7) and full combustion (Figure 18.8) regenerators. In partial combustion applications, the ESP is installed downstream of the flue gas cooler and subsequent orifice chamber. In full combustion applications, the ESP is installed after the CO boiler. The ESP can also be installed with a TSS and is typically the preferred choice of equipment when emissions are required to be less than 0.5 lbs/1000 lbs of coke burned.

The ESP can also be used to collect catalyst fines prior to a Wet Gas Scrubber (WGS). This eliminates the need for the high pressure drop across the WGS required to achieve low particulate emissions and reduces the need for solids removal from the WGS blow down (Figure 18.9). In installations with an SCR (Selective

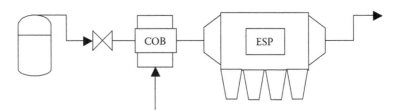

FIGURE 18.7   ESP on partial combustion unit.

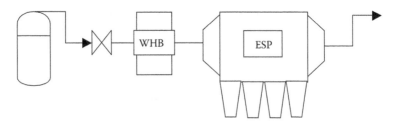

FIGURE 18.8   ESP on full combustion unit.

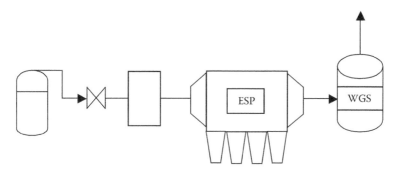

FIGURE 18.9   ESP with scrubber.

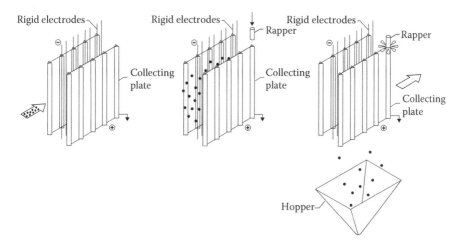

FIGURE 18.10   ESP particulate collection fundamentals. (With permission from Hamon.)

Catalytic Reduction for NO_x control) the ESP has been placed both before and after the SCR.

The ESP operates with a low pressure drop making it suitable for installation on existing units with pressure constraints such as CO boilers. The typical ESP will require less than 2″ w.c. additional pressure drop, which is typically less than that of a TSS or WGS.

18.4.2   ESP PARTICULATE COLLECTION

The process of particulate collection in an ESP is shown in Figure 18.10 and described in general terms below.

1. FCC flue gas laden with catalysts fines enters the ESP in a uniform flow pattern between parallel collecting plates into an area of high voltage.
2. Once exposed to the high-voltage field, the fines are bombarded by a flow of negatively charged ions from discharge electrodes obtaining an overall negative charge.

3. These negatively charged particles migrate to the (positively grounded) collecting plates where they are captured. The current flows to ground leaving the particles behind. The particles are continuously recharged through the negative ions and electrons created by the corona discharge. Complete discharge results in particle reentering the gas flow after collection.
4. The captured catalysts fines are transported down the collecting plate using a mechanical device known as a rapper, which accelerates (moves) the plate normal to its surface and assists gravity in moving the collected particles to the bottom of the plate. It is critical to maintain the electrical field and charge on the particles during rapping to minimize re-entrainment of dust.
5. When the particles reach the bottom of the collection plate they fall into a hopper for subsequent evacuation into a collection device such as a dumpster or are pneumatically transported to a storage silo for storage and disposal. Because of their fine nature, these particles are not typically reintroduced into the FCCU for reuse.

A Four Field Hamon Research-Cottrell Inc. Refinery ESP with top inlet plenum and weather enclosure is shown in Figure 18.11.

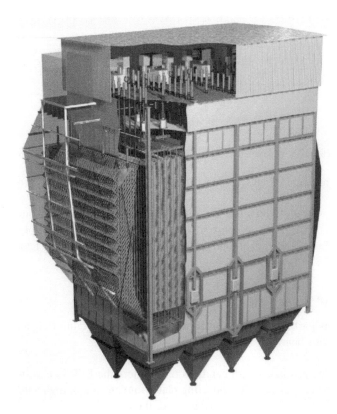

FIGURE 18.11 Cut away view of four field Hamon Research Cottrell ESP. (With permission from Hamon.)

18.4.3 PARTICULATE COLLECTION THEORY

The function of an ESP is to remove particles from gaseous streams. This is accomplished by passing the gas between a pair of electrodes—a discharge electrode at high potential and an electrically grounded collecting electrode. The potential difference must be great enough so that a corona discharge surrounds the discharge electrode. The secondary voltage at which measurable secondary current is observed is called the corona onset voltage. After reaching the corona onset voltage, electrons and negatively charge ions will enter the gas passage. Under the action of the electrical field, gas ions formed in the corona move rapidly toward the collecting electrode and transfer their charge to the particles by collision with them. The electrical field interacting with the charge on the particles then causes them to drift toward and be deposited on the collecting electrode.

If the gas flow through the precipitator corresponds to viscous flow, near 100% collection efficiency can be achieved in a precipitator of finite length. Unfortunately, this does not happen because the gas is in turbulent flow. As a result, the collection efficiency is an exponential term influenced by several variables. The following is a general collection efficiency relationship for ESPs based upon the Deutsch-Anderson equation [4]:

$$\text{Efficiency} = 100 \times (1 - e^{(-A \times w/Q)}),$$

where
    A = collecting electrodes cross-sectional area
    w = particle drift velocity
    Q = gas flow rate

This equation shows that for a given collection efficiency, the precipitator size is inversely proportional to particle drift velocity and directly proportional to gas flow rate. Increasing the gas density (migration velocity is a function of gas viscosity) by reducing its temperature or increasing the pressure will reduce the precipitator size. However, theory does not account for gas velocity. This is a variable that influences particle re-entrainment and the drift velocity. This typically requires an ESP design at lower velocities than predicted in theory.

ESP particle removal efficiency is exponentially related to the ratio of collecting plate area available to the gas volume treated. This ratio is called the Specific Collecting Area (SCA). Small changes in this ratio result in big changes in particulate removal. This ratio is modified directly by the empirical value for migration velocity. Increasing the gas volume by increasing temperature or throughput will reduce particle collection efficiency. Losing an electrical field to a ground or other mechanical problem reduces the size of the ESP causing particulate emissions to increase.

18.4.4 PERFORMANCE AND CRITICAL PARAMETERS

The performance of the ESP is dependent on several factors that include treatment time, temperature, particle size distribution, and resistivity. The single most important parameter is temperature, which is directly related to resistivity. These critical parameters are detailed below.

18.4.4.1 Particle Size

There are three charging mechanisms present in an ESP, diffusion, field, and a transition between diffusion and field. The larger particles, greater than about 1.5–2 microns, tend to accept charge rapidly to saturation. In this particle size range, field charging is the dominant mechanism. In an ESP, almost all of the particles greater than 10 microns are removed in the first field. Smaller particles less than 0.5 microns do not accept charge, but comingles with the charge. Diffusion charging is the dominant mechanism for the very fine particles. This Brownian type movement toward the collecting plate is slower than field charging. The particles between about 0.5 and 1.5 microns do not have a dominant charging mechanism and as a result present the lowest migration rate. Very fine particles, less than 1 micron in diameter, provide small cross sections or targets for negative ions. As such, it is more difficult and requires more time for fines to capture a sufficient number of ions to attain a maximum charge. Also, fine particles travel to the grounded collecting plates in a more random motion since the fines are in a size range where gas molecules affect them. A significant increase in the population or number of small particles may require the precipitator to be larger in size to achieve the required performance efficiency than for the case of a larger size distribution of particulate.

18.4.4.2 Particulate Loading

High particulate loading interferes with particle charging by suppressing, to a degree, the corona, and thus negative ion generation from the emitting electrodes. This effect is seen on the first and second inlet electrical fields of the precipitator. This effect becomes more significant if the higher particulate loading has a high population of fine particles. The typical dust burden at the outlet of an FCCU is comparable to the second or third field of a utility boiler. The low dust burden avoids problems associated with current suppression due to excessive space charge.

18.4.4.3 Operating Temperature

FCCU ESPs operate at temperatures ranging from 450 to 800°F (230°C–430°C) with the optimal level being 600°F (315°C) and above. Resistivity is directly related to the catalysts temperature. Catalyst resistivity is measured in Ohm-CM. While each manufacture of FCC catalyst may have a slightly different composition, it is generally made up of 99% oxides of silica and alumina. At lower temperatures, both silica and alumina have properties that make the catalyst resistant to the acceptance of electrical charge required for the collection process to occur. Not being able to accept an electrical charge results in poor collection efficiency. As can be seen in Figure 18.12, the resistivity at 400°C is two orders of magnitude lower than at 250°C. This results in PM collection at 400°C being much better than at 250°F. At this lower temperature, the ESP may be as much as 30% larger to achieve the same performance as the higher temperature.

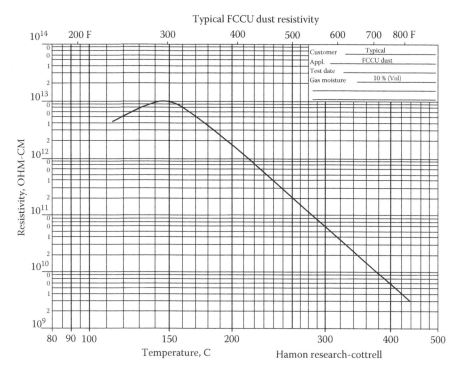

FIGURE 18.12 Typical resistivity curve For FCC catalyst. (With permission from Hamon.)

Two types of conduction mechanisms occur in an ESP—volume conduction and surface conduction. The conduction referenced occurs in the dust layer on the collecting plates. Particles arrive at the collecting plate comingled with or carrying charge. Charge is also continuously imparted to the dust layer by the electrodes. In the dust layer, charge is constantly conducting to ground. The rate at which this conduction occurs is a function of the particle resistivity. The curve shown in Figure 18.12 shows a typical relationship between gas temperature and catalyst resistivity. The peak dust resistivity occurs at a temperature of approximately 150°C. The portion of the curve to the right of this peak is called the volume conduction phase. The portion of the curve to the left of the peak is called the surface conduction phase. When the gas temperature is below 150°C, acids and other condensables in the gas stream can accumulate on the surface of the catalyst. In this temperature range, current flows over the surface of the particle through the condensate. For this reason, the composition of the particle is not as significant to resistivity as is the presence of condensables. At temperatures above 150°C, volume conduction becomes dominant. The presence of condensables on the surface of the particle becomes less likely as the gas temperature increases. At these elevated temperatures, conduction occurs through the particle. The composition of the particle becomes a major influence on the dust resistivity. Catalyst is predominantly silica and alumina that exhibits very high resistivity at the mid-range operating temperatures. As the temperature continues to increase, the resistivity of the constituents of the particle decrease improving conduction.

When dust resistivity is in the elevated range, the rate at which charge dissipates to ground is less than the rate at which new charge is being applied to the dust layer. This creates a situation where charge accumulates on the surface of the dust. At some point the charge will find a weak spot in the dust layer and discharge rapidly to ground. This discharge results in a phenomenon called "Back Corona."

18.4.4.4   Treatment Time

Treatment time is the measure of the time in seconds that the gas is in contact with the energized electrical field. It is determined by the cross section of the ESP, the length of the precipitator, and the gas volume treated. Adequate treatment time is critical when high efficiency removal of fine dust is required. The diffusion charging mechanism is slow, requiring more time to move particles from the gas passage onto the surface of the collecting plates. The treatment time will vary depending on the outlet emission required, the flue gas temperature, particle resistivity, particle size destruction, and inlet loading. Over the years, the treatment time has increased from 10 seconds in the 1970s, to as much as 25–30 seconds in recent installations requiring ultra low emissions (<0.005 gr/acf).

18.4.4.5   Gas Velocity

Typically, gas velocity (the velocity at which the gas travels through the treatment zone) is kept under 3 ft/sec. When emission requirements were less stringent, gas velocities operated between 3 and 5 ft/sec. The velocity at which gas moves through the ESP is a factor in determining the level of reentrainment of dust. High-velocity units will sweep dust off of collecting plates and out of hoppers. Re-entrainment of collected dust is a major cause of noncompliance. Nonuniform gas distribution can also result in excessive velocity in portions of the unit. This high-velocity section can result in channeling of flow through a small number of gas passages leading to reentrainment.

18.4.4.6   Aspect Ratio

Aspect ratio is the ratio of the collecting plate length (sum of all fields) divided by the collecting plate height. For example, a precipitator with 4 fields @ 9 feet with 30 feet high collecting plates would have an aspect ratio of $4 \times 9/30 = 1.2$. An aspect ratio of 1.2 or greater is considered to be the industry minimum required for high performance. Aspect ratio becomes more critical as the particle size decreases and the required particulate removal efficiency increases. More time is required to move particles from the upper portions of the gas flow into the hoppers.

18.4.4.7   Inlet Gas Flow Uniformity

Inlet gas uniformity as measured by the inlet root mean square (RMS) should be kept less than 20%. Nonuniform gas flow or maldistribution will result in poor collection, excessive reentrainment, or hopper sweepage. The precipitator is designed to have very little flow at the bottom of the collection plate to allow the particles to drop freely into the hopper without entrainment back into the gas stream.

18.4.5 THE EQUIPMENT

18.4.5.1 Collecting Plates

Collecting plates are categorized in two groups, strip collecting plates and single piece collecting plates. The pediment type has been single piece collecting plates with vertical stiffeners called welded unitized collecting panels. A common brand is the G-Opzel. These stiffeners allow for plate highs up to 36 feet. This type of panel is regularly utilized at 42 foot lengths with limited distortion that allows interelectrode spacing to be accurately maintained. Plate lengths range up to 14.5 feet.

18.4.5.2 Plate Spacing

With the advent of rigid discharge electrodes in the late 1970s, plate spacing has changed. Until approximately 1980, most units were designed with weighted wires and had spacing of 9–10″. With the advent of the rigid electrode, the plate spacing has increased to 12–16″ with the most predominate being 12″ spacing.

18.4.5.3 Collecting Plate Height

The industry standard collecting plate height currently is 36–42 feet. This maximum height has been specified by most owners to eliminate concerns that include inter-electrode clearances, plate bowing, and reentrainment resulting from large rapping forces associated with large and tall collecting plates.

18.4.5.4 Discharge Electrodes

An ESP works by applying a high voltage to emitting electrodes, which produce points of corona emission along the length of the emitter. The location and number of these corona discharges depends on the emitting electrode design. The corona discharges are ionization points, which release electrons from gas molecules. Other gas molecules forming negatively charged ions that are attracted to the positive polarity collecting plates capture these free electrons. Particles of dust entering the precipitator capture these negative ions and become themselves negatively charged to a degree depending on the number of ions captured. The negatively charged particles are attracted to and migrate to the collecting plates where they are captured. The migration velocity of the charged particulate is proportional to the applied precipitator voltage squared. The higher the migration velocity of the particles, the shorter the distance the particles must travel in the direction of flow in the precipitator before capture. This determines the size of the ESP required for a specific collection efficiency. It is critical that a high-voltage level be maintained in the precipitator for optimal charging and collection. The process and dust characteristics define the voltage at which a field will operate. A voltage control seeks the maximum secondary voltage at which the field will operate, which is a dynamic condition.

Discharge electrodes have evolved over the past several decades from a weighted wire type unit (0.109″ diameter wire) with a lower tensioning weight and internal insulators to a rigid mast type electrode ranging from 1–2″ in diameter with no internal insulators and are virtually unbreakable. Figure 18.13 shows a 2″ diameter rigid electrode and G-Opzel collection plate commonly used in the refining industry.

FIGURE 18.13   Discharge electrode and G-Opzel collecting plate. (With permission from Hamon.)

18.4.5.5   Transformer-Rectifiers

Transformer-rectifiers and the associated controls deliver the power to discharge electrodes. The 480 volt single phase power is supplied to a transformer-rectifier control where the incoming power is modulated via SCRs (silicon controlled rectifiers) to provide the transformer primary voltage. The transformer steps up the incoming voltage of approximately 400 volts to a secondary voltage of 65,000–80,000 volts. The secondary voltage limit specified is a function of the gas passage width. The secondary current limit defined for the transformer-rectifier is a function of the size of the field that it serves. This can range from 450 mA to over 1 amp (1000 milliamps). The high voltage is rectified from AC to DC and then directed to the discharge electrodes.

18.4.5.6   High-Frequency Power Supplies

The high frequency or switch mode power supplies are becoming more prevalent on ESPs. A T/R set is a pulsed DC output device that operates with a peak to average voltage ratio of 1.2 to over 1.4. Sparking in an electrical field occurs at the peak secondary voltage. A switch mode power supply operates at a peak to average voltage ratio of about 1.03. The output of the switch mode unit is near DC with respect to the wave form. That allows the electrical average secondary voltage to operate near the peak secondary voltage that a conventional T/R set would generate. Field strength and charging rate both benefit when the average secondary voltage can be increased without increasing the spark rate of the field.

18.4.5.7   Rapping System

The collecting plates and discharge electrodes require periodic cleaning to remove collected particulate. This collection process is continuous. As such, devices have been developed to clean both the discharge electrodes and the collecting plates on a continuous basis. The two types of systems used are internal rotating hammers and external magnetic impulse gravity impact rappers, commonly referred to as MIGI rapping. These are generically termed electromagnetic impact rappers. The MIGI rapper is the most common type of rapping as it allows the operator to have control over both the amplitude, frequency, and number of impacts of the force applied to the discharge electrode and collecting plate. The rotating hammers have a fixed amplitude and a variable frequency for a group of collecting plates or discharge electrodes.

18.4.6   GAS CONDITIONING

18.4.6.1   Ammonia Conditioning

Ammonia can be injected into the flue gas prior to the ESP to modify the catalyst resistivity and agglomerate the fine particles and improve the units collecting efficiency. Ammonia is most effective at operating temperatures lower than 550°F. Injection rates are typically 10–50 ppm of which > 50% of the ammonia is adsorbed on the catalyst fines and the remainder of the injected portion shows up as ammonia slip in the flue gas. Ammonia is good for ESP performance when used in moderation. Excessive injection can cause a degradation in performance due to collecting plate build-up.

18.4.6.2   Steam/Water Injection

Many local regulatory agencies require the ESP to be energized whenever the FCC main blower is in operation. Some units have found that steam/moisture injection is useful at temperatures lower than 400°F as a form of resistivity modification. This is because the unit is in the surface conduction phase of the resistivity curve. At this temperature, the moisture improves surface conduction of the particles.

18.4.6.3   NO$_x$ Reduction Synergy

Marathon has one FCC unit that uses an ESP with NH$_3$ injection. The unit was originally operated without NH$_3$, but experienced a decrease in performance due to lower Ecat sodium content and poor resistivity of the fines. The system was modified to inject anhydrous NH$_3$ as shown in Figure 18.14.

Stack testing was competed at ~45 ppm and ~90 ppm NH$_3$ injection. PM performance improved as expected with opacity decreasing from ~10% (0.75 lb/Mlb) to ~7% (0.45 lb/Mlb). However, an unexpected benefit was a dramatic reduction in NO$_x$ from >70 ppm to <30 ppm. The results are shown in Figure 18.15.

This information was used to develop performance curves to show expected PM emissions with NH$_3$ and without. Figure 18.16 summarizes the PM performance curves with two levels of NH$_3$ injection.

Other units in industry, but not all, have seen a similar response. Figure 18.17 shows the observed NO$_x$ response with continued NH$_3$ injection.

Marathon has not observed a visible plume with use of NH$_3$ injection. The majority of the injected NH$_3$ is either reacted or absorbed on the catalyst fines. Stack test data measured <3 ppm NH$_3$ slip with 90 ppm NH$_3$ injection at the inlet.

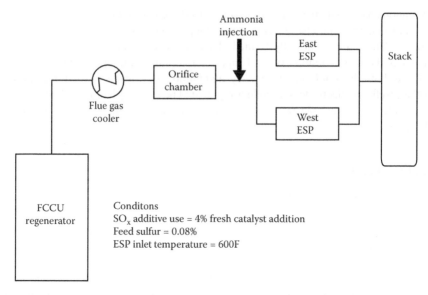

FIGURE 18.14   ESP configuration with NH_3 injection.

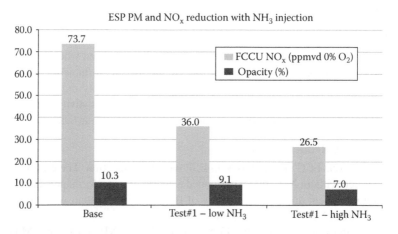

FIGURE 18.15   ESP PM and NO_x reduction with NH_3 injection.

18.4.7   ESP RELIABILITY

To maintain acceptable performance of the ESP, it is necessary to understand the mechanisms responsible for performance degradation and how to mitigate via monitoring and design. The following are typical factors that contribute to performance degradation of the ESP:

- Reduction in TR power levels
- PM accumulation on collecting plates
- Fouling and corrosion on insulator surfaces leading to electrical degradation

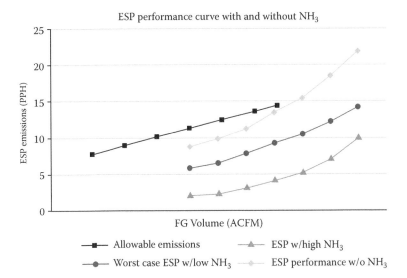

FIGURE 18.16   ESP PM emission curves with NH₃ injection.

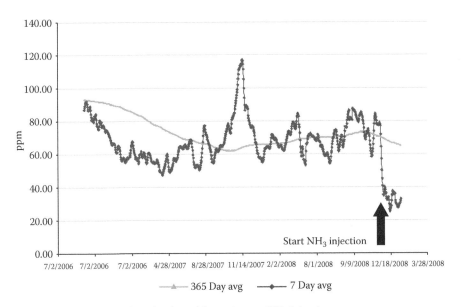

FIGURE 18.17   ESP NOₓ reduction with continuous NH₃ injection.

- Poor gas flow distribution
- High level of fines in hopper
- Improper control settings
- Rapping system degradation
- Process gas conditions
- PM fines electrical resistivity

Most FCC units target FCC turnarounds at 3–5 year intervals. This requires the ESP to be fully functional over that time frame. Some electrical degradation is expected with time. However, performance can be maintained by careful monitoring and a good maintenance program including regular cleanings. To assure high reliability, most design specifications consider the following options:

1. Supply two ESPs capable of 70% flue gas treatment that allows the FCC to be brought to a lower rate and one ESP can be taken out of service for any required maintenance.
2. Supply one ESP with a redundant or spare mechanical/electrical section, this approach protects the FCC from being brought down should some degree of internal fault occur due to insulator breakage, full hopper, and so on.
3. Newer ESPs with rigid electrodes and unitized collecting plates provide improved reliability compared to weighted wire units. Designing a unit with multiple independent electrical sections allows a portion of the ESP to be taken out of service while remaining online.

It is noteworthy to mention that many older ESPs are undersized as FCC unit throughputs have increased substantially without similar debottlenecking efforts on the ESPs. Many of the undersized ESPs are over-worked operating at higher flue gas flow rates, rapper frequencies, and collection voltages relative to their initial design conditions. Thus, higher maintenance costs and poorer reliability are characteristic of these units. A larger ESP with modern technology and redundancy may be required to achieve reliability and performance targets.

18.4.8 ESP Safety

ESPs represent a safety concern due to sparking of the electrodes being an ignition source. During upset conditions, hydrocarbon can be present in the FCC stack. If the ESP is operating, the hydrocarbon can be ignited. The results of a refiner survey completed in 2001 were presented at an NPRA meeting [5]. This survey included 114 FCC units and found 52 have or did have ESP units. Of these, a total of 20 units had experienced a fire or explosion representing 40%.

Several refiners have taken steps to improve ESP safety. These actions include the following:

SIS: Safety instrumented shutdown (SIS) systems have been installed on several units. These systems automatically remove feed and hydrocarbon from the process to place the unit in a safe state when a set of conditions that represent an unsafe operation is observed.

De-energize: The ESP is de-energized through automatic controls when the following process situations exist:

• SIS trip: The ESP is de-energized when the SIS trips and pulls feed.
• Start-up: Start-up conditions represent a risk for upset conditions. The ESP is typically not commissioned until after the unit has reached steady-state.

However, some refineries are required to operate the ESP whenever the main air blower is energized due to local regulations. These units have subsequently developed procedures to allow the ESP to operate at low voltages to minimize the risk of fire or explosion caused by high-hydrocarbon carryover in the presence of high-oxygen levels.

- External heating: Supplemental heat is required during start-up and potentially turndown conditions to establish proper regeneration kinetics. A direct fired air preheater and torch oil in the regenerator are typically used. The ESP is de-energized whenever these fuel sources are in use.
- CO boiler upset: The ESP is de-energized during an upset of the COB.
- High CO/low O_2: Poor regeneration conditions can result in hydrocarbon leaving the FCC stack. These conditions typically result in high CO and low excess O_2 emissions. The ESP is automatically de-energized when these conditions are seen since they are surrogates for potential hydrocarbon sources.
- Operator intervention: A manual shutdown can be activated by an operator if the potential exists for an ESP fire or explosion.

18.5   WET GAS SCRUBBERS (WGS)

18.5.1   WET GAS SCRUBBER

The WGS is an excellent device for removing PM with efficiencies typically >90%. Most scrubber designs rely upon pressure drop to reduce PM. This technology was previously discussed in the chapters on SO_x and NO_x technologies. A scrubbing system is a device that can potentially reduce SO_x, NO_x, and PM in one system.

18.5.2   WGS WITH WET ESP

Wet electrostatic precipitators (WESP) are used for removal of liquid contaminants such as sulfuric acid mist, aerosols, and particulate matter. The acid mist and aerosols are typically formed in a WGS by condensation of SO_3. Unlike dry precipitators, wet precipitators do not require rapping to remove the dust. The collected mist and particulate matter form a liquid film that runs down a vertical collecting plate. In some cases, a continuous spray of liquid is used to prevent solids deposition on the collecting plates.

The WGS/WESP combination consists of a spray tower and a WESP, arranged in a common casing with the stack mounted directly on top. The arrangement of the WESP and the stack on top of the scrubber minimizes the footprint area required, allowing the equipment to fit into many existing plants. Moreover, this design eliminates interconnecting ductwork and changes in gas flow direction with typical design pressure drop less than 3″ H_2O overall. A typical WGS/WESP configuration is shown in Figure 18.18 (GEA Bischoff design).

For scrubbing of sulfur dioxide and a large portion of catalyst fines, a countercurrent spray tower with a multiple stage nozzle system is typically applied. The WESP is applied after the scrubber section to remove the residual particulate matter with a

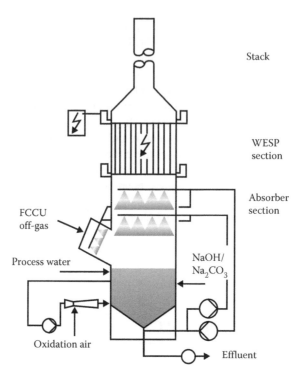

FIGURE 18.18 Typical WGS/WESP configuration. (With permission from GEA Bischoff.)

high efficiency and low pressure drop. In addition, the extremely fine sulfuric acid
aerosols that pass through the scrubbing system are also captured in the WESP.

The WESP consists of a bundle of hexagonal tubes forming the collecting electrodes, with the discharge electrodes suspended down through the vertical axis of
each tube. The discharge electrodes are spiked rigid tubes. They are supported by
an upper frame and held firm in the center of the collecting tubes by a lower guide
frame. A typical arrangement is shown in Figure 18.19.

The high voltage supply is connected to the upper frame. Both frames are suspended from ceramic insulators mounted in lateral compartments outside the main
vessel. Those compartments are constantly purged with heated air from individual
blowers to avoid infiltration of humidity, dust, or acid mist from the process gas to
the insulators. The collecting electrode tube bundle is supported by a ring and support beams in the casing and is electrically grounded to the casing. To address concerns about reliability, selected critical components in the system have been specially
modified for the FCC application. These modifications include increased robustness
in construction, installed spares, and the ability to repair or replace critical equipment during operation. A typical WESP configuration is shown in Figure 18.20 (GEA
Bischoff design).

Dust particles, water droplets, and sulfuric acid mist (and if present, ammonium
salt aerosols) are electrically charged in the same way as in the dry precipitator.
The negatively charged particles are collected on the positive collecting electrodes.

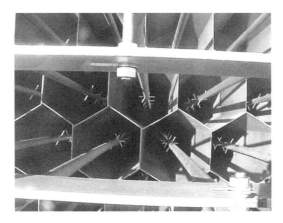

FIGURE 18.19 WESP collecting electrode tube bundle with discharge electrodes. (With permission from GEA Bischoff.)

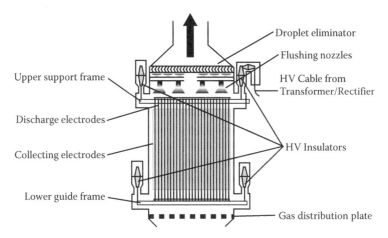

FIGURE 18.20 Typical WESP configuration. (With permission from GEA Bischoff.)

Collected water and sulfuric acid mist form a falling film of condensate on the surface of the collecting electrode. The condensate and captured particles fall from the WESP, back into the scrubber sump. The required high voltage is supplied by a transformer-rectifier (T/R). The T/R sets are controlled by an automatic, microprocessor based, precipitator voltage control system that enables limitation or cut-off of the short-circuit current within the shortest possible time, and ensures the highest possible voltage relative to the gas conditions. A second T/R with controller is installed as a standby unit. The liquid condensate film on the collecting electrodes normally provides a good, constant self-cleaning of the surface. To prevent the formation of a dust layer on the collecting surface, a flushing system installed above the electrical field can be activated in certain time intervals. The basic operating principals of a WESP are shown in Figure 18.21.

Wet ESPs have been used extensively in other industries. To date, there are three units installed on FCC units in North America. In all cases, SO_3 and PM emissions

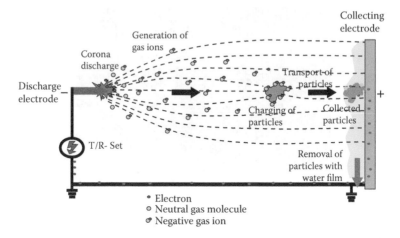

FIGURE 18.21   WESP operating principals. (With permission from GEA Bischoff.)

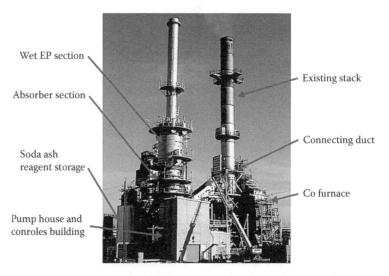

FIGURE 18.22   Commercial WGS/WESP application by GEA Bischoff. (With permission from GEA Bischoff.)

were reasons for selection. In one case, the stringent SCAQMD regulation 1105.1 was met using a WESP. Vendors that have applied a WESP to an FCC include Hamon (ExxonMobil Scrubber) and GEA Bischoff. Figure 18.22 is a picture of an FCC application by GEA Bischoff.

18.6   BAGHOUSE

A fabric baghouse is a common technology in the utility industry for removal of PM. These devices can achieve very high PM reductions since they are a barrier filter. However, the bags are prone to tear, do not react well to temperature excursions, or high-moisture applications, and typically require high maintenance. Often, the operating temperature of the FCCU precludes the use of a fabric filter. There is one commercial unit using a baghouse for PM control on the whole flue gas. Several other baghouse units are used as the FSS for TSS underflow.

18.7 SUMMARY AND CONCLUSIONS

In the 1950s, a typical FCC unit operated with >10 lb/1000 lbs coke PM emissions with a resulting stack opacity of 30–50%. Environmental regulations and technology improvements have reduced PM emissions by an order of magnitude for current FCC units. Several methods to control PM emissions from FCC units are available including cyclonic devices, electrostatic precipitators, WGSs, wet ESPs, and baghouses. All have been able to demonstrate compliance with environmental regulations and consent decree requirements.

ACKNOWLEDGMENTS

The following individuals and companies are acknowledged for their input and permission to use information and data: Neil Dahlberg, Vice President, Hamon-Research Cottrell; Patrick Walker, Technical Fellow, UOP; Scott Evans, Technical Manager, Clean Air Engineering; Charles Leivo, Sales Manager, GEA Bischoff, Inc.; Bob Taylor, Consulting Engineer, GE Energy.

REFERENCES

1. EPA Web site, http://www.epa.gov
2. Rosin, P., Rammler, E., and Intelmann, W. *Z. Ver. Dtsch. Ing.*, 76 (1932): 433.
3. Walker, P. and Sexton, J., Multiple Stage Separator Vessel, U.S. Patent #7,547,427, 2009.
4. Davis, W. T. *Air Pollution Engineering Manual/Air & Waste Management Association*, 87–89. New York: John Wiley & Sons, 2000.
5. Wardinsky, M. *Reducing the Risks of Fires and Explosions in FCCU ESPs*. NPRA Q&A P&P Session, October 13, 2009.

Index